PROTEIN SECRETION

Tell all the truth but tell it slant,
Success in circuit lies,
Too bright for our infirm delight
The truth's superb surprise;

As lightning to the children eased
With explanation kind,
The truth must dazzle gradually
Or every man be blind.

EMILY DICKINSON

PROTEIN SECRETION

A CRITICAL ANALYSIS OF THE VESICLE MODEL

Stephen S. Rothman
University of California
San Francisco, California

A WILEY-INTERSCIENCE PUBLICATION
JOHN WILEY & SONS
New York • Chichester • Brisbane • Toronto • Singapore

To
Doreen, Jennifer, and Peter

Library of Congress Cataloging in Publication Data:

Rothman, S. S. (Stephen S.)
Protein secretion.

"A Wiley-Interscience publication."
Includes index.
1. Proteins—Secretion. 2. Biological transport.
3. Cell membranes. I. Title.
QP551.R85 1985 591.19'245 85-12252
ISBN 0-471-07976-6

Printed in the United States of America

10 9 8 7 6 5 4 3 2 1

Preface

Picasso said that the creation of new art forms can come about only as the result of the rejection of older esthetic values. That is, in art, creation is in the first instance an act of rejection or a negative, rather than an affirmative event, contrary to the popular notion of such things. The issue is much the same in science, and new perceptions of nature—discoveries and inventions—come about, or so it seems, most often as the result of questioning older perceptions, and not merely as a continuing positive accumulation of new knowledge, as is so often thought. Indeed, it can be argued that such rejection is an absolute prerequisite to all fundamental shifts in our view of the universe around us.

Rejection often starts modestly in the form of unanswered questions or anomalous observations that current models or paradigms cannot explain in a satisfactory fashion. Sometimes, although certainly not always, as the result of such problems a dominant view of a system is challenged, and evidence is sought not merely in its support but in an attempt to test its predictive power in a questioning manner. Under such circumstances defenders of a paradigm may be constrained to try to make the anomalous fit and to answer the unanswered within the borders of their construct.

This book is about such a situation. It concerns questions that the author and others have raised about the current paradigm for the movement of proteins across biomembranes, and that paradigm's harmony with nature. As I shall discuss in what follows, the paradigm originated in and has been bolstered by a widely held belief that protein molecules pass neither through the substance of biological membranes nor through channels within them. In great part as a result of this belief, models have been developed to explain the transport of proteins by processes that do not require their movement through membranes individually.

It is these models, which we can call "vesicle transport," and the accompanying belief in the essential impermeability of membranes to proteins, that my colleagues and I have questioned. At first our questioning was tentative, but as we explored the issues, testing predictions of the vesicle model as we went, our original questions were not satisfactorily answered, and our uncertainty grew.

As a result, we realized two things. The first was that much of the evidence that had been presented in support of the vesicle paradigm gained its strength as evidence from the a priori assumption that biological membranes were impermeable to proteins. If this support was removed, then the evidence became much more ambiguous and uncertain. The second realization was that if our questions about the paradigm were correct, then we needed to develop an alternative theoretical framework in which to explore protein transport. The obvious alternative, and one that had been in my mind from the beginning, was to apply a traditional diffusion-based construct. Indeed, in my first paper on the subject, regarding protein secretion by the exocrine pancreas, in *Nature* in 1967, I used the term "protein transport," I believe for the first time to describe such processes. This, instead of descriptive terms for particular hypothetical vesicle mechanisms as had become traditional, was an attempt to focus attention on the fact that the secretion of protein should be viewed in the most general manner and not as a subject apart from the study of biological transport otherwise.

In a diffusive model one does not make a priori assumptions about whether a membrane is permeable to a substance, but instead the kinetic and thermodynamic characteristics of the system are experimentally determined and from that knowledge models are developed in an attempt to explain what is observed. If the membrane is permeable to the substance of interest, one starts with simple diffusion as the explanation of parsimony, and rejects it only when and if the evidence requires more complex constructs, and even then, unnecessary complexity is eschewed as we move to the next required level of complexity.

Our particular diffusion-based hypothesis for protein transport, known as the equilibrium theory, has three important characteristics that recommend it as a satisfactory theoretical perspective from which to try to understand the nature of these transport processes. The first is that it is a true alternative to the paradigm. This gives us the power of choice in attempting to account for observations in terms of theoretical models. The second characteristic is that it is readily testable. The third characteristic is that it is a deep or general theory which provides the broadest framework upon which to build an understanding of transport processes, being based, as it is, on general properties of matter.

Applying a diffusive model to protein transport means rejecting the assumption that proteins cannot pass through or otherwise cross membranes individually. It is important to mention that one need only reject the *assumption* that this is true. We need not assume that membranes *are* permeable to proteins by diffusive processes. This is matter for experimental exploration and evidence.

The belief that proteins cannot cross membranes in any kind of general sense was, and is, deep-seated and dates back to the early days of cell theory and our definition of cells themselves. As a result, my colleagues and I encountered much resistance to this idea from others, as well as within ourselves, despite that fact that the evidence that membranes are impermeable to proteins in general was not only not compelling, but totally absent.

In the 20 years since the ideas discussed in this book first saw the light of day, two events may have had an important impact and deserve special note. The first was that the paradigmatic view of protein transport became even more firmly established when a Nobel prize was awarded to George Palade, a scientist whose work and ideas were central to its development. The climate after such an affirmation is less receptive to a different opinion. The second, and somewhat paradoxical, event was the growing realization that individual protein molecules did indeed cross biological membranes, and that this was perhaps a very common occurrence; indeed it was even required by the paradigm itself. I say that this is "paradoxical" because as the vesicle paradigm became more and more firmly ensconced as the manner of means whereby protein transport occurred, its underlying premise, that membranes are impermeable to proteins, was shown to be incorrect even if one attempted to explain such transport solely within paradigmatic borders.

The subject is approached here in great part by considering how cells secrete their protein products. The book has been divided into three sections. The first brief section, *general considerations,* sets the stage for what follows. In it I define terms and present the context in which I will consider the vesicle paradigm. In the second section, *the vesicle paradigm* is described briefly and then considered critically in regard to certain of its aspects, in particular, its underlying assumptions (namely, impermeability), the proposal for filling secretion granules with product, the exocytosis model for secretion, and the signal hypothesis. In the third section, *testing the paradigm,* I discuss a variety of experimental tests of the paradigm, other observations that bear on its adequacy as a general model for secretion and protein transport, and also consider these observations from the viewpoint of an equilibrium or diffusion-based theory of transport. Although I find the general discussion at the beginning helpful for what follows, some readers may wish to go right to the experimental issues.

When I have been faced with a choice between a discussion of experimental details and a more accessible consideration of the question aimed at the nonexpert, I have tried to choose the latter. The primary literature is there to be examined and considered in greater detail for those who wish to do so. I have not written this book solely for those who are actively engaged in studying protein transport, but for a much larger group of people who have an intellectual curiosity about the subject, from biologists working in other areas, to students of biology, and even the curious layman who has the necessary background.

The current paradigm for protein secretion, known variously as the vesicle model, the cisternal packaging–exocytosis model, the exocytosis model, the segregation model, and so forth, is known, not only to professional biologists, but to many college, high school, and, dare I say, grade school students who have been exposed to cell biology. In considering the subject I will assume some knowledge of this model and the character of the evidence that has been adduced in its behalf. The information that I present may not be sufficient in itself if you are not familiar with these ideas, and I suggest that in that case you read one or more of the review articles listed in the bibliography at the end of Chapter 6, which, although often written with different audiences in mind, present the model from a similar perspective that has changed little over the years. These articles have been written by its inventors and should give the reader a sense of how they view its nature, origins, and the evidence in a way that I probably cannot convey as well, or as accurately. However in the final analysis, an informed opinion about the paradigm is not possible without the required search of the primary literature and a thorough and critical reading of the studies that form its experimental basis.

There is one theme in the book that goes beyond the specific subject matter. It has to do with the way in which we go about developing hypotheses, models, constructs, and theories. Although imagination is crucial to the development of hypotheses, it cannot be given free rein in science. This despite the fact that history has taught us that chance, error, and erroneous hypotheses have all played important roles in scientific progress. It seems to me that this is more difficult in biology than in the physical and chemical sciences, because it is generally agreed at the outset that we are dealing with an inordinately complex system. As a result, it often seems reasonable to imagine a complex mechanism at work, and for this reason rather complex models are at times put forth seemingly without sufficient thought to or exploration of alternatives, and evidence is merely sought in their behalf. When positive experimental evidence can be obtained, it affirms the correctness of the model. However, when such results are critically analyzed, they are often found to be weak and uncertain as evidence in support of the investigator's particular dei ex machina, as opposed to other potential explanations. I believe that protein transport, in particular, the secretion of protein by cells, is a case in point.

In a recent lecture at Berkeley, Salvatore Luria, one of the fathers of the great 20th century achievements in molecular biology, made this comment, which I paraphrase: "A scientific hypothesis should never, even tentatively, be accepted as being true, unless it has been exposed to tests that are capable of proving it false." I believe that the particular vesicle theory that I consider in detail here is a hypothesis that has not been subjected to, nor in some cases is even capable of being subjected to, tests that might falsify it. Yet it has not merely been accepted tentatively, as unwise as that might be, but is so widely believed to be true that textbooks at all levels discuss it as established fact. The alternative hypothesis, the equilibrium theory, that I discuss in this book, was in the first instance, and remains at its core, an attempt to develop potential falsifying tests for the vesicle hypothesis. In my view, it is primarily on this basis that its usefulness as a theory should be judged.

Questioning views that are widely held, whatever their factual basis or intellectual justification, is rarely appreciated. If my criticism appears to be tendentious in places, the reader has my apologies. It is my desire to have the issues viewed fairly on their merit, not hyperbole.

Acknowledgments

I dedicate this book to my family, my wife Doreen, and my children Jennifer and Peter, without whose nourishing love and shared experience I could not have overcome the obstacles during these years. I would also like to dedicate it to the memory of my father, Abraham Rothman.

I owe my deepest gratitude to my students and scientific collaborators, who have experienced with me not only the exhilaration of discovery, and the other personal rewards of science, but its dark side as well. "How I wish that somewhere there existed an island for those who are wise and of good will," Einstein importuned.

I would like to thank the many people who have offered encouragement and support, particularly those who gratiously offered to read the book in manuscript form and offered their valuable comments and criticism. I would like to thank Dr. Thomas Ermak for the micrographs in Chapter 4. I would also like to thank Stanley Kudzin and the staff of John Wiley, whose expertise and professionalism have been invaluable.

STEPHEN S. ROTHMAN

San Francisco, California
January 1986

Contents

PART 3 TESTING THE PARADIGM

PROTEIN SECRETION

PART 1

General Considerations

Before a theory explaining a certain process can be tested, the process must first be known.

ALBERT EINSTEIN

CHAPTER 1

Two Processes

I will discuss two related biological processes—the secretion of organic products by cells and the transport of one of these products, proteins, across biological membranes. Our understanding of each of these processes has become entwined with that of the other. This intermingling has historical, intellectual, and experimental roots whose basis and character will form an important and recurring theme throughout this book.

1.1 SECRETION

Secretion can be broadly divided into two types. The first is secretion in which substances, mostly fluid and electrolytes, move across polarized tissue surfaces, particularly epithelia, from one extracellular compartment to another, namely, from blood or the internal milieu to bodily cavities or the external milieu. Its opposite is absorption. The second type of secretion is the release or efflux of organic compounds from cells that manufacture or accumulate them.

Only this second type of secretion will concern us here and my use of the term "secretion" will refer exclusively to this process. Such secretion processes are widely seen, diverse in terms of the biological products secreted, and imposing because of their important place in biological function. They account for most, if not all, communication between cells, as in the secretion of neurotransmitters and hormones; in the secretion of digestive enzymes, they account as well for the ability of organisms to digest food, a function central to the survival of all animal species.

When we say that a substance is secreted by a cell, we imply the existence of an underlying biological *purpose* or *function* for the process, such as neurotransmission or digestion. Sometimes a purpose is assumed even though it is unknown, that is, it awaits discovery, or why else would the substance be secreted? Of course, although the release of a contained substance from a cell may serve some biological function, its release in and of itself does not demonstrate it, and the activity may be purposeless. A more convincing inference about function can be made when an externally applied stimulus, particularly a "physiological" one such as a neurotransmitter or hormone, is found to increase the rate of release of a specific substance. In this case a function is implied by the *response,* although even here increased release may merely reflect a passive, dependent, and purposeless sequel to another process.

Thus, secretion is often thought of as a special process in which the transport of a substance is thought to serve a particular function or purpose, and while this inference may sometimes be justified (that is, be more than teleology) and sometimes is not, nevertheless, secretion is often set apart in this way from biological transport as a general phenomenon in which the mere transport of a molecule tells us nothing of purpose.

The term secretion is also used as a synonym for the model that attempts to explain the transfer of "secreted" molecules across cell membranes, namely, the vesicle paradigm. Although this model is complicated and varies in its details from system to system, the central and final step in the construct, the step that is thought to lead directly to the release of the product from the cell, is common to them all. It proposes the fusion of the membrane of intracellular vesicles, which contain or are thought to contain the products that are secreted, and the cell membrane; the result is a hole produced in both membranes through which vesicle contents leave the cell and enter the extracellular environment. This model, commonly called exocytosis, will have an important place in the discussion and analysis that follows.

Thus, the study of secretion is frequently set apart from the general study of biological transport in two important ways. First, there may be a purpose, known or assumed, for the process—a particular raison d'être. And second, release may be thought to occur in a particular manner. In this way, ideas of *purpose* and *mechanism* have often formed an a priori assumptive foundation upon which knowledge of particular secretion processes is built. Unlike the general study of biological transport, our understanding of secretion is not usually derived from a description of the events as free from special, as opposed to general, mechanistic, and functional hypotheses as possible, but is often developed in just such a context. Indeed, preconceived notions of mechanism and function have often taken the place of an actual or real description of the system under study, undertaken *without* presumed particular purposes or presumed mechanisms of transport. The question of whether the proverbial cart—models of mechanism and hypotheses of purpose—has been placed before the horse—a full and unbiased description of the transport events as they occur—shall appear frequently in what follows. I will try to separate the concept of the *process secretion* from specific hypothetical models that attempt to explain how it is carried out, as well as from particular functions in which it may be involved.

1.2 PROTEIN TRANSPORT

Protein transport, the second process that will concern us here, includes examples of secretion, such as peptide hormone and digestive enzyme secretion, as well as the more general topic of the permeability of cells and biological membranes to large molecules which includes transport processes that we would not characterize as "secretory"; for example, protein transport across intracellular membranes (mitochondria, chloroplasts, peroxisomes, and so forth), the uptake of peptidic bacterial toxins by host cells, and the absorption of proteins across intestinal epithelium.

Protein transport and the secretion of organic cell products are processes that exist at different levels of logical organization. In the case of secretion, we consider different classes of molecules being transported out of cells by a process called secretion with whatever functional or mechanistic trappings we might wish to attach to that word, whereas protein transport is the movement of a particular class of molecules across biological membranes in general whatever orientation, location, mechanism, or purpose such transport might have.

As I have said, our understanding of each of these processes has had an important bearing on how we have come to view the other. This is primarily because of a particular experimental focus over the course of a century on *how cells secrete proteins.* Because the secretion of protein is an example of both sets (formally, the class "protein secretion" forms a subset at the intersection of the two larger sets "secretion" and "protein transport"), our view of the secretion of organic molecules has been greatly influenced by our understanding of how cells secrete the specific class of molecules, proteins, and similarly, our knowledge of protein secretion has been the substrate on which our understanding of protein transport has developed.

1.3. DIGESTIVE ENZYME SECRETION BY THE PANCREAS

I will approach the two general topics, secretion and protein transport, by means of a particular member of the subset protein secretion. One reason for this choice is that, as protein secretion has been the process upon which much of our knowledge of both secretion and protein transport has been built, our knowledge of protein secretion has in great part been derived from this specific case, *the secretion of digestive enzymes by ducted glands, in particular the mammalian pancreas.*

CHAPTER 2

The Constructed Hypothesis

During this century, reductionism has come into full flower not only as the philosophical underpinning of biology, but in a sense as its central experimental methodology. Based on the idea that biological forms appear to be made of the same stuff as the rest of the material world and obey the same laws, reductionism as an experimental strategy argues that we can understand their workings through knowledge of their parts, ultimately at the fundamental level of the substituent molecules and atoms. From this atomistic knowledge the nature of the native or whole system can then be inferred. This experimental reductionism and the philosophical system that proposes that biological law can be "reduced" to more general natural laws are not one in the same as is so often thought. Nonetheless, the experimental approach has been strikingly successful. We have made great progress in defining the contents of cells and organisms both anatomically and chemically. What would probably be agreed upon as the major achievement of 20th century biology, the determination of the structure of DNA and the details of the genetic code, is no doubt the brightest star in the firmament of experimental reductionism in biology. Nevertheless, and in no way demeaning the importance of these achievements, merely knowing, however intimately, the parts of a system provides us with only half of the information that we must have in order to understand how the intact system works. If our models are to be satisfactory, then we must also have a substantial independent description of the workings of that intact system, a four-dimensional space—time continuum, into which the presumed parts must fit. To describe a system we must have knowledge of its parts, obtained from atomistic scientific investigation *and* an accurate and equally rigorous description of the system's behavior in its native state, that is, how the parts interact with each other as part of the observed natural continuum.

Trying to construct a process solely from knowledge of its parts, or presumed parts, is somewhat like trying to solve a four-dimensional puzzle in which the pieces can be fit together in numerous ways, without knowing whether you have all of the pieces, or do not have extra pieces mixed in from another puzzle, and most importantly without knowing the nature of the picture that you are trying to construct. To contradict this view, one might point to the elucidation of the structure of DNA. After all, didn't the genetic code fall out of our knowledge of this structure? But what could one have inferred about the function of the DNA molecule, knowing its structure completely, without the work of geneticists over the course of the prior 75 years defining the nature and behavior of the gene? That is, considerable knowledge of the whole genetic system fortunately *preceded* the discovery of the structure of DNA.

The great changes that have occurred during this century in our view of the physical world came about in great part because mechanical Newtonian models could not account for certain observations, particularly those related to the nature of the physical continuum. If we believe that we can explain the workings of biological systems in physical terms, then we must understand both the parts and the whole, that is, the molecules, reactions, and organelles that comprise the whole, *and* the way in which the native system functions. To do otherwise is to ignore one important aspect of the physical world, while focusing solely on another. Physicists have tried to explain both the parts (matter) and the continuum (the context in which matter exists) within a single theoretical framework. The issue is the same in biology.

2.1 THE ESSENTIAL DUALITY

As a physiologist, I have witnessed throughout my career what have always seemed to me to be fruitless arguments that should have ended long ago about the relative appropriateness of in vitro versus in vivo investigation. Some physiologists argue that one learns nothing (or precious little) from in vitro, or worse yet, cell-free, studies because we can only learn from the whole system with all of its natural functions in place, and that all else is artifact. On the other side, there are those who just as vigorously argue that the complexities of the whole system in vivo are beyond rational analysis, and that "rigorous" science can only be carried out on parts of systems and cells.

In order to understand, we must know of what the system is comprised and how the isolated parts behave, as well as about the intact process. Awareness of this necessary duality seems to be lacking when mechanistic models are developed solely or primarily from knowledge of the parts of a system and their isolated behavior. I will call this a "constructionist" approach to model building in which the researcher constructs a model for an intact system based on his or her knowledge of its parts in the absence of a substantial *independently* arrived at description of the intact system. Of course, it is only knowledge of its presumed parts, since they can only be presumed in this case. There is nothing wrong with constructing mechanistic hypotheses based on knowledge of a system's parts when there is also knowledge of the native system with which to confront a constructed model, as in the case of DNA, or, even in the absence of such knowledge as a starting place for further investigation and an ordering of thoughts—a working hypothesis, if you will. However, hypotheses constructed in the absence of an adequate understanding of the native system can pose difficulties that go beyond the substantial uncertainties of trying to put together an extremely complex and unknown puzzle.

2.2. THE ASSUMPTIONS ON THE CONSTRUCTED HYPOTHESIS

When a model is not derived from the known characteristics of a natural system, it can only be constructed on the basis of assumptions about that system; that is, how we think it should behave. These assumptions may either be a matter of convenience or belief: convenience—assumptions simply made to facilitate the experimental process—or belief—in that certain views are held as being generally applicable, either in the absence of evidence, or on the basis of evidence from other systems. In experimental biology such assumptions are rarely akin to assuming that some universal theorem holds, such as energy and matter being conserved. Nor are they similar to mathematical assumptions that help define a theoretical system without necessarily judging its claim to being true. In the absence of significant knowledge of the native process, that is, its kinetic and thermodynamic characteristics, assumptions used in constructing hypotheses and models for biological processes are likely to be based on evidence collected in other systems that is deemed applicable to the system of interest; or even sometimes on beliefs that are not supported by substantial experimental evidence or other proof at all. In the latter case, we often only have our intuition or feelings as a guide. While in the former, we must necessarily also hold (assume) that the observation in the other system is correct, that the interpretation given in that setting is correct, and that the analogy is apt. For these reasons, such assumptions are often weak and clouded by chains of uncertainty, which are then reflected in the constructed hypothesis.

2.3. KNOWLEDGE GAINED BY TESTING A CONSTRUCTED HYPOTHESIS

Thus, constructed hypotheses cannot be tested against knowledge of an intact process; that is, we cannot ask whether or not our model is consistent with the known behavior of a natural process, because its behavior is unknown to us. But could we not argue that while this criticism of the constructed hypothesis may have some merit, it should not pose any serious problems for the hypothesis used as a tool in the design of experiments? Indeed, isn't that really the purpose of our hypothesis—to permit us to learn about the workings of the natural process in an orderly way? Therefore, shall we not inevitably learn about these things as we proceed with experimentation? Although we may learn about the intact process by testing our hypothesis, there is no guarantee that this will be an outcome. Testing a system in the context of a particular model may only provide information about the system's natural behavior *if* the proposed model turns out to be correct; that is, the experiments may add to our knowledge of the system's real behavior if and only if our hypothesis is correct, and not regardless of whether or not it is correct. Thus, it is possible to test a particular hypothesis in many different ways and yet be essentially as ignorant of the native process as we were initially. This is because our experiment is not designed to tell us about the system, but about our hypothesis.*

2.4 SPECIFIC VERSUS GENERAL HYPOTHESES

But even if this were not the case, and our tests clearly provided information about the real system, we might still have substantial problems if our hypothesis were a relatively specific and complex mechanistic construct at the outset.

When complex hypotheses are proposed for processes in the absence of substantial knowledge of the system's characteristics in the native state, tests of specific aspects of the hypothesis (subordinate hypotheses of the larger model) can at best only provide us with an extremely limited view of the natural continuum. That is, what we learn about it is necessarily derived from a selected and limited sampling of its behaviors as seen through the myopic eye of our hypothesis. The broader or more general our hypothesis, the less restricted our probing and the more we can hope to learn from it. The more specific and particular our predictions, the less we will learn about the system overall. This may seem somewhat perverse in a "reductionist" context since the latter type of experiment often appears to be more rigorous, getting to the essence of "mechanism" because of its concrete and detailed character. One may even talk of molecules or parts of molecules and their role in the larger process. But nevertheless it seems to me that the

*Consider the following process A that is transformed into another Z by an event or series of events. Of the conceivable $a \cdots n$ different series of potential transforming events, we hypothesize that A becomes Z by means of a particular series b. If, however, we are wrong and it is transformed in another way, let us say by means of x, then we may learn nothing about how A becomes Z by studying our hypothesis that b accounts for this conversion.

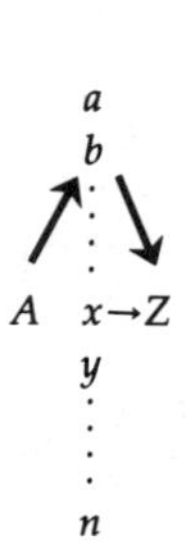

more general our hypothetical statement, the more valuable the experiment and the broader the resultant understanding of the natural system.*

2.5 LOGICAL EMPIRICISM

Might we not still argue, that even with the shortcomings that I have just discussed, if our hypothesis fails a test, then it will be rejected and replaced by a new one; and that with time and experimental test we shall nevertheless hone in on the true nature of the system? This approach to experimental science, often called logical empiricism, rests to a considerable degree on the belief that the hypothesis will be rejected as a result of a negative test, and will be replaced by another that will then similarly be tested against reality. Hypotheses are usually derived from a scientific paradigm deeply embedded in an area and based on a wide range of evidence, assumptions, and related beliefs. A negative test of such a construct, by means of a particular hypothetical prediction derived from it, often, and perhaps usually, leads, at least initially, to the rejection of the negative observation, not the paradigm.

Moreover, the likelihood that a construct will be rejected by a negative test becomes increasingly remote the more specific and complex the model. In this case, negative evidence is often readily explained by means of ad hoc or auxiliary hypotheses that represent minor modifications of the comprehensive construct, and is often attributed to the investigator's own, as yet, incomplete understanding of the details of the construction. This allows one to hold on, sometimes with great tenacity, to the model's general concept despite negative evidence. On the other hand, the more general a hypothesis, the more likely its rejection will be required by experimental test, and hence its greater value.

*Let us again consider the model in the footnote on p. 11. We can call our hypothesis b a specific hypothesis because if it is wrong, of the n potential explanations for the conversion of A to Z, we are left with $n-1$ other hypotheses to consider still and while we may have learned something about the transmutation of A it may be trivially related to the actual events. On the other hand, a more general hypothesis, let us say one that encompasses behavior that we would expect of $n/2$ of the potential systems, will teach us about the nature of the system more effectively, and appropriate tests of our more general hypothesis will include or exclude half of the prospective total.

2.6. POSITIVE TESTS OF CONSTRUCTED HYPOTHESES

What happens if tests of the hypothesis are positive? It is of course in the nature of a hypothesis that if experimental support for it is obtained, its claim on reality will be strengthened. Sometimes so are the assumptions that gave rise to it in the first instance, however circular this reasoning may be. Thus, as the researcher gains support, however tentative, for his or her particular hypothesis, the model as a tool in the design of experiments, gives way to a specific construct that attempts to describe real processes concretely, and the claim may even be made that the proposed process or mechanism actually occurs as envisioned. Making this leap may be problematic, however, particularly if we have used the most common logical methodological tool in experimental biology, weak inductive inference, to test our hypothesis.

In this case, we make a prediction about the behavior of a system based on our model, a hypothesis deduced from it, if you will, and then test this hypothesis. If the prediction is proven correct, then we infer that our general hypothesis is correct as well. However, this result really only permits us to say that "if our hypothesis is correct, then the following should happen, and it does." It does not rule out alternative explanations for the same phenomenon along which path the truth may lie instead. It is rare, if it ever occurs in the world of experimental biology, that we are capable of saying "if and only if our hypothesis is true will the following occur."

Nevertheless, when a series of tests of this sort are carried out and a chain of positive answers provided, the tentative quality of the original hypothesis can be lost in a most elating affirmation that the investigator's ideas were correct. Such lines of evidence can be very deceiving as tests in the real world since observations that are consistent with false hypotheses are rather common occurrences. Thus, a list of such observations, however long, can never alone prove that our hypothesis is true, because it does not provide a unique solution for the observations, nor consider one explanation against another. This problem is accentuated when our model has been constructed without an independent description of the natural continuum against which to test it.

This does not mean that we should discard experiments whose design leads to inferences of the weak inductive variety. Indeed, it is hard to imagine experimental biology in their absence. And even if models derived from such weak inferences cannot be considered as satisfactory approximations of reality, since science is an unending search for better and better approximations, they may be useful for the moment whatever their shortcomings. The problem occurs when we lose sight of the fragility of hypotheses supported by this type of evidence alone; once again, particularly when our hypothesis is for a system for which we do not have substantial knowledge of the intact process other than as viewed from within the framework of our specific hypothesis.

2.7. THE GENERATION OF A CONSTRUCTED PARADIGM

In the absence of an independent description of the natural process, our hypothesis and our description of the process may be made of the same cloth. The parts of the system that we describe are only those that our hypothesis leads us to, and sometimes not even that. The more specific and elaborate our hypotheses become, the more likely we are to go wrong, because our models are easily modified in trivial ways to account for negative observations without the whole construct being seriously challenged. Thus, in the absence of independent knowledge of the behavior of the whole native or natural system, we cannot properly test such hypotheses in their own right, or in comparison to others.

SELECTED REFERENCES

1. Kuhn, T. S. *The Structure of Scientific Revolutions*, Univ. of Chicago, Chicago, IL, 1962.
2. Popper, K. R. *The Logic of Scientific Discovery*, Harper and Row, New York, 1965. (Reprinted).

CHAPTER 3

What's in a Name or Anatomic Reductionism

As I noted in the Preface, until relatively recently, the term "protein transport" was not generally used to describe the movement of proteins across membranes. Instead, descriptive terms for particular hypothetical transport mechanisms were, and to a considerable degree still are, used. These terms, usually constructed from Greek or Latin roots in the classical tradition, evoke in the knowledgeable reader a specific and complex series of events. They have included pinocytosis (cell drinking), its reverse (reverse pinocytosis), emiocytosis (cell vomiting), exocytosis (to remove from the cell), and endocytosis (to bring into the cell). Even when common nouns are used, the word "transport" is usually replaced by others such as "transmission," "extrusion," "sequestration," "internalization," "externalization," or "translocation." "Translocation" has become popular recently, and one talks of "cotranslational translocation," which means the transport of a peptide chain across a membrane as it is being synthesized; and "posttranslational translocation," which means that the protein is transported after it is synthesized. This latter terminology of course refers to all transport that occurs in the life of a protein after synthesis. Why this special and somewhat arcane terminology? Why haven't we talked about the membrane transport of proteins in the same way that we talk of the movement of other molecules and atoms across membranes?

3.1 THE SMALL AND LARGE OF TRANSPORT

The term "transport" refers to a physical event, not to a particular mechanism or purpose. When we talk of the transport of a substance across a membrane, we are not saying how it crosses, or why, but simply describing a phenomenon. This, as we have already discussed, is different from talking of "the secretion of digestive enzyme by exocytosis," in which case we are not only saying that protein is transported out of the cell, but that it leaves in a particular way (exocytosis) that is purposeful (secretion).

This having been said, however, the word "transport," as it has been commonly applied to the movement of molecules and atoms across short distances such as biological membranes does imply something about mechanism at the "microscopic" level. That is, the movement of the transportable substance occurs as a result of its interactions with nearest neighbor molecules and atoms due to the kinetic energy of the individual molecules themselves. It is these interactions that are thought to account directly for the movement of a particular molecule within a phase or from phase to phase.

The transport of large organic molecules such as proteins has been considered sui generis; that is, as a system apart from this general perception of how molecules move across membranes. This is because such transport has been considered, until very recently, a "macroscopic," rather than a "microscopic" process. For example, the secretion of macromolecules is thought to occur as the result of the release of hundreds of thousands or even millions of molecules from the cell at once through holes in the plasma membrane produced by exocytosis. In this case, microscopic interactions between individual permeating molecules and the plasma membrane would not occur, and their transport would instead occur en masse as the result of the integrity of the membrane barrier or boundary being compromised temporarily by exocytosis. Characterizing such transport in "microscopic" or general transport terms would be inappropriate. Even when models have been considered in which the protein is thought to interact with the membrane directly (binding), as in certain pinocytotic or endocytic models, transfer across the barrier is still thought to be the result of its removal. Hence, even though interactions between the permeating molecule and the membrane may occur, they do not directly account for the passage of the molecule across it. It is this line of thought that makes it possible to say that membranes are impermeable to proteins and at the same time know that protein transport across them is a process that occurs throughout animal and plant kingdoms.

3.2 THE ELECTRON MICROSCOPE

Placing protein transport as an independent subject has in part been the result of historical circumstances that have led to the process being studied predominantly from an "anatomical" or visual perspective. Cell biology, the child of classical anatomy and histology, came into full flower after the Second World War with the development of the electron microscope, which made it possible to look into the cell and describe with some apparent precision intracellular parts and organelles. With the increased power to resolve small objects that this instrument gave us, cells that had merely seemed to be sacs of nuclei-containing protoplasm, were found to be far more complex and visually interesting. Many new features were observed, and structures only guessed at in the light microscope could now be confirmed at the ultrastructural level.

Like many of their predecessors, anatomists of ultrastructure were not content to just describe what they observed. They also wanted to know what these objects did. Attempts to answer this question often followed a particular line of thought. If a cell was known to carry out a particular function, then it was reasonable to think that structures found within them might be related to that process in some way or another. This line of reasoning was further validated by the widely held, although often unstated, view that cells of eukaryotes, particularly higher forms, are differentiated to the ultimate degree; that is to say, in their differentiated state virtually all of the cell's activities subserve one particular or special function such as muscle contraction, secretion, etc. If one could indeed assume that contraction was what a muscle cell was up to, then it made good sense to try to relate muscle ultrastructure to this function.*

*Such a conclusion may seem quite reasonable for skeletal muscle for example, based not only on its structure and known function, but on its most common location in the organism (i.e., connecting joints). However, we should be aware of what we are saying when we draw this conclusion. First, we are saying that there are no other as yet undiscovered functions that a given cell performs. Since we in great part direct our efforts to learning about known and

One could seek clues about function from the form and location of particular objects in cells. Considering muscle, as detailed knowledge of the ultrastructure of the striated muscle fiber became known, it was natural to ask how the structures running parallel to the long axis of the cell (thick and thin filaments) might be related to the contractile process that occurred along this axis. Thus, by assuming that a particular form is related to a particular process, and by then trying to guess what its role in that process might be from its shape and location in the cell, it was possible to develop specific models for processes at the ultrastructural level that could then be tested.

Thus, the greatly increased resolution that the electron microscope gave to cell biologists led to the development of a great many new hypotheses that attempted to explain how particular cellular functions are carried out by trying to relate observed structures to known functions. This was certainly not a new pattern of hypothesis development, but nevertheless, the increased focus of our vision had led to the development of functional models based on structures whose existence was made known by being able to examine the system at a higher magnification, closer to the molecular level. Thus, experimental reductionism had another success. The technological advance that the electron microscope embodied led to a great increase in our knowledge of the cell's structures, and this in turn was food for our imagination as we tried to relate these structures to the known functions of cells.

3.3 THE ATTENDANT PROBLEMS

Even though the visual resolution of forms inside of cells was increased by several orders of magnitude, being able to resolve even certain single molecules, the material that we examined in this way was static and related to the living state in often unclear and ill-defined ways. Fixation, staining, dehydration, and other treatments required for electron microscopy alter

---

obvious functions, other processes, even ones of importance, unknown and perhaps not so obvious, may escape our view. In any event, such a negative assumption based on what we do not know does not seem very reasonable or useful. Second, we are saying that the evolution of the cell from the multipotent bodies of microorganisms to a single-function body is complete in today's eukaryote. While we may be well down this path, our understanding of the nature of irreversible processes and evolution does not, I believe, permit us to conclude that the trip is over.

samples in ways that we often cannot explicitly predict or characterize. What we see is a generally agreed upon distortion of reality that we hope, with some reason, bears a resemblance to the real thing. Of course, this is true of all scientific theories, but unfortunately when we look at "pictures" of "real" cells, the fact that the images are merely hypotheses themselves can easily be lost.

But perhaps more to the point here than the accuracy of the images is the fact that they are invariably static. We cannot actually follow continuous (dynamic) processes with this technique. The scientist looks at his or her material and from it imagines how the images may be related to actual occurrences in the living system. At best we can examine samples of material taken at different points in time and then attempt to "reconstruct" actual occurrences as we think they might occur in real space-time. It is sometimes thought that this is wholly analogous to following cytologic changes in living cells. Unfortunately, it is not, unless of course we have independent evidence that such changes occur in living cells in the first place. If we have an hypothesis that a cytologic change of a certain type occurs and observe electron microscopic images that are consistent with this hypothesis, then we have provided evidence in its support (weak inference), but have not described the events involved in the natural process independent of our hypothetical construction. This approach to the discovery of physiological mechanisms when used in isolation has the important limitations that were discussed in the last chapter.

3.4 THE IMPORTANCE OF A DESCRIPTION OF THE PROPERTIES OF A SYSTEM INDEPENDENT OF SPECIFIC MECHANISTIC HYPOTHESES

As I have said, the term "transport" is descriptive of a phenomenon, and the study of transport in biology is in the first instance the delineation by direct measurement of the kinetic and thermodynamic characteristics of a particular transport process. How fast does transport occur? How is rate related to the concentration of the transported substrate? How do alterations in the structure of the permeating molecule alter its rate of transport? And so on. It is on the basis of these measurements that we can *then* build models that attempt to incorporate the observed properties of the system and will be useful in predicting its responses. These models may also be used to characterize the underlying mechanisms in specific molecular terms. Only with such knowledge of the system's properties in hand, can we have confidence that the isolation (chemical or anatomical) of what we think are parts of the system indeed relate to the natural processes in a way similar to what we imagine. That is, we can only have this assurance, if we know how the system should behave however it is put together; i.e., if our initial description of the system is accurate and broad. In this case, our "reductionist" or "atomistic" model can be compared with the actual behavior of an intact system. Thus, our approach to the study of molecular transport has at its center the actual physico-chemical description of the system from which reductionist or atomistic science attempts to determine its detailed structure (Figure 1).

Thus, in examining a transport process we must first describe the actual or "real world" properties of the system before developing detailed models at the molecular or other "reducing" level. To do so in the absence of this prior knowledge in a sense presumes that the investigator has devised an hypothesis that is adequate and that other possibilities have not escaped attention; in essence, we assume that the hypothesis is correct, at least in general terms, before we even test it. In this case, our hypothesis and the description of the system may become confused with each other. I believe that this has occurred in the study of protein transport wherein the vesicle hypothesis and what is thought to be a description of the system have in some instances become virtually indistinguishable.

3.5. THE GENERATION OF STRUCTURE-FUNCTION HYPOTHESES

The presence of particular visual images within cells and knowledge that a certain function is carried out by a cell permits us to propose or assume a connection between the two; but it, of course, does not demonstrate one. In some cases, as with muscle contraction, considerable knowledge of the

---

1. The discovery of a process or event.
2. A description or characterization of that process or event in real space-time (kinetic, thermodynamic, etc).
3. Attempts to identify and isolate the parts of the system.
4. Reconstruction or other tests to demonstrate whether the isolated parts can be reassembled or otherwise can account for the behavior of the intact system in terms of a particular model.

Figure 1 A sequence for the investigation of a biological process or event.

function (or physiology) of the intact process from a dynamic and thermodynamic point of view existed prior to electron microscopy studies that attempted to correlate structure and function. Therefore, a model constructed from forms seen within the cell could be based on and tested against substantial knowledge of the system's actual behavior. Whatever limitations this approach may have for muscle contraction, attempts to make similar inferences about the relationship between the structure and functions of the acinar cell of the pancreas has been far more problematic, as we shall see.

PART 2

The Vesicle Paradigm

Mere description is so difficult because one believes that one needs to fill in the facts in order to understand them. It is as if one saw a screen with scattered colour-patches and said: the way they are here, they are unintelligible; they only make sense when one completes them into a shape.—Whereas I want to say: Here **is** the whole. (If you complete it, you falsify it.)

LUDWIG WITTGENSTEIN

CHAPTER 4

Assumptions

When the electron microscope permitted the first detailed look into the world of the cell, the acinar cell of the pancreas was particularly interesting. As with striated muscle, it was also quite natural to wonder how the forms seen in the microscope might be related to the cell's ability to perform a particular function, namely, the secretion of large quantities of protein, and to then build a model for secretion based on these forms and their location in the cell.

The most striking structures were the spherical zymogen granules that were visible in the light microscope (Figures 1 and 2). But in addition, one could now see what the Golgi stain did stain: discontinuous stacks of membrane-bound cisterns, the Golgi complex (Figures 2 and 3). And the staining characteristics of the basal portion of the cell could be attributed to small, roughly spherical structures, ribosomes, that we now know are involved in protein synthesis (Figures 2 and 4). Most of the ribosomes appeared to be attached to numerous parallel flattened membrane saccules, the rough-surfaced endoplasmic reticulum or RER (called rough due to the attachment of the ribosomes to the membrane's surface) (Figures 2 and 4).

A microscopist looking at this bewildering array of subcellular detail might well attempt to develop an hypothesis for protein secretion that would relate these structures to each other and to protein synthesis,

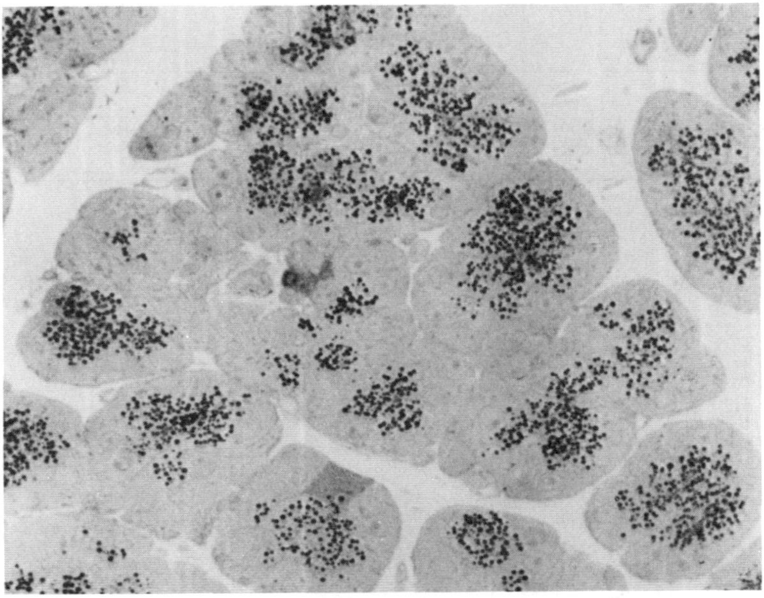

Figure 1 Pancreatic tissue from an adult rat. Large regions containing zymogen granules surround duct lumina. Peripheral cytoplasmic regions contain nuclei. Magnification, 530×.

intracellular transport, storage, and the release of digestive enzyme proteins from the cell. Zymogen granules, predominantly located at the cell apex, contained digestive enzymes and their disappearance had been associated with secretion in the past. Furthermore, it was thought that proteins were synthesized in association with ribosomes that were localized primarily in the basal or blood-facing portion of the cell.* Thus, new protein had to get from the ribosomes at the cell's base to the zymogen granules at its apex, and from there into the duct lumen. But how did this occur? How might the objects seen under the microscope be involved in the movement of this material? What led to the disappearance of granules? What was the role of the RER and the Golgi complex, assuming that they had one? The pattern of hypothesis development flowed from questions of this nature.

The task then appeared to be to relate structures to events and to overcome the limitations of inferences based solely on static electron

*While the generalization that the RER (and associated ribosomes) is at the base of the cell and zymogen granules at its apex is more or less true, it is only a quantitative approximation. Considerable amounts of RER are normally found at the cell apex—more noticeably so in cells with few granules—and zymogen granules are sometimes found throughout the cell.

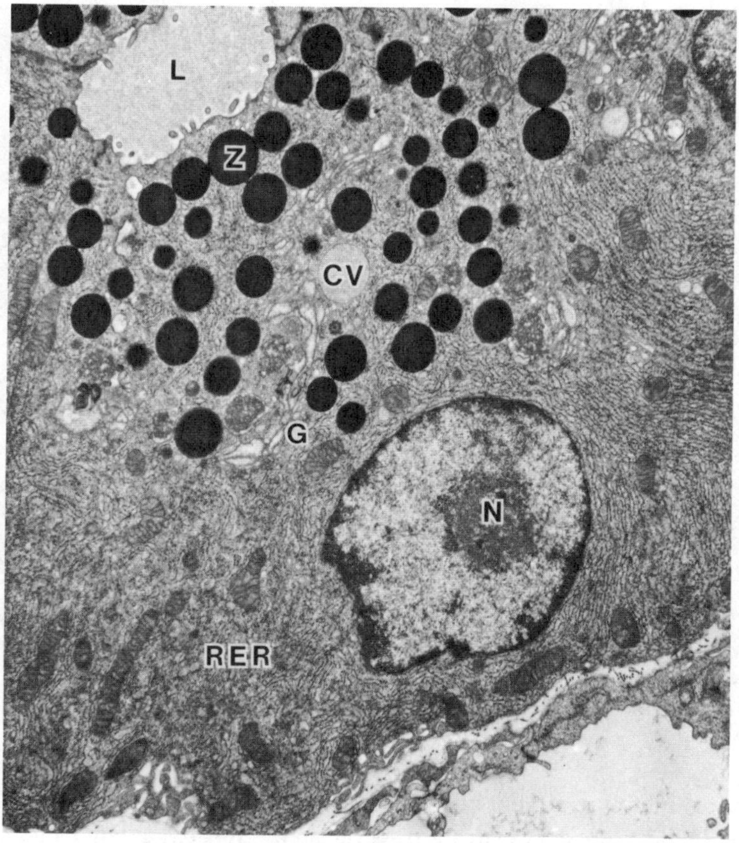

Figure 2 Acinar cell from pancreas of a young weaned rat (3 weeks old). Zymogen granules (Z) cluster around the lumen (L) at the cell apex. Rough endoplasmic reticulum (RER) fills the base of the cell. CV, condensing vacuole; G, Golgi complex; N, nucleus. Magnification, 10,000×.

microscopic images. But against what information could a putative structure-function model be tested? Unlike the muscle fiber, there was not much in the way of kinetic or thermodynamic knowledge about the intact secretory system. In muscle, as I have said, our "reductionist" or "atomistic" hypothesis must in the end account for the knowns of contraction: i.e., its physiology, its various heats, the length-tension relationship, the time-course of the process, its latency, etc. Although this fit may have yet to be satisfactorily achieved, what is required of a successful model is relatively clear.

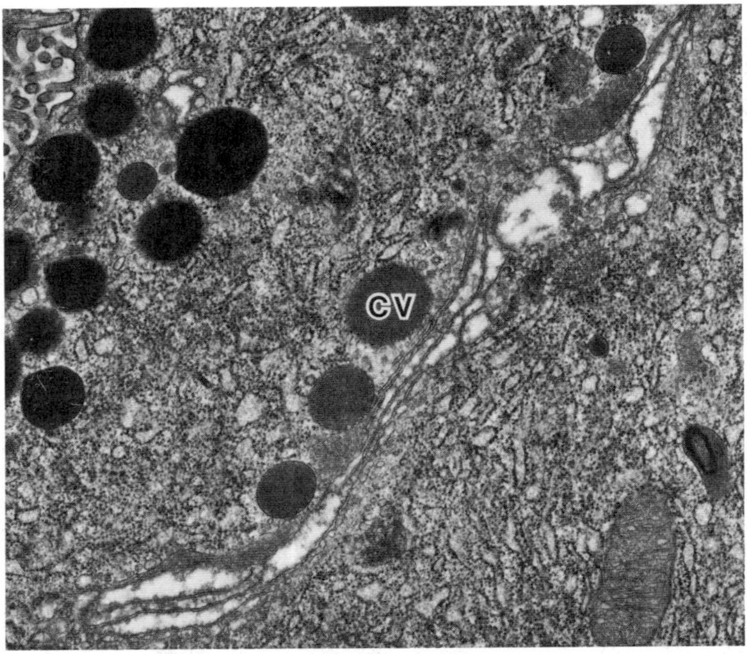

Figure 3 Golgi complex of acinar cell from secreting rabbit pancreas in organ culture. Condensing vacuoles (CV) lie along concave surface of the membrane lamellae. Magnification, 22,500×.

Unfortunately, a similar body of knowledge against which a model could be tested did not exist for protein secretion by the acinar cell. Indeed, little was actually known about the transport process beyond the fact that the enzymes were secreted into the extracellular space (the duct system), that a variety of different proteins were secreted, that the concentration of protein in secretion was variable, and that the zymogen granules that contained these proteins appeared to disappear when secretion was stimulated.

Thus, a model that attempted to describe the mechanism of protein secretion could not be much more than a construct from the imagination, based on assumptions about the system; a structural model for a physiological process evoked by static images in the absence of much information about the kinetics and thermodynamics of the intact process that it hoped to explain. Step 2 in Figure 1 of the last chapter was essentially missing, viz., "a description or characterization of (the) process or event in real space-time." It has been from this perspective that our understanding of protein transport has evolved.

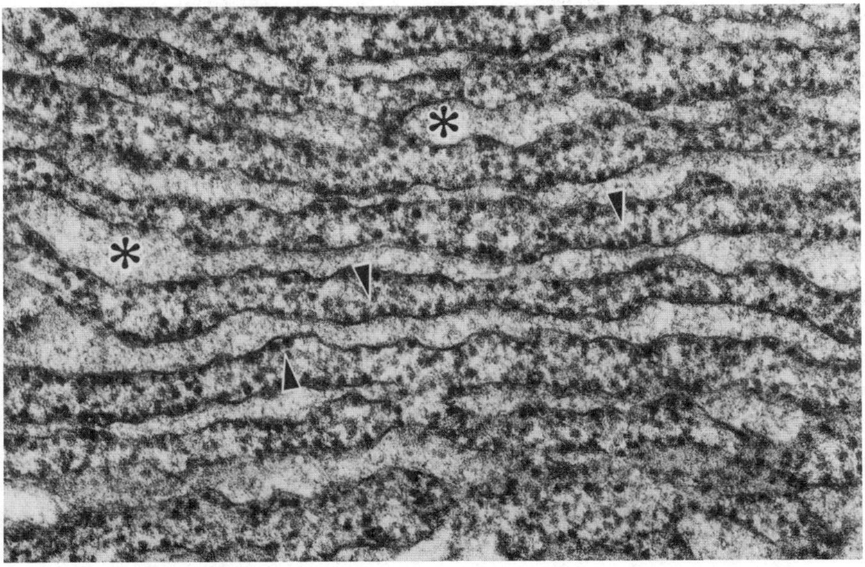

Figure 4 Rough endoplasmic reticulum of acinar cell from rabbit pancreas in organ culture. (*) Intracisternal space; (►) attached ribosomes. Magnification, 60,000×.

4.1 THE VESICLE PARADIGM

Thus, the vesicle or cisternal packaging–exocytosis model for protein secretion by the pancreatic acinar cell is an example of a constructed hypothesis; one built to a considerable degree on a series of assumptions about the system and knowledge of its presumed parts in the relative absence of a foundation of accurate knowledge of the intact system whose behavior it seeks to describe. The knowledge of secretion that we have gained since its proposal more than 25 years ago has for the most part been limited to information gathered by testing, by weak inductive inference, one aspect or another of this particular model.

The model can be summarized briefly as follows. New protein chains are "sequestered," or segregated, within the cisterns of the endoplasmic reticulum as they are being synthesized on ribosomes attached to the membrane surface of this structure. The secretory products contained within RER cisternae are then moved to the Golgi region of the cell in small vesicles thought to bud from the RER and subsequently fuse either with Golgi membranes or the membranes of filling zymogen granules, called condensing vacuoles (Figure 3), wherein the small vesicles release their contents. Secretion granules fill in this way and eventually form mature zymogen granules that are capable of fusing with the plasma membrane of the cell in the exocytosis event leading to the final and irreversible release or secretion of product.

Palade and Jamieson (1977, Ref. 4, Chap. 6) claim that the following is known or established:

"Transport along the secretory pathway is: (a) vectorial, i.e., it proceeds regularly in the direction of secretion granules (apically) in exocrine cells; (b) irreversible, i.e., there is no evidence of back flow; and (c) functionally discontinuous, i.e., the secretory products are moved from one compartment to the next in series by discrete membrane containers (vesicles or vacuoles) and the pathway behaves as if provided with locks whose opening requires energy and involves membrane fusion-fission."

4.2 TWO ASSUMPTIONS

However substantial or weak, clear or ambiguous, one considers the evidence that has been offered over the years in support of the paradigm, the issue rested in the minds of many, and perhaps still does, on our assumptions about the system. If we can assume that a mechanism similar to the one that has been proposed must exist, then whatever weaknesses the evidence may have as convincing proof dissolves into insignificance, because we really have little choice in the matter.

The vesicle hypothesis has in great part been based on, and has its origins in, two particular assumptions. If we accept either of them, then we have a priori excluded the only apparent general alternative to it: a diffusive or diffusion-based transport process in which proteins move individually, interact directly with and move through membrane elements in order to effect transport. That is, if we believe these assumptions to be necessary and sufficient, then the existence of a vesicular system in which proteins are moved en masse is established without the need for direct evidence of its existence at all.

4.3 THE FIRST ASSUMPTION

The major assumption is that proteins cannot cross biological membranes individually; that is, biological membranes present an absolute barrier to this class of molecules, and therefore other means must exist to account for their known movement across them. This assumption was thought to be quite compelling during the period in which the paradigm was being developed.

The cisternal packaging–exocytosis paradigm was itself based on another paradigm, the then-current hypothesis for the structure of biological membranes, the Davson-Danielli model, which proposed that biological membranes are comprised of a continuous lipid bilayer to which proteins are appended at the surface, binding to the polar groups of the lipids, particularly phospholipids, facing the two aqueous environments (cellular and extracellular fluid) bordering the bilayer. Proteins (at least globular and so-called random coil proteins) were seen as having considerable, and essentially invariant, surface charge that prohibited their entrance into the resolutely nonpolar center of the membrane. Hence, charged macromolecules could not enter a membrane, no less cross it. This view of impenetrability was further supported by the fact that many secretory proteins were water-soluble globular proteins that were insoluble in many bulk nonpolar solvents. Thus, the idea that membranes were impermeable to proteins, left us to find other means to explain their known movement across them, hence vesicle mechanisms.

4.4 SEMIPERMEABILITY

Our ideas about biological membranes and their permeability are in great part derived from the concept of "semipermeability" first introduced by Nägeli in the middle of the 19th century. Semipermeability simply meant that membranes were permeable to some substances, water in particular, but not others. In this regard, the cell membrane has been thought to have two barrier functions. The first is to *permit,* or even facilitate, the entrance of needed substances, such as water and nutrients, to which it is permeable. The second is to *prevent* the entrance of other (harmful) substances or the loss of important cellular components such as proteins and nucleic acids, to which it is impermeable.

At first biological membranes were thought to be impermeable to most substances. For many years, even the sodium ion, whose entrance into cells is now not only well established, but considered crucial to cellular function, was not thought to enter cells. At first indirectly, and then most notably with the advent of isotope tracer techniques, it was shown directly that even though the rate of sodium permeation was rather slow compared with some other atoms and molecules, it did enter cells at a measurable rate, and that over time the cell would, if permitted (in the absence of countervailing processes), come into electrochemical equilibrium with external sodium.

In fact, as experimental data accumulated it became clear that cells were permeable to a very wide range of substances, even ones that they might never see in the normal course of events (such as those made in the laboratory), or that might be harmful to them (such as urea is to red blood cells, destroying the cell just because it enters). Thus, the permeability of cells was a general phenomenon, and whether or not a cell was permeable to a particular substance did not necessarily reflect either its usefulness or its potential danger. Whether a molecule did or did not cross the barrier simply depended upon its physical and chemical characteristics in relation to those of the membrane.

It was also realized that the concentration of a substance in a cell was determined by the algebraic sum of its rate of entrance and exit. That is, substances did not merely enter *or* exit, but, as with simple diffusion, permeability was bidirectional, and intracellular concentration was determined by the relative rates of both influx and efflux (or outflux). In addition, for nutrients such as sugars and amino acids and products of cellular metabolism such as organic acids, cellular concentration was also determined by a third factor, the rate of utilization or production. In this case, the observed intracellular concentration of course reflected a balance between influx, efflux, and rate of utilization or production.

Despite the fact that the sodium ion enters cells, its concentration in many cells, notably those that have been widely studied, such as the red blood cell, nerve, and striated muscle, is very low in comparison to extracellular biological fluids (N.B.: A central reason why the membrane was thought impermeable to it). This presented a paradox. How could sodium both enter the cell from a high-concentration external medium and yet remain at a low concentration within the cell at the steady state? This problem was resolved when it was demonstrated that the cell could maintain a dynamic disequilibrium in which the rate of sodium efflux for a given concentration was greater than its rate of entrance; hence, the ability to maintain a low internal concentration at the steady state. That is, the cell's sodium content constantly changes due to influx and efflux, but nevertheless a low concentration is maintained by differences in the respective transport "affinities" (permeabilities) that makes it "easier" for sodium to exit than enter the cell. Thus, it became clear that the cell had "special" processes that could produce and maintain nonequilibrium steady or stationary states across the membrane for sodium, and, as it was discovered, numerous other molecules as well.

In this way, our view of the presence or absence of particular substances in cells has gradually evolved from one in which the plasma membrane was seen as being permeable to some molecules, but not others, to one in which the presence or absence of a substance in a cell (or whether it is present at a high or low concentration) in itself tells us nothing about whether or not it crosses the cell membrane or if it does, at what rate. The idea that the cell's contents of a molecule reflects a distribution between cell and environment in which the observed steady-state concentration is determined by a continuous, dynamic balance between influx, efflux, and utilization or production when that occurs has become the paradigm for the great majority of biologically significant substances.

Nevertheless, despite the realization that the membrane is permeable to a wide range of substances, that this permeability is bidirectional, that the degree of permeation varies broadly for different molecules and membranes, and that nonequilibrium steady states can be maintained in biological systems, the dichotomous view that a membrane is either permeable to a substance or impermeable to it has remained with us over the years. However the list of substances thought to be absolutely excluded has gradually decreased as our knowledge has grown. The most prominant members of this dwindling list of excluded molecules are macromolecules, particularly proteins and nucleic acids. Thus, although the older view of absolute impermeability has gradually changed to a more relativistic, statistical, and dynamic conception, it has not changed for proteins. Although as we shall see later, it has been nibbled away at the edges. And even though it is clear that intracellular proteins, for example, exist in a dynamic steady state, their concentration is thought to be solely due to a balance between the rate of their manufacture (synthesis) and subsequent breakdown (by enzymatic degradation) within a cell—the possibility of their exit or entrance is generally excluded a priori.

4.5 A PORE FOR EVERY MOLECULE?

Some membranes were found to be permeable to many water-soluble molecules, including water itself, at rates that were seemingly substantially greater than predicted for their transport across a lipid bilayer by solution and diffusion. This posed another paradox. If the membrane is formed by a continuous lipid bilayer, then why should the rate of transport be greater than predicted for diffusion through such a medium, at least for molecules for which no special mechanism appeared to exist to accelerate or facilitate movement, such as was apparently the case for water? In an attempt to resolve this problem, it was proposed that the membrane was perforated by a mosaic of water-filled pores or channels that transected the bilayer and through which water and other water-soluble substances passed.

Might not proteins exit or enter cells through such channels as well? The conventional wisdom has been "no." When pore size has been estimated from the observed rates of passage of test molecules across specific membranes, it has frequently (although not always) been found to be significantly smaller (4–10 Å radius) than the radius of most proteins (about 10–30 Å for most globular protein monomers),* and thus proteins have been thought to be excluded.

In a sense this was reassuring. If the pores were large enough to accommodate proteins, then how could the cell maintain its integrity? Wouldn't crucial intracellular proteins be lost? Indeed, wouldn't the existence of large pores in the cell membrane simply be incompatible with life? Isn't our concept of the cell membrane as a semipermeable barrier in part that it maintains the life of the cell by preventing the loss of many crucial substances, particularly nucleic acids and intracellular proteins, and in this way helps keep the entropic wolf from the cell's door? Isn't the view that the cell is impermeable to proteins embedded in no less than our understanding of the purpose of the cell membrane, and in a sense even our definition of the cell itself? It goes beyond models of membrane structure.†

Regardless of our beliefs about the nature of cells, the purpose of their limiting membranes, or our models of membrane structure, how justified is the assumption that proteins cannot pass through channels or pores, if such they be, or pass through the cell membrane in other ways? Might the factors that determine the steady-state concentration of cellular proteins include a transport component as well? I will now consider this idea, first in terms of the pore concept and then in more general terms.

The permeability of a membrane to a series of water-soluble test molecules drops sharply as the size of the test molecule increases, following functions that appear analogous to the Stokes-Einstein equation in which the diffusion coefficient, D, or the ease with which a substance moves through a medium, decreases as the radius of the particle increases.‡ This

* Based on crystal radii.

† Indeed, does not such a definition of the cell retain a remnant of 19th century vitalism?

‡ $D = k\,1/r$ where D is the diffusion coefficient, r is the radius of the diffusing molecule, and k is the proportionality constant which includes temperature, Avagadro's number, the gas constant, the viscosity of the solvent, and a geometric constant for the assumed spherical structure of the particle.

analogy is thought to fail however as we consider molecules of increasing size such as moderately sized organic compounds, or biological macromolecules such as proteins. In diffusion, although the rate of a particle's movement decreases as its size increases, in dilute solution at temperatures in the ambient range, the limit of zero mobility is not approached, and specifically for molecules of the size of biological macromolecules, mobility is substantial in solution (in water), even though it may be magnitudinally less than for smaller molecules (Figure 5). The permeability (P) of biological membranes, on the other hand, is thought to reach a particular discrete limit at which the "mobility," or permeability, of molecules of greater than a certain specified size can be said to be zero (Figure 5). That is, the movement of molecules larger than the size of the theoretical pore spanning the membrane would be absolutely, and not merely relatively, restricted.

This view is derived in great part from studies that estimate a small pore size by measuring permeability decreases with increasing molecular radius, weight, or volume for a series of chemically homologous compounds. However, such measurements do not establish absolute limits for pore size. This conclusion is an extrapolation from the data; it is not demonstrated by it. Indeed, pore size estimates are statistical averages obtained by a combination of experimental and mathematical techniques that permit the isolation of a value for average pore radius from the area component of the overall proportionality constant that relates solute flux across the membrane to concentration difference. Because the value is an average, the conclusion that the calculated pore radius is, for example, 5 Å, does not permit us to conclude that *all* pores are 5 Å or that 5 Å is the maximum size of a pore. If we assume for convenience that the distribution of pore size is normal and that the standard deviation around the mean value of 5 Å is a modest ±100%, then on a purely statistical basis we would estimate that 0.5% of the pores would be four times the average radius, or 20 Å. A pore of this size would be able to accommodate many protein monomers. Thus,

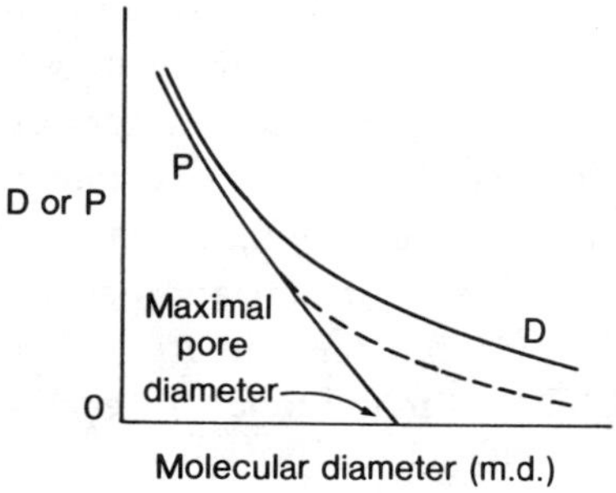

Figure 5 The relationship between the free diffusion coefficient (D) or the membrane permeability coefficient (P) and the diameter of a test molecule (m.d.). Two scenarios are shown for P versus m.d.; one with an absolute limit to permeability (——) and another with a limit that only approaches zero (-----).

estimates of pore radius do not permit us to conclude that there is an absolute limit to the size of permeating molecules. Such a limit may or may not exist, but we can not set it at less than the total surface area of the cell itself from such measurements. We can however say that if the standard deviation of pore size around the mean is rather small, then large pores, if they exist, would be relatively few in number and would only account for a trivial percentage of membrane conductance and total pore area. That is, if such pores or analogous structures exist, and proteins can pass through them, then we would expect the rate of protein passage across the membrane to be very much less than that for small molecules.

But what is the importance of such a conclusion? For example, if we ignore potential differences (for example, due to diffusivity or electric field) in the rate of passage through a small (5 Å radius) pore that can accommodate a sodium ion (23 daltons) or a large pore (20 Å radius) that can accommodate a particular 25,000-dalton protein, then for a given concentration gradient, 1000 small pores would move the same mass as one large pore, even though they would nevertheless account for over 99% of the total pore area. For the acinar cell, which moves large quantities of protein across the relatively small surface area of the cell's apical membrane, one can calculate that no more than about 0.001–0.01 of 1% (10^{-4}–10^{-5}) of the surface area need be occupied by pores large enough to accommodate proteins in order to account, not only for modest rates of protein secretion, but the maximal observed rate (N.B.: Based on the assumptions that free diffusion through the pores applies and that only one protein molecule can occupy it at a time).

Thus, although there is no disputing the fact that membranes are less permeable to large molecules than small ones, this does not necessarily permit us to conclude that there is an absolute limit to permeability, and that there is not a flux, even though small in relative terms, of proteins into and out of cells that occurs in the normal course of events and that is in the end analogous to the Stokes-Einstein relationship. The question then is whether or not membranes are indeed absolutely impermeable to molecules of greater than a certain size by measurement, at least within the limits of our capability to make such measurements. That is, is the decreasing permeability seen with increasing molecular size due to the absolute exclusion of molecules of greater than a certain size (is the reflection coefficient actually 1.0), or is the exclusion merely relative (the reflection coefficient approaches but does not reach 1.0)? Since the existence of pores has primarily been inferred from measurements of the permeability of membranes to certain molecules, it certainly seems reasonable to place limits on its permeability (to proteins) in the same way by attempting to measure it directly.

In this regard when a protein is said to "leak" from or into a cell, it is often assumed that this is because the cell membrane has been damaged or compromised in some fashion or another that permits the release or entrance of the protein. Whether or not this is the case in one particular circumstance or another, this inference is often drawn with no evidence that the cell membrane is damaged at all, other than its permeability to the macromolecule itself. Of course this is circular reasoning, and is based on the belief that the membrane is normally absolutely impermeable to such molecules. In this case any exhibited permeability must perforce reflect damage and nothing more. Although it certainly is true that if one "damages" the cell membrane, at the limit by removing it, that most of the contained molecules will diffuse or otherwise move away; and while it is also clear that an unnatural permeability of membranes to proteins can occur, for example in the release of hemoglobin from red blood cells during hemolysis, we cannot assume that whenever a protein enters or exits a cell (other than by vesicular means) that it is unnatural. This must be proven by evidence in each case, and cannot be assumed a priori, any more than we can assume that a membrane is permeable to a substance without testing our hypothesis.

4.6 ONE CELL, TWO CELLS, THREE CELLS, ALL CELLS

A central goal of scientific work is to develop models that successfully predict the behavior of a system. We also ask that our models explain as wide a range of behavior as possible, ideally incorporating all systems that fall into a particular category. Thus, it is not surprising that studies on the permeability of one particular cell type lead us to generalize about all cells, even though we know full well that permeability can only be formally defined in terms of each individual molecular species and each individual cell type whose membrane it crosses.

Particular cells have been studied because they are thought to be interesting and also often because they have provided an experimentally accessible system for the study of permeability. We then generalize to all cells from these particular examples. The red blood cell is perhaps the premiere example of an accessible system, and along with nerve and muscle, has provided, and continues to provide, the source material for many of our generalizations about cell membranes and permeability.

When considering the possibility that proteins cross cell membranes through polar channels of some type or another, the choice of many of the cells from which these generalizations have been drawn seems unfortunate. Of course, they were not chosen to learn about protein transport. For example, the plasma membrane of red blood cells, nerve, and muscle may be among the least "leaky" (smallest pored?) of cell membranes. The red blood cell, half a cell, unable to replace its important cargo hemoglobin by new synthesis, would be unable to carry out its functions effectively if it lost a substantial portion of this protein during its approximately 120-day life span. Thus, one would not be surprised to find that the red blood cell membrane helps prevent such leakage. The membranes of muscle and nerve must propagate an electrical signal, and to do so must have a high transmembrane resistance; once again these are not the most propitous choice of cells in which to look for protein permeability. This is not to say that these cells are not permeable to proteins, but rather that we might expect them to be *relatively* impermeable as cells go.

Thus, to assume a priori that the pancreatic acinar cell, or other cells, whose major known function is to transport or secrete proteins, is as impermeable to large molecules as the red blood cell or nerve would be inappropriate; this is a generalization based on an abstraction of the structure of biological membranes made at the wrong level. But of course such a generalization is commonly made, and our view of the permeability of all cells to large molecules has been in great part derived from a limited, and perhaps poorly chosen, selection of particular cases. Even without considering cells actively involved in protein transport as a specific group, a wide range of estimated average pore radii is seen in different cells or in cells in different functional states and these vary by over an order of magnitude, from about 2 to 30 Å. But the important point here is that if we wish to examine the permeability of cells to proteins, then we should study those that are actively engaged in such processes, not those that do not appear to be. In recent work, Mélèse and I have shown that the acinar cell is not only permeable to molecules that often do not enter cells readily (e.g., sucrose, mannitol, and inulin), if at all, but that its permeability to these molecules is substantial and moreover increases greatly during stimulant-induced augmented protein secretion (for further discussion of these studies and references see Chapter 15). Compared with red blood cells, for example, that usually can be stored in isosomotic solutions of sucrose or mannitol, the acinar cell appears to be quite different. If the red blood cell is the cell from which we draw our paradigmatic generalizations about membrane permeability, then the acinar cell can be viewed as being "anomalous." But which cell is really "anomalous" and which is "typical?"

4.7 PERMEABILITY AS A CONTINUOUS PROCESS

To this point we have considered the possibility that pores of sufficient size might exist in the cell membrane to accommodate some proteins, and we have concluded that we can neither exclude this possibility a priori, or on the basis of experimental estimates of pore size. We have however thus far considered pores as the only potential modus operandi for proteins to cross membranes individually. We have not questioned the Davson-Danielli concept of the membrane in which proteins are viewed as being excluded from its lipid core, and are thus of course excluded from passing through it. Curiously, this latter idea (proteins cannot pass through the membrane's core) remains widely held today, even though the former (proteins are excluded from the membrane's core) is no longer held and is clearly incorrect. There is little question that many different proteins are embedded in and transect the membrane's lipid bilayer both individually and in groups. The current paradigm for the structure of biological membranes, the fluid mosaic model, proposes that membrane proteins are suspended *in* a liquid crystal lipid bilayer in which there is considerable motion of lipids and proteins both in the plane of the membrane and across it.

If proteins are embedded in the membrane and transect the bilayer, then at least some, and perhaps all, membrane pores might be polar faces within transecting protein molecules or between groups of such membrane proteins. If so, these channels would not necessarily be particular structures designed to function as pores and they would in all likelihood be nonuniform, perhaps of variable size, and would not display regular geometry. This does not mean that special pore structures, such as those traditionally conceptualized, do not exist. That remains uncertain. But rather that there are probably a variety of pore-like structures in membranes. The existence of a diversity of polar channels means that some of the assumptions that have had to be made in order to estimate pore size are probably incorrect. For example, pore size estimates require the assumption that the number of pores per unit area is constant or unchanging and that pores are regular geometric figures, in particular right circular cylinders [pore radius is calculated by assuming that total pore volume, $V = n\,\pi/2\,r^2\,l$, where n equals the number of pores, r is the radius of the individual pore (really the radius of the average pore), and l is the length of the pore]. Both of these assumptions are simplifications without clear experimental support, and indeed if we take our present ideas about the structure of the membrane seriously, then certainly not all, if any, membrane pores would fit this simple construct of uniform, invariant, regular cylindrical proes. Thus, not only do estimates of pore dimension not provide a single fixed value for the radius of such structures, but we can be relatively sure that pore size is nonuniform, even if we can not be certain about the extent and form of this variation.

The idea that the "pore" may really reflect a variety of different structural elements leads us to question the pore concept itself. Is it satisfactory, or perhaps more correctly, necessary to explain the permeability of membranes to water-soluble substances? This has been a central point of controversy. It has been suggested that the membrane should be viewed, both on the basis of its structure and permeability characteristics, as a polymeric lattice and not a bilayer punctuated by special porous channels. It should be remembered that the concept of the pore was introduced to explain the anomalously high permeability of water to what was thought to be a lipid layer. That is, the idea of pores was proposed to explain why water permeability was greater than predicted for its diffusion through a lipid solution. However, we now know that even pure lipid bilayer vesicles show such "anomalous" permeability to water, and they contain no pores in the usual sense. Therefore, we must ask whether or not the passage of water-soluble substances across membranes is due to their movement through fixed polar channels at all, uniform or nonuniform, or whether a more general construct, in which permeability is viewed as a statistical event for the membrane as well as the permeable solute, and pores are merely a particular spacing and order of membrane lipids and proteins, offers a more predictive model? That is, does permeability reflect the local kinetic and thermodynamic instability of a particular geometric organization of membrane structure? The lipid portion of the membrane is after all not a bulk lipid solution, even if we choose to use that simplification in our attempts to estimate permeability, but an extremely thin liquid-crystal approximately two molecules thick.

How does the application of a lattice or polymeric model affect the idea of macromolecular permeability? Can we predict protein permeation within the context of a polymeric structure? A statistically-based lattice model proposes that permeability should be a probabilistic function of the size of the permeant and no specific discontinuous limit can be set. Therefore protein permeability would be expected, although its magnitude might be trivial. In a sense the question remains the same however we wish to think of the organization of the membrane. Is there a statistical relationship between permeability and molecular dimensions in which permeability decreases as size increases, but whose limit (zero permeability) is not attained in practice? If this is so, is the magnitude of this permeability biologically meaningful? Alternatively, is there a clear absolute or finite limit to the size of molecules that can cross these structures?

4.8 THE STRUCTURE OF PROTEINS AND THEIR PERMEABILITY TO BIOLOGICAL MEMBRANES

We have been discussing protein permeability in the simplest possible terms, considering the proteins as uniform, roughly spherical objects of varying radii, and asking "if it could be that the membrane is a sieve of some sort or another that proteins can pass through?" But proteins cannot realistically be considered as a single entity that varies only in size. Proteins are of course an extremely, indeed the most, diverse class of biological molecules in terms of physical and chemical characteristics, as well as size and shape. Thus, the concept that the permeability of a membrane to a molecule must be established for the particular molecule of interest, as well as for the particular membrane, holds no less for proteins than for other groups of compounds.

Variations in protein structure are crucial to the expression of their unique differentiated functions. For example, the serine proteases, a series of enzymes apparently related by common ancestry that are often of roughly the same size, shape, net charge, and contain large segments of homologous amino acid sequences, nevertheless display very different enzymatic specificities. Knowing that "small" differences in structure can account for very different and distinct enzymatic behavior, can we properly assume that analogous and equally specific differences do not exist between protein molecules that are related instead to their ability to cross particular membranes? For example, if the rate of loss from a cell of a particular cytosolic protein is very low, might this low rate of loss be due to a particular structural adaptation that helps retard loss, rather than to an absolute impermeability to such molecules? Such characteristics are important determinants of permeability at the glomerular membrane of the nephron. Perhaps the particular molecule is especially large, charged in a certain fashion, tends to form large polymers, is oddly shaped, increases the viscosity of its immediate environment, and so forth. Conversely, might not proteins that leave the cell, or enter membranes, have properties that make these events more likely?

Thus, particular *physical* characteristics may have a role in determining whether or not a protein is more or less able to cross a membrane. But this is only half of the story. The transported molecule or the membrane itself may be *chemically* modified to effect transport. Indeed, as it has become clear that proteins do cross membranes individually, chemical transformation has been the most seriously considered mechanism. I believe that this is because it permits one to maintain the notion that membranes are impermeable to proteins as a general rule, and yet explain the existence of such permeability as special instances. Proposals for chemically mediated transport generally fall into two major categories: (1) the splitting of covalent bonds in the peptide chain of the protein being transported itself in order to produce movement and (2) the separation of associated peptide chains into subunits or the formation of peptide chain complexes to produce movement. Although there is evidence for a variety of different mechanisms, I will not develop a discussion of these processes here, with one important exception, the signal hypothesis of Blobel and Dobberstein (See Chapter 7 for references.) This transport model has particular significance because it attempts to provide a single coherent mechanism to explain all of protein transport within the context of the cisternal packaging–exocytosis paradigm.

4.9 THE MINIMALIST POSITION

The vesicle model was built in great part on the assumption that some "external" process, in particular a vesicular transport process, is required to account for the known movement of proteins across membranes, because membranes are impermeable to water-soluble molecules of this size. However, proteins were found to be synthesized in association with ribosomes that were either suspended in the cytoplasmic matrix or were attached to the cytosolfacing surface of the membranes of the RER (Figures 2 and 4). In either case, wouldn't new protein be released from the ribosomal complex into the cytoplasm of the cell? And if so, how could one explain the filling of the zymogen granule without proposing that individual protein molecules cross the membrane of these structures? There was no morphologic evidence of a vesicle uptake system from the cytoplasm at this location. Therefore, contrary to its own assumption that membranes were not permeable to these proteins, the vesicle model seemed to require just such a permeability.

As I have already described, the solution to this paradox was sought by proposing another fate for newly synthesized proteins. It was proposed that secretory proteins were only manufactured on ribosomes attached to the membranes of the RER, and that this permitted the transfer of new protein directly into the cisternal spaces of the endoplasmic reticulum wherein they would be held (sequestered) until they were moved by vesicular processes to other, distal compartments. In this way new protein would never enter the cytosol. This of course did not really deal with the paradox. It merely proposed that individual proteins crossed a different membrane.

Thus, not only is the a priori assumption that membranes are impermeable to proteins questionable in terms of our knowledge, or lack thereof, of the structure of biological membranes, but even if we make this assumption and attempt to construct a vesicular model for transport, we find ourselves required to propose that individual protein molecules must cross at least one membrane in order to gain exit from the cell. Thus, we have the curious situation that a model built on the concept that membranes are impermeable to proteins requires such a permeability (I consider this thought further in Chapter 7).

It is from this perspective that minimalist models for protein transport have been developed. The original hypothesis, the signal hypothesis of Blobel and Dobbestein, as well as latter-day derivatives of this hypothesis, argue that even though secretory proteins cross the membranes of the endoplasmic reticulum individually, that it is a special, indeed a unique, case and we can still maintain the notion that membranes are impermeable to formed proteins, and hence that vesicular mechanisms are still required for their transport across other membranes.

In its original form the signal hypothesis restricted the passage of secretory proteins across membranes to the following conditions: (1) passage normally only occurs across the membranes of the RER; (2) this transport occurs only once in the life history of a particular protein; (3) it occurs only for the elongating peptide chain prior to the development of secondary and tertiary chain geometry; (4) it occurs in one direction only from the cytosolic to cisternal face of the reticular membrane; (5) it requires the formation of special pores in the membrane—temporary or fugacious pores; (6) it proposes that a hydrophobic peptide sequence at the amino end of the elongating peptide chain is necessary for its passage; and (7) it requires that this amino acid sequence be split from the rest of the peptide chain once it enters the cisternal spaces of the endoplasmic reticulum. This original list must be extended by the addition of many modifications and exceptions as we consider further details of the model and the now extensive literature on the subject.

I will discuss the signal hypothesis at much greater length in Chapter 7. It is hard to imagine a more restrictive (or more complex) view of protein transport. This series of hypotheses permits one to accept the idea that individual proteins cross this membrane, and yet retain the notion that membranes in general are nevertheless impermeable to them. It is interesting that implicit to the idea of splitting the aminoterminal sequence is the thought that if the proteins were not modified in this way, they might still be able to cross membranes. Thus, in this sense the system is viewed as actually being adapted to protect itself against protein permeability of a more general type.

Since its original proposal in 1974, the signal hypothesis has generated a great deal of research, which is the sign of a successful hypothesis. However, it seems clear from the present vantage point that this attempt to squeeze protein transport into a rather tight corset has failed. Each new study brings new exceptions to the original hypothesis, and new complexity to its latter-day variants, so that it may be fair to say that exceptions to its rules appear as numerous as examples of them. These studies have however made it clear that individual proteins, secretory and nonsecretory, cross and enter numerous membranes in a rather substantial variety of ways. There have been other attempts to formulate general models for protein transport, some of which appear superior to the original signal hypothesis because of their more general nature (accounting for a greater variety of the observed cases). Nevertheless, these ideas still focus on the view that we must seek *the* special mechanism that accounts for such processes, in much the same way as the signal hypothesis. The idea that a unique, special process accounts for all protein transport, by reassuring us of its special nature, helps bolster the belief that we need not consider protein transport in more general phenomenological terms in the same fashion as we do other (nonmacro) molecules.

In seeking common elements in the different types of protein transport processes that have been described, we should consider the possibility that as diverse and complex a group of molecules as proteins might display a variety of transport modes, from the simplest diffusive mechanism to complex chemical processes, as has been found for the transport of much simpler small organic molecules. It certainly would not be surprising if the diversity of protein structure and function, a fundamental characteristic of living systems, were reflected in the transport behavior of this group of molecules.

4.10 SUMMARY ON IMPENETRABILITY

Thus, we find the original notion that the membrane is impermeable to proteins as a class not only an unjustified a priori assumption, but that even if we make this assumption, we are in any event, forced to propose protein permeability, if only as exceptions to a general rule of impermeability.

Because the current paradigm for secretion is constructed on this a priori assumption, we must often ascribe a particular meaning to a given observation when viewing the system through the paradigm's eye even though there are other potential explanations that we might otherwise consider. For example, when granules "disappear" during active secretion, this can only be attributed to their loss from the cell in toto, since their contents, in particular their protein contents, cannot leave the granule across its membrane. Observations that granule size decreases during active secretion, made for the first time 100 years ago, must either be in error or must be unrelated to their disappearance. Based on the belief that vesicular mechanisms are necessary to account for protein transport, a complex series of hypotheses, that in turn have required numerous subsidiary and ad hoc hypotheses, such as the signal hypothesis, were put forth with some confidence that such mechanisms or ones quite similar were required by what seemed to be the reasonable assumption of protein impermeability. And it was in great part for this reason, that it was thought unnecessary to learn about the behavior of the natural continuum "secretion" independent of a particular secretion model, and pointless to put the hypothesis that the relevant membranes might be permeable to proteins to experimental test. The assumption of impenetrability seemed good and adequate. Among other things, the fact that this model was constructed on the assumption that another hypothesis, the Davson-Danielli model of the biological membrane, was correct, was insufficiently appreciated.

4.11 THE SECOND ASSUMPTION

There was another reason to think that newly formed digestive enzyme would be separated from the cytoplasm of the cell at all times. This was to protect the cell from the potentially damaging effect of certain of these proteins. It was known that many digestive enzymes have the potential for destructive hydrolytic activity, and indeed in certain pathologic conditions, the relatively inactive precursor forms of proteolytic enzymes, such as trypsinogen and chymotrypsinogen, undergo substantial autocatalytic activation that presumably leads to cell death and organ disease. Trypsinogen activation appears to be the main danger since the active form of this enzyme, trypsin, initiates the activation cascade for other proteolytic enzymes, although the cell does contain a cytoplasmic trypsin inhibitor protein that binds and inhibits the active enzyme and that presumably helps protect against this possibility.

Since so much of the cell's contents are membrane-bound compartments, it was thought that the cell might be protected from these dangers by the separation of its digestive enzymes from the cytoplasm within membranes to which they were thought impermeable. And thus it was assumed that "sequestration" was necessary to protect the cell from its own potentially dangerous products.

However, this assumption is merely that, even though it has often been stated as the reason why there must be an absolute separation of secretory product from the cytoplasm. It is a teleological assumption in that it not only assumes that the compartments in which the proteins are sequestered are safe from destruction, but assumes that their purpose is to provide this protection. Potentially dangerous enzymes (proteolytic enzymes, phospholipases, etc.) are present however in virtually all cells; cells that function normally in their presence. Are they all afforded protection by membrane partitioning? Protection can be, and probably is, produced in numerous ways, including phase separation, the presence of inhibitors in the cytoplasm, the rigors of substrate specificity, the accessibility of substrate, the reactive environment (pH differences, and so forth), as well as the ability of the cell to repair whatever damage may be unavoidable or inevitable. The idea of obligatory sequestration for secreted proteins comes primarily from work on the pancreas, where the question of "protection" is at least one to think about. However, the intracellular transport of many, if not all, secreted proteins, is viewed similarly, even though many of these proteins pose no obvious threat to the cell. Thus, the teleological assumption of cytoplasmic exclusion, derived from the study of digestive enzyme secretion, has been applied to other protein-secreting cells, even in the absence of a teleological basis for making this assumption.

In the end, the idea that membranes act as a protective barrier cannot be assumed. It is a hypothesis that proposes a particular means whereby the cell might protect itself in a potentially damaging situation. Of course, as a hypothesis it requires evidence in its support.

4.12 A FINAL COMMENT ON ASSUMPTIONS

The purpose of this chapter is to convince you that an a priori assumption of either protein impenetrability or a protective role for cellular membrane-bound compartments is not warranted, and that such assumptions should not form the foundation upon which we attempt to learn about how proteins are secreted or cross membranes in general. Whether or not a particular protein can cross a particular membrane molecule by molecule, or whether secretory proteins are kept in special membrane-bound compartments to protect the cell from potential damage are matters to be dealt with by experimental test.

SELECTED REFERENCES

1. Singer, S. J. and G. L. Nicolson. The fluid mosaic model of the structure of cell membranes. *Science* **175**:720–731, 1972.
2. Stein, W. D. *The Movement of Molecules Across Cell Membranes*, Academic Press, New York, 1967.
3. Wallach, D. F. H. and A. S. Gordon. Lipid protein interactions in cellular membranes. *Fed. Proc.* **27**:1263–1268, 1968.

CHAPTER 5

Exocytosis

As I have already noted, there are thought to be natural processes by which invaginations in the cell membrane carry molecules into cells within vesicles formed from the invaginating membrane and molecules are also thought to be moved out of cells, or secreted, in an analogous fashion, usually envisioned today in the form of the exocytosis model. Both of these concepts are examples of a wide range of processes, hypotheses, and constructions that we call "vesicular transport," and that have in common the following notions: (1) membrane-bound vesicles are either formed as the result of transport, as in invagination processes, or preparatory to its occurrence, as in the formation of secretion granules prior to exocytosis; (2) transport is not driven by the kinetic energy of the transported molecules, but by an external force; (3) transported molecules do not cross the membrane by passing through its substance or through channels in the membrane; (4) transported molecules do not move individually, but en masse or in bulk; (5) the membrane barrier must be removed for transport to occur; (6) this involves the macroscopic separation of adjacent portions of membrane from each other; and (7) this schism in the membrane occurs either during the insertion of additional membrane, or the deletion of membrane segments.

5.1 AN EXISTENTIAL HYPOTHESIS

What kinds of evidence might be of value in helping us evaluate such an hypothesis? In the most general sense, we can state that all specific or particular vesicular hypotheses, as well as vesicular hypotheses in general, are strictly existential, as opposed to universal, hypotheses. An existential hypothesis is one that cannot be proven false in any general sense by any single case for which we reject the hypothesis. For example, if we are unable to demonstrate the presence of a particular vesicular mechanism in one cell or another, who can say that it will not be found elsewhere? In the case of a universal hypothesis, a single negative case requires rejection of the hypothesis for all possible cases. For example, we only need one exception to the law of conservation of energy and mass to require its rejection. Since existential hypotheses cannot be rejected in general from a particular false experience, likewise they cannot be proven true in general from a specific example or case.

With this idea in mind we can make the following statements about vesicular models:

1. Proof for the existence of one type of vesicular mechanism does not bear upon the existence of any other type of vesicular process. This is a matter for independent proof. I would add here, only because I have heard it expressed, that agreement that there is the possibility that a process might exist, based on the existence of another similar or analogous process, has no bearing on its actual existence. It only means that we should not reject the idea out of hand.

2. Even if a particular vesicular process can be shown to account for the transport of a molecule or group of molecules in one cell type, nevertheless, the existence of the same process in another cell in pursuit of the transport of different substances must be independently proven.

3. Proof that a process occurs under a particular set of experimental circumstances does not necessarily mean that it occurs as a natural process. "Real" processes can be produced by the investigator's experimental system or design, and therefore the mere existence of an event does not indicate that it occurs naturally or accounts for transport under normal circumstances. There must be specific evidence to support this contention.

4. Even if we can show that a particular vesicular process occurs naturally and transports a molecule of interest, we cannot presume that it

exists to perform the physiological function that we envision for it without evidence that our hypothesis is true in this regard as well.

5. Even if we can demonstrate that a particular vesicular mechanism occurs naturally and transports the particular substance of interest, we cannot assume that it accounts for its movement quantitatively without specific evidence concerning the quantitative aspects of its transport.

5.2 THE FALSIFICATION OF HYPOTHESES

Although the existence of some vesicular processes seems clear, for example, the phagocytosis of food by protozoans, evidence for many is quite fragile and is at times no more than an assertion by an investigator that it occurs. The evidence for most such hypotheses is *not* the *direct* observation of its occurrence in real space-time, as in phagocytosis where one can observe a particle being engulfed by a living cell in the light microscope. Usually it is indirect; observations that are consistent with such a mechanism, but that do not rule out alternative explanations for what is seen, nor that demonstrate its existence directly by measurement or observation of the natural continuum. The importance of this latter point will become clearer as we consider the evidence for exocytosis below.

For the moment, let us consider a point of scientific methodology that concerns the difference between "seeking evidence in support of a hypothesis," and "testing a hypothesis." Frequently, investigators "seek support" for a particular hypothesis or point of view, but do not actually "test" it. That is, one possible outcome of an experiment is not that the hypothesis will be rejected. At best "support" will not be forthcoming. We say that if our hypothesis is correct, then we would expect to observe this or that. However, the statement that the absence of the particular occurrence provides negative evidence for our hypothesis cannot be made. When this approach is used, all hypotheses are in a sense correct to one degree or another, and differ only in their relative degree of corroboration; that is, how much support exists for the particular hypothesis. Its weakness can perhaps best be appreciated by considering the real possibility that two totally antithetical hypotheses have been proposed to explain a particular phenomenon and that they are equally supported by different corroborative observations. Applying this reasoning, we would conclude that both hypotheses (a hypothesis and its logical antithesis) are simultaneously true, or that "truth" itself fluctuates between one reality and another depending upon a continuously changing balance of evidence. Thus, in order to determine satisfactorily to what degree a hypothesis reflects the real world, we should do more than seek support for it. We should attempt to test it, against other hypotheses, even antithetical ones, as well as in its own terms, as either being true or false. It is more powerful to be able to say that our hypothesis has been substantiated or corroborated by numerous tests of its validity, *its falsification being rejected*, than merely asserting that there is accumulated evidence that supports or is consistent with our view. We must be willing and able to prove our hypothesis false in order to prove it true: a more difficult task than it might seem at first glance.

We know intuitively that some experimental tests of hypotheses are more powerful than others. Such differences derive from the ability of a particular experiment to really test our hypothesis; that is, from the degree to which it is capable of proving our hypothesis false. Thus, an experiment that has a high falsifiability quotient, if you will, is a stronger test and hence a better experiment than one that has little potential for proving our hypothesis false. In addition to tests, hypotheses themselves can be more or less powerful depending in part upon how amenable they are to being proven false.

Some hypotheses are so vague or otherwise flawed that they are virtually impossible to prove false. Of course, such a hypothesis cannot likewise be proven true, even though one might amass considerable evidence in its support. Thus, to the degree that falsifying tests are not carried out, or are not possible or practicable, we cannot show that our hypothesis satisfactorily accounts for the observed phenomenon regardless of how much evidence we might marshall in its behalf. Only when a hypothesis is adequately falsifiable, and when we have performed numerous tests that could have, but did not, lead to its falsification, do we have the right to accept it as a satisfactory approximation for the moment.

This issue of the "strength" of tests and hypotheses should be borne in mind as we move deeper into particular experimental issues and ideas, and it is in this context that I would like to discuss the evidence for the exocytosis model. First I will consider the evidence for this hypothesis in general terms, and then I will consider the evidence, such as it is, in our particular system of interest, the exocrine pancreas. In addition to the issue of falsification, we should also keep in mind the existential nature of the exocytosis hypothesis, that is, the need to prove its existence system by system, and cell by cell. We cannot generalize from one cell type to another, nor accumulate evidence for the process in general by adding that obtained in one system to that obtained in another.

5.3 EVIDENCE OF FORM

Although evidence for exocytosis varies in regard to both type and degree of support from cell type to cell type, the preponderance is of one particular kind that we can call "evidence of form"; that is, observations of particular static shapes in cells that are viewed as being consistent with the occurrence of exocytosis in real space-time.

In recent years there have been some attempts to prove the existence of such processes in cell-free systems by demonstrating that fusion of granule and plasma membrane can occur under certain conditions. I will not consider this issue here, except to point out that such experiments may demonstrate that fusion can occur between two membranes, but that the relationship between such an observation and the existence of exocytosis in an intact cell is tenuous.* Perhaps we can agree that it is possible under a variety of conditions for biological membranes to fuse with each other. The issue is not this capability, although there are questions concerning the potential for its occurrence in certain specific situations *in situ,* but its relationship to the existence of a natural process called exocytosis. Perhaps someday conclusive evidence for exocytosis will come from such a purely "reductionistic" approach, whereby one attempts to prove the existence of exocytosis by elucidating its underlying detailed mechanism, but this has yet to come to pass, and for the reasons I discussed in Chapter 2, I wonder if it ever can.† This having been said, I will now consider several examples of evidence of form.

5.4 THE PRESENCE OF SECRETION GRANULES IN CELLS

Secretion granules are found in many, but not all, cells that are known to secrete proteins and other organic compounds such as neurotransmitters. It has been established that many of these granules contain the product or

*Often these studies even fail to prove conclusively that fusion of membrane has actually occurred. Sometimes intermembrane binding and actual fusion are not distinguished or distinguishable. In some studies, the movement of material from one group of vesicles to another is assumed to indicate fusion, when movement independent of such an event has not been ruled out.

†Such an approach "stacks" hypothetical constructions—a construction for a particular mechanism of exocytosis layered on the exocytosis construct itself.

products that are secreted, although it has not been established that all of their contents are secreted in any case to my knowledge.*

The mere existence of granules in cells can be taken as evidence of exocytosis. The reasoning is that "granules are involved in exocytosis where it is thought to occur, and therefore their presence, particularly if they contain a product that is secreted, is *prima facia* evidence for exocytosis in a cell." Of course, the presence of granules suggests the possibility that exocytosis might occur, but it provides no *evidence,* of even a weak variety, that it does, even in the most granule-laden cell.

If we were to argue that the presence of granules in cells is prima facia evidence for an exocytosis process, then wouldn't their absence be evidence that the hypothesis is false? Many cells that are known to secrete large quantities of proteins, such as the parenchymal cells of the liver, which contain few if any true secretory granules, are still proposed to carry out transport by an exocytosis mechanism similar to, if not identical with, that proposed for cells with granules. Their absence is merely explained by ad hoc hypotheses; either there is little storage of product within the cell and hence there are very few granules, or the rate of granule turnover is so rapid that few granules are required to account for the observed secretory rate, or both, and so forth.

Of course, the presence of granules in cells is evidence for exocytosis in the same way that the presence of the sun in the sky or a rock on the ground is evidence that it moves. This is not to say that granules are not involved in such processes in some cells or under some circumstances, nor that their presence in cells does not suggest the possibility or hypothesis that such a process might exist there, but their presence is simply not evidence for one type of transport process or another. The presence of granules in cells is evidence for and proof of the presence of granules in cells, and nothing more.

5.5 THE DISAPPEARANCE OF GRANULES

The number of granules in micrographs of secretory cells may decrease when the secretion of a product contained within them is augmented, although such a decrease is not always seen. This disappearance, as I have already mentioned, is often taken as evidence for exocytosis. "How else could they disappear?" it would be asked.

*To the contrary, internal structural elements have been described for many secretion granules.

The disappearance of granules during augmented secretion was first shown for the acinar cell of the pancreas by Rudolf Heidenhain in the mid-19th century. Heidenhain found that under certain conditions of augmented secretion the number of granules in the cell, as viewed through the light microscope, appeared to be diminished. Of course, this observation in itself did not necessarily mean that granules left the cell in toto. Heidenhain was aware of this because both he and his contemporaries, Kühne and Lea, had also observed that the size of granules in the tissue decreased simultaneously. These size changes suggested an alternative to the release of granules in toto; that is, a decrease in size to their actual, or merely apparent, disappearance within the cell.

I say apparent disappearance, because if granule size is decreased, even if their number remains unchanged, micrographs may well display fewer granules. For example, the zymogen granules of the pancreas average a little less than 1 μm in diameter in the fasted state, which is close to the theoretical resolution of light microscopy (the wavelength of visible light being $\sim$0.5 μm), and if granule size decreases by half when secretion is augmented,* then the resolution of the light microscope starts to become limiting; and small, although still present granules, may not be seen in the microscope and may appear to have disappeared.† The same effect occurs at the level of electron microscopy, although for different reasons. In this case it is not that resolution is limiting, but that the probability of sampling a particular granule in any given tissue section decreases as size decreases [number/unit volume = number/unit area $\times$ (mean diameter)$^{-1}$; DeHoff and Rhines, 1968]. Ermak and I have found that large apparent changes in granule number may "disappear" when the probability of sampling granules of various sizes is taken into account (Table 1).

Therefore, the apparent disappearance of granules can be explained in ways that include but are not limited to their loss from the cell *in toto.* And for this reason, we cannot argue that such "loss" or "disappearance" is evidence in favor of a particular mechanism of secretion, any more than we can argue that the presence of granules in cells is evidence for such a process. As with their presence, the disappearance of granules can only be proof of their disappearance.

*After feeding fasted rats, average granule diameter was found to be 0.65 μm and granules as small as 0.1 μm in diameter are frequently seen in electron micrographs of actively secreting cells (Ermak and Rothman, 1981).

† In addition, during active secretion, granules are sometimes clustered and occupy a smaller percentage of the cell volume giving the appearance of a reduction in number. However, this conclusion may be erroneous, particularly if the size of individual granules is also decreased.

TABLE 1. The Effect of Granule Size on Estimates of Granule Number[a]

Time Period[a] (Days Postcoitum)	Mean Granule Diameter (μm)	Number of Granules/Volume (Arbitrary Units)	
		Uncorrected	Corrected for Size
17	0.55	1	1
20	1.22	3	1.36

[a]Modified from study of Ermak and Rothman (1983) on prenatal development of the zymogen granule in the rat pancreatic acinar cell.

We can consider the negative case here as well. If secretion can be greatly augmented without a decrease in real granule number, does this observation falsify the exocytosis hypothesis? That is, do we conclude that since secretion occurs in the absence of a change in granule number, that the exocytosis hypothesis is false? Proponents of the hypothesis would be quick to point out that constancy of number might merely indicate that the rate of formation of new granules is able to keep pace with their rate of loss from the cell. A decrease in granule number is proof for exocytosis in the same way that the absence of a change in number is proof that it does not occur.

5.6 THE EXOCYTOSIS OR OMEGA FIGURE

The most frequently cited evidence for the existence of exocytosis is the exocytosis or omega figure. It is an outline of the surface membrane of the cell that has a form that is consistent with the prior occurrence of exocytosis; that is, it could be a geometric sequelae to exocytotic fusion. These forms are called omega figures because they are shaped like the Greek letter omega (Ω), the circular portion being thought to represent the membrane of the secretion granule fused to the cell membrane and the arms of the figure the cell membrane proper. Omega figures are seen from time to time, more frequently in some secretory cells than others, and are sometimes very evocative of the idea that an exocytosis-like event has occurred, although in many cases they are not very convincing. However, as with the presence of secretion granules in cells or their disappearance from them, regardless of the visual impact that a particular form may have, the mere presence of the omega figure does not indicate that exocytosis has taken place. Although it is true that if exocytosis had occurred such a geometric figure might well be seen in the outline of the cell for a period of time, and hence its existence is consistent with such a process, the prior occurrence of exocytosis is only one possible explanation for its presence. Often the outlines of omega figures cannot be or are not distinguished from chance or other geometric variations in the outline of the cell surface; or they may be real enough, but the result of an artifactual, rather than a natural, fusion of membranes, for example, produced by the fixation of tissue or other steps in the preparation of tissue samples for viewing. Thus, it is an hypothesis that the omega figure is related to the prior occurrence of a natural process of exocytosis. As with the other evidence of form, omega figures can only be proof for the existence of omega figures.

The weakness of the omega figure as evidence for a dynamic process is compounded by the fact that they are usually sought by investigators; that is, they tend to be infrequent occurrences that are discovered as the result of a search. If they are found, then the exocytosis hypothesis is thought to be supported by the observation, but if they are not, the search continues by increasing the size of the sample surveyed or by looking for better circumstances and conditions. Thus, the negative observation (the absence of omega figures) does not lead to the rejection of the hypothesis, but to a continued and perhaps expanded search to validate it. In the end we can only view the omega figure as significant evidence for exocytosis if we *assume* that exocytosis is the reason for its presence, and of course to assume that is to assume the existence of exocytosis in the absence of evidence.

If we wish to show that the omega figure is more than a chance, artifactual, or other occurrence either unrelated or only indirectly related to natural secretion, then we must provide a substantive connection between such figures and the secretion of product. First and foremost, if the omega figure is related to secretion, then its frequency should be related to secretory rate. If the secretory rate increases so should the frequency of omega figures. Indeed, the frequency of omega figures would be expected to increase proportionately. Although this correlation would certainly help place the observation of the omega figure on a firmer scientific footing, we would still have to demonstrate in addition that the observed frequency of such an event accounts quantitatively for the amount of material that is secreted; that is, the fusion of so many granules leads to the release of so much product. Unfortunately, even if we were able to show this, we still could not, despite the correlation and obvious intuitive connection, say that the omega figure proves the existence of exocytosis because we still would lack direct evidence that such a process occurs in real space-time and causes these forms.

Perhaps we can appreciate the weakness of such evidence better by considering the negative case once again. When an increase in the frequency of the omega figure is *not* observed when the rate of secretion is increased, as seems to be the case for the acinar cell of the pancreas and many other cells in which exocytosis is thought to occur, it is sometimes argued that exocytosis is too rapid for us to "catch." But of course, if there are "real" omega figures in the unstimulated state, then we should see proportionate changes with increased rates of secretion however fast the process may be. Thus, if the omega figure is in fact a sequel to secretion by exocytosis, then shouldn't an increased rate of secretion necessarily lead to an increase in their observed frequency? If this is not seen, then would we not be required to reject the hypothesis? There are numerous instances where on this basis, rejection would appear to be required by the available evidence. However, it is not generally forthcoming.

The reason for this is that even in the absence of such a correlation, we cannot necessarily reject the exocytosis hypothesis, because exocytosis might either occur without a discernable omega figure, or the duration of its appearance might be reduced when the rate of secretion is increased and so forth. Thus, these figures, even when they increase in frequency during augmented secretion, are proof for exocytosis in the same way that their absence, or the absence of changes in their frequency, is proof that such an event does not occur.

5.7 THE SURFACE AREA OF THE CELL MEMBRANE

Another deductive hypothesis derived from the exocytosis construct proposes that the surface area of the plasma membrane should increase if the rate of exocytosis is increased, because additional membrane is being inserted into the plasma membrane of the cell. Such increases have been observed, although this is not by any means a universal occurrence in cells thought to secrete in this fashion. At first glance, this type of evidence may seem to make a "dynamical" connection that much of the evidence discussed above lacks. However, as with the other evidence of form, an increase in surface area of course does not necessarily indicate the prior occurrence of exocytosis. There may be other explanations, most obviously, an increase in the volume, and hence the surface area, of an actively secreting cell. The case for a dynamic connection is more convincing if large disproportionate increases in surface area occur, particularly if localized to a particular cell surface, as appears to occur when HCl secretion from the oxyntic cell of the stomach is augmented with stimulants. Thus, if we can show a change in surface area due to alterations in the rate of secretion we still need to be able to ascribe it to our proposed mechanism, as opposed to other causes, either unrelated, indirectly related, or related to secretion in another way. Also, as with the other examples that we have discussed, when increases in surface area are not observed, the hypothesis need not be rejected. Many cells that are thought to secrete by exocytosis do not show this effect, and our hypothesis can be saved by the ad hoc hypothesis that the rate of membrane reuptake by the cell is equal to the rate of its insertion, and hence a steady-state pertains. Of course, this kind of reasoning allows us to argue that a decrease in the surface area of a cell coincident with an increased secretory rate is also evidence for exocytosis; an increase in the rate of exocytosis leading to an even greater increase in the rate of membrane uptake into the cell.

5.8 PARTICLES AND PITS

Patterns of particles, which are probably membrane-associated proteins, are sometimes observed in freeze-fracture images of membranes taken from cells thought to secrete by exocytosis and have been proposed to be evidence for its occurrence. The particles (proteins) would be directly involved in or otherwise related to fusion between granule and cell membrane. This proposal is analogous to the other evidence of form that we have discussed. In itself it is merely an hypothesis for the function of the particles and is not evidence for exocytosis. This confusion between hypothesis and evidence is further compounded when investigators themselves order independently obtained micrographs that contain different patterns of particles in an attempt to convey the impression of a sequence of events occurring in real space-time, as in the formation and subsequent dissolution of rosette or linear patterns of particles. Of course, this is merely a visual display of the investigator's hypothesis since he or she has no idea of how the different forms are really related to each other, if indeed they are related at all; no less if they are related to secretion and in the way proposed.

If fusion of granule and cell membrane takes place and produces a hole in both membranes, as the exocytosis hypothesis proposes, then these holes should appear as pits in the membrane or in half membranes viewed in freeze-fracture replicas. This observation is quite similar to the omega figure and the same types of issues apply. Their presence may be due to events other than the proposed secretory mechanism, and their absence can be explained by ad hoc assumptions concerning the rate of the secretory process. And their discovery, like that of the omega figure, is usually the result of a search for an infrequent form.

5.9 GENERAL DISCUSSION OF EXOCYTOSIS—SUMMARY

The weakness of evidence of form and the ambiguity and confusion that it produces between what is evidence and what is hypothesis too often goes unappreciated. I hope the examples that I have given are sufficient to show both the kind of "evidence" that has been mustered in support of the exocytosis hypothesis and the deep problems with it.

One cannot emphasize enough the need to correlate such cytologic observations with the presumed functions in a rigorous and quantitative manner. However, the central weakness of essentially all of the evidence that has been brought to bear on the exocytosis hypothesis rests inescapably on the fact that the secretory process itself is never measured or followed in real space-time independent of the hypothetical construct of exocytosis. It is for this reason that it is so easy to propose ad hoc hypotheses to account for negative results, and it is also for this reason that evidence and hypothesis are so easily confused with each other. Perhaps this can be highlighted by considering an experiment that attempts to study the release of acetylcholine at the cholinergic synapse. It is widely, although not universally, believed that the exocytosis of acetylcholine-containing vesicles occurs at the synapse and accounts for the transmission of the impulse. However, this process, as all other proposed "exocytoses," has never actually been observed in living cells. In the past, even evidence of form in support of the exocytosis hypothesis at the cholinergic synapse was rather weak. Synaptic transmission occurs very rapidly and perhaps, it was thought, the exocytosis event was too rapid to be seen by the usual approaches and special measures would have to be taken to "see the process" as it occurred. In the particular experiment at hand, the investigators proposed to stop exocytosis in midstream by rapidly freezing the tissue and then looking for evidence of its occurrence, most importantly, pits in freeze-etched membranes. They were able to find them under certain circumstances (that is, in the presence of an exogenous chemical that "helps" increase their apparent frequency) and from these observations concluded that they had provided direct, even kinetic, evidence for exocytosis in this system. But what kind of evidence was it? What was stopped in midstream? It was not exocytosis at all, but rather the "hypothesis" of exocytosis. That is, a known process was not stopped; indeed, the investigators sought proof of its existence. Thus, a hole in a membrane cannot be proof that exocytosis occurs as a natural process, although it is certainly consistent with it. The hole may be due to other unrelated or artifactual causes that we have no way of knowing.

As discussed earlier, the problem is simply that we cannot ask whether or not the *secretion process* is behaving as our mechanistic hypothesis predicts, either kinetically or thermodynamically, because we are not actually following that process. Even if we confirm a particular prediction of our hypothesis, we cannot say with certainty that the observation is related to the transport process that we think we are investigating, because its behavior has only been characterized in terms of the hypothesis itself. The question that we must ask then is what, if anything, our observations tell about the secretory process independent of the truth or falsehood of our hypothesis. That is do our observations have an existence as descriptions of the events of secretion independent of our hypothesis? For evidence of form in general, the answer is negative and observations are only descriptive of secretory events *if* the hypothesis is correct.

5.10 EXOCYTOSIS IN THE ACINAR CELL OF THE EXOCRINE PANCREAS

However weak this evidence may seem, it should be noted that it is the unusual case (if one exists) in which all types of evidence of form can be shown for a single proposed exocytic process. For many years the pancreas was cited as the system in which the evidence for exocytosis was most prepossessing, and indeed it has been used to argue for the existence of such processes in other cells where the evidence was not thought to be as secure.

20 years ago when I first became interested in the subject of protein transport, the idea that exocytosis accounted for the secretion of the digestive enzymes by the pancreas was already well established, and

textbooks and monographs assured us, as they still do, that this was the way in which things happened. I should perhaps point out that the process was called "reverse pinocytosis" at that time; but no matter, the proposed mechanism was essentially the same. However, a search of the primary literature for evidence left one feeling that something was missing. The evidence appeared to be threefold: (1) the cell contained secretion granules, zymogen granules, which were known to contain the enzymes that were secreted; (2) these granules seemed to disappear from the cell during augmented secretion under certain circumstances; and (3) the omega figure had been reported.

The following is a brief dialogue that occurred between Prof. Neaverson-Hokin and myself after I had presented a paper on the subject of protein transport by the pancreas at a symposium on exocrine glands in 1968. I think that it sums up the situation quite well. Prof. Neaverson-Hokin asked, "If (I had) really examined evidence that reverse pinocytosis occurs at all in the pancreas?" I responded weakly, "The evidence for reverse pinocytosis is based on two sorts of things (being terribly precise). One is electron micrographic pictures (of omega figures)* which are . . ." She interrupted me, "Is it pictures or picture?" I responded, "Pictures—there are several." But her point was made. There may have been more than one picture, but not much more, and the published evidence was certainly minimal. Proponents of the exocytosis theory argued that such figures were seen frequently, but this was merely hearsay, and the frequency of their occurrence was really unknown and apparently unstudied.

In the years since this exchange, the cisternal packaging– exocytosis paradigm has become even more generally accepted as *the* mechanism for protein secretion by the pancreas, and exocytosis is believed by most scientists in this and related areas to account for the final step in the transport process. Yet the evidence for its existence has remained essentially unchanged. Several new omega figures have been published since 1968, but little else has been forthcoming. There does not appear to be an increase in the occurrence of such forms in actively secreting cells. There does not appear to be a change in the surface area of the cell membrane or its apical surface. Freeze-fracture evidence of pits or particles has yet to be published to my knowledge. The fact remains that evidence for the existence of exocytosis in the pancreas, no less evidence that such a mechanism accounts quantitatively for the natural process of secretion during the digestion of a meal, is weak at best.

*The other was a picture suggestive of granule to granule fusion within the acinar cell.

5.11 SUMMARY

The evidence that can be brought to bear in support of the exocytosis model from a visual perspective is wholly comprised of static images of the sort that I have discussed that are consistent with the existence of such a dynamic process. Some of these observations are strengthened by functional correlations of one sort or another, although most are not. One cannot prove that they are not related to such events, but neither can the hypothesis be affirmed. If visual evidence is what we seek, then clear proof for exocytosis will have to await our ability to look at living cells at the resolution of the electron microscope. In any event, even accepting the existence of exocytosis and a role for it in the normal process of secretion, that it solely accounts for secretion is another matter altogether, needing of evidence to establish that particular fact. At the moment, certainly in pancreas, one must say that only the most fragile evidence exists to support the notion that exocytosis occurs at all, however reasonable such an event may seem, no less that it accounts wholly for the secretion of organic products by cells.

SELECTED REFERENCES

1. DeHoff, R. T. and F. N. Rhines. *Quantitative Microscopy,* McGraw Hill, New York, 1968.
2. Ermak, T. H. and S. S. Rothman. Zymogen granules of pancreas decrease in size in response to feeding. *Cell Tiss. Res.* **214**:51–66, 1981.
3. Ermak, T. and S. S. Rothman. Increase in zymogen granule volume accounts for increase in volume density during prenatal development of pancreas. *Anat. Rec.* **207**:487–507, 1983.
4. Heidenhain, R. Beiträge zur Kenntnis des Pancreas. *Pflügers Arch.* **10**:557–632, 1875.
5. Heuser, J. E., T. S. Reese, M. K. Dennis, Y. Jan, and L. Evans. Synaptic vesicle exocytosis captured by quick-freezing and correlated with transmitter release. *J. Cell Biol.* **81**:275–300, 1979.
6. Rothman, S. S. and J. J. L. Ho, eds. *Nonvesicular transport,* John Wiley and Sons, New York, 1984.
7. Satir, B. The final steps in secretion. *Sci. Am.* **233**:28–37, 1975.

CHAPTER 6

Filling the Secretion Granule

Knowledge that secreted proteins are synthesized in association with ribosomes found predominantly in the basal region of the acinar cell, are then stored in substantial quantities in apically placed zymogen granules, and are finally released across the cell's apical membrane into the duct lumen led to the conclusion that new digestive enzyme moves as a wave from its synthetic site of origin at the cell's basal pole to its apex. In this sense, the product can be said to follow a "vectorial" path. Although this anatomical polarization is not always as striking as usually envisioned, it is nevertheless true that the great preponderance of ribosomes are found basally and the majority of secretion granules are seen at the apical pole, and of course the product eventually ends up, at least in substantial part, in the most apical site of all, the duct lumen. Thus, we can say with some assurance that as new protein travels through and out of the cell, we would initially expect to find it predominantly at its basal end, and over time at more apically placed sites, to and through the region of the zymogen granules, on its way out of the cell. Although, on the average, such movement of new protein appears to occur, "how" is another matter, and research efforts for over a quarter of a century have been directed at trying to find out in just what way or ways this transfer takes place. This, as I think you will agree as you read on, is still a matter for considerable speculation.

6.1 THE FILLING OF THE GRANULE

For the purposes of our discussion here I will assume, as the paradigm does, that all new protein enters the cisternae of the rough endoplasmic reticulum (RER) and that all cellular membranes are impermeable to these proteins once they have been so "sequestered." How would zymogen granules fill from this source? How would new digestive enzyme get from the internal spaces of the RER to the zymogen granules if the membranes of the RER are impermeable to them?

It is widely, although not universally, believed that the major subcellular membrane-bound compartments of the acinar cell are discontinuous. That is, the interior spaces of the RER are a separate compartment from the Golgi cisternae, which in turn are separate from the interior of secretion granules or condensing vacuoles, which are separate from each other as well. If this is the case, a special modus operandi would have to exist to move the newly sequestered protein from the compartment in which it finds itself (the RER) to its eventual storage repository (the zymogen granule). As I have already pointed out, the cisternal packaging-exocytosis model proposes that "vesicular" processes account for this movement. Indeed, how else could movement occur? Vesicles that contain digestive enzymes are thought to be formed from the membranes of one compartment and their contents then transferred to another as a result of the fusion of the membrane of the vesicle and the membrane of the recipient compartment. In this way the movement of protein from the cisternal spaces of the RER to the zymogen granule would be achieved, and the absolute barrier that the membrane poses to the passage of protein broached.

What specific events have been proposed to account for this transfer? Unfortunately, the model, even now, some 25 years after it was first put forth, is not clear in this regard. This of course makes description difficult and experimental test regarding ambiguous points wholly problematic. However, here is what I believe to be the most commonly described scenario.

6.2 THE FORMATION OF SMOOTH-SURFACED VESICLES FROM THE RER

The removal of secretory protein from the cisternal spaces of the RER is thought to occur in small vesicles, often called smooth-surfaced vesicles, of about 400 Å internal diameter, formed, by budding, from the membranes of the RER. Budding is thought to take place at the end (apical end only?) of individual RER stacks at areas of the membrane that are denuded of ribosomes. The only evidence of which I am aware for such an event is the occasional appearance of narrowings at the poles of individual membrane stacks in electron micrographs that could reflect the beginnings of such a process. Such visual images are, of course, not very compelling evidence, as we have already discussed, and might well reflect other things, including static narrowings that have no functional significance whatsoever.

Even if we allow for the existence of budding, the proposal that protein is transported in this way still needs to be proven. That is, proof of budding does not prove that secretory protein is moved in this fashion, no less that such a process accounts for its movement quantitatively. In any event, there is no proof that such a process occurs beyond this evidence of form (narrowings), and the hypothesis that transfer occurs as the result of the budding of ends of RER stacks remains an essentially unsubstantiated hypothesis. There is no proof (1) of a dynamical kind that budding takes place; (2) that the small smooth-surfaced vesicles seen in these cells are derived from the membranes of the RER; or (3) that these vesicles in fact contain digestive enzymes.

6.3 COMPLEXITY AND THE BUDDING OF THE RER

As I have already discussed, the complexity of an hypothesis has a great bearing on its adequacy as a theoretical framework upon which to develop an understanding of a system. While a system may seem complex to us as we approach it, we must nevertheless try to avoid complexity, to the degree that we are able, in the development of our hypotheses. As we have discussed, this among other reasons, is because a complex hypothesis, as evidenced by the number of subsidiary hypotheses that must be proposed, may be difficult, if not impossible, to prove false.

With this thought in mind, the budding hypothesis can be our point of departure for thinking about the considerable complexity of the cisternal packaging-exocytosis model. This single step or subsidiary hypothesis of the larger model requires the following subsidiary hypotheses of its own:

1. The existence of a specific force to produce the budding of the membranes of the RER.
2. A means of restricting expression of that force to certain locations on the membrane [only at the (apical) poles of RER stacks].
3. A means of excluding ribosome binding sites from budding sites.
4. The replacement of membrane at a rate equal to the rate of budding so that at the steady-state the amount of RER membrane is constant.
5. And, of course, to produce the above effects, the existence of parts of RER membrane specifically (chemically) modified to carry them out.

There are other subsidiary hypotheses that we could add, such as a mechanism for selecting those intracisternal proteins destined for secretion granules from others that are not. This list is not given to be complete, but to point out that what may seem a simple proposal at first glance, the budding of membranes, is really quite a complex idea, at least in this particular setting. The level of complexity becomes even more profound when we consider further details, such as what additional processes and mechanisms are required to account for the replacement of membrane at a rate equal to the rate of budding (I will touch on this below). The important thing to keep in mind here is that these complex ideas are not being considered because our knowledge of the kinetics and dynamics of the process requires such complexity. That would be justifiable. But rather, because our *assumption* that membranes are impermeable to proteins requires it.

6.4 SOME CALCULATIONS

To some the absence of evidence for budding may represent something of a challenge. That is, they would seek the missing evidence. And if they were successful in finding something in the parts of the cell or its behavior that is consistent with this hypothesis, might conclude that they have provided evidence in support of the idea. And such evidence it might well be, but without knowledge of how it fits with an actual kinetic and thermodynamic description of an intact, known process of budding, it might be fragile evidence indeed. In this light and before moving on to the next step in the model, we should consider the complexity of the budding hypothesis from a dynamic point of view; that is, what would be required of such an event if it accounted for the transfer of enzyme, based on what we know of secretion.

Many of the calculations that follow merely reflect the geometric fact that as a sphere becomes smaller the ratio of surface area to volume becomes increasingly large, and hence the smaller the vesicle, the smaller the mass carried per unit diameter or per unit membrane surface area. As I have said, the vesicles that are thought to be formed from the RER are of about 400 Å internal diameter, and therefore have an internal volume of some 3×10^{-17} ml. How many molecules of secretory protein would each such vesicle carry? The proteins in the zymogen granules are predominately digestive enzymes, and their concentration there has been estimated for the rat and rabbit to be between 4–9 *mM* for an average mixture (Ho and Rothman, 1983). The concentration of secretory protein in the cisternal spaces of the RER can be estimated by comparing the specific activity of digestive enzymes in the RER (in microsomal fractions of tissue homogenates, in point of fact) to that in the zymogen granule. For example, microsomal specific activity is about one-fifth of that in the zymogen granule in the rat (Rothman, 1970), and therefore, the concentration of these proteins in the RER can be estimated at about one-fifth of that in the zymogen granules or about 0.8–1.6 *mM*. Much of the enzyme in microsomes appears to be adsorbed to the membrane or associated with the ribosomes and thus a better, but still conservative, estimate would be about half this value (see Chapter 7). By this calculation, if the concentration is the same in the small vesicle as in the cisternae, then each vesicle would contain $1.3–2.6 \times 10^{-23}$ moles or approximately 8–16 protein molecules per vesicle on the average. For ease of calculation, let us assume that there are 10 secretory protein molecules within each formed small vesicle.

At a steady-state in which protein synthesis and secretion are equal, the rate of vesicle formation (that is, the number of vesicles × the content of each vesicle) would equal the rate of protein secretion. If this balance or steady-state can be maintained at moderate rates of protein secretion, let us say 25% of a maximal rate of about 10–20 mg/hr/g of tissue (based on observed rates in rat and rabbit), then about $3–7 \times 10^{-7}$ moles or $1.9–4.4 \times 10^{17}$ molecules of protein (using an estimated average molecular weight for the secreted digestive enzymes of 30,000 daltons) would have to be packaged in small vesicles each hour per gram of tissue. From this we can calculate the number of vesicles containing 10 protein molecules apiece that would have to be formed per unit time. For each cell (with a volume averaging about 1000 μm^3 or 1×10^{-9} ml), 1.9–4.4 $\times 10^8$ molecules would be synthesized and secreted per hour on the average. At 10 molecules per vesicle, 20–40 million vesicles would have to be formed per cell per hour. This would require an amount of membrane equal to 170–330 times the surface area of the cell's plasma membrane (using cell surface area = 600 μm^2 and small vesicle surface area = 5×10^{-3} μm^2). However one might quibble with the particular numbers that I have chosen to make this estimate, we must propose a remarkably active process of membrane turnover or recycling, to my knowledge uniquely so among cells.

We can also ask a related visual question. If we assume that the narrowings observed in RER sections reflect an event premonitory to budding, then how frequently should we see such forms given a rate of 20–40 million vesicles being formed per hour per cell. Converting to seconds, this would be approximately 5,000–10,000 vesicles/cell/sec, or an equal number of narrowings in every cell each *second.* Does the observed frequency of narrowings agree with this calculation? If we estimate that substantial amounts of RER would be seen in 75% of 0.08-μm-thick electron microscopic sections through a 10-μm-thick cell, or in 94 sections through each cell, then we would expect to see about 1/100th of the total number of narrowings in each RER-containing cell section or about 50–100 narrowings or forming vesicles in each and every section if such an event was of the order of one second in duration. This is clearly not observed.

One might argue, as has been regarding exocytosis, that the process occurs too rapidly for us to be able to observe more than a small fraction of the total number of events. For example, for a millisecond event, a very fast biological occurrence of about 10–20% of the duration of a fast action potential, assuming a proportional reduction from a 1-sec event, only 5–10 narrowings would be seen in each cell. Even this frequency may be greater than what is actually observed, although it has not been quantified.* Of course, we could posit an even faster process if necessary. With constructed

*The number of forms, such as narrowings, that are seen of those that occur is really dependent upon the relationship between the rate of the presumptive process (for example, budding) and the time required to "fix" the cell in a particular state. If fixation occurs more rapidly than the event, then they should all be seen.

hypotheses, such as the budding hypothesis, a low frequency of a form can be taken as evidence of the rapidity of the process, without any real knowledge of its existence, no less its rate.

6.5 THE MOVEMENT OF SMOOTH-SURFACED VESICLES

Despite these problems, let us assume that budding occurs and accounts for the transfer of protein out of the RER quantitatively. Next we would ask, how do these vesicles move? The model proposes that they "migrate" (like swallows to Capistrano?) to the Golgi region of the cell where they empty their contents into the lumena of waiting condensing vacuoles and perhaps into Golgi cisternae as well (N.B.: Certain sugars are thought to be added to sugar-containing secretory proteins in the Golgi complex. If this occurs within the Golgi cisterns, then we must either propose vesicle budding at this locus or that condensing vacuoles are formed from these structures).

If the evidence for budding is weak, then it is nonexistent for the movement of vesicles. There is no experimental evidence that the small vesicles seen in the acinar cell even exhibit Brownian motion, although one assumes that they do, no less directed movement, since they have only been seen statically under the electron microscope. Indeed, with some exceptions, there is little evidence for the directed movement of granules or vesicles thought to be involved in secretion processes in general. Sometimes differences observed in their location in the cell from time to time or situation to situation are offered as proof of movement. For example, when there are fewer zymogen granules in the acinar cell, they tend to be apically placed, and the inference may be drawn that they have moved to the apical surface from other locations. However, from electron microscopic images of fixed or otherwise nonliving tissue, movement cannot be observed, only imagined. The fact that the secretory product moves is also sometimes viewed as being tantamount to proof that vesicles, thought to contain that product, also move. Of course, even assuming that one can demonstrate that the latter is contained in the former, the movement of the vesicle and its contents are separate matters, unless one can show that the contents cannot leave the vesicle and move independently (see Rothman, 1975, for a discussion of this question). And of course we should not assume this without explicitly testing the proposition.

None of this is to say that vesicles do not move, but rather that direct evidence for movement has been lacking. Moreover, even if we could demonstrate movement, it alone would certainly not validate a vesicular transport hypothesis. We would also have to demonstrate that vesicle movement is directed and occurs at a rate that is quantitatively appropriate for the process. It is a particularly telling fact I believe that there is no convincing evidence that vesicles move in situ, no less in a directed and quantitatively appropriate manner, when such movement is central to the definition of vesicular transport hypotheses—that is, the directed movement of vesicles acts to transport product from place to place. Given this fact, I find the spate of studies describing the role of intracellular tubules and filaments in such movement premature. Perhaps the hope is that once the details of the mechanism become clear, so will the existence of the process it carries out. But of course one would then have to ask, "What process?"

6.6 CONVERGING ON THE CONDENSING VACUOLE

Let us put aside the problems with a vesicular mechanism up to this point and assume that the smooth-surfaced vesicles do contain digestive enzymes and somehow reach the condensing vacuoles at a satisfactory rate, fuse with them, and release their contents within the vacuolar lumen. Now new problems arise. We must assume that this fusion is a specific event, that is that specific membrane structures exist to promote fusion between these particular membranes, as well as prohibit fusion between the vesicles and undesirable cell components; for example, small vesicles should not fuse with each other, nor condensing vacuoles with each other, nor either with fully formed zymogen granules, nor should any of these fuse with mitochondria, at least not to an appreciable degree. Compared with some of the complexity that we have already encountered, this may seem to be a rather insubstantial problem. Lipid bilayers certainly seem capable of fusing with each other, and a couple of proteins might well take care of the issue of specificity.

Nevertheless, however simple or complex one may think these ideas, there is no significant body of evidence, either anatomic or dynamic, describing such events. The fusion of smooth-surfaced vesicles to condensing vacuoles is, at the least, an uncommon image in electron micrographs. We can of course again argue that the events occur too rapidly to be seen at a substantial frequency.

6.7 THE DYNAMICS OF FUSION BETWEEN SMALL AND LARGE VESICLES

Now let us take a more detailed look at the idea that small vesicles fuse with the condensing vacuole. What would we expect to see if such an event occurred? We can use our fact about secretion, a known rate of protein secretion, for help. We have already calculated that about 5,000–10,000 vesicles should be formed per cell per second in order to account for observed rates of secretion and thus an equal number of fusion events should occur at the condensing vacuole each second. Thus, vesicles formed from the extensive membranes of the RER, occupying some 20% of cell volume, now converge on a minor cell pool, the condensing vacuole which occupies about 2% of the volume of the average cell. From stereological measurements we can estimate that the total volume of the condensing vacuole pool is about 15 μm^3. If we consider the average condensing vacuole to be a little less than 1 μm in diameter or about 0.5 μm^3 in volume, then we would have approximately 30 condensing vacuoles per cell. As I have said, small vesicles would fuse with the condensing vacuoles at a rate of about 5,000–10,000/sec, or for a population of 30 condensing vacuoles, about 200–400 fusion events per second per vacuole. And yet such events are rarely, if ever seen. Even if we assume a very fast process, such images should still be seen regularly, and to my knowledge they are not.

Perhaps what would be required can be appreciated in another way. If the completely filled condensing vacuole (now a zymogen granule) contains protein at an average concentration of 5 *mM* and if the average granule volume is about 0.5 μm^3 (for granules of about 1 μm average diameter), then each zymogen granule would contain about 1.5×10^6 protein molecules. If 10 molecules were deposited at the condensing vacuole per small vesicle fusion, then 150,000 such events would be required to fill each granule. The surface area of one small vesicle is about 5×10^{-3} μm^2, and thus, about 750 μm^2 of membrane or an amount of membrane about equal to that enclosing the cell would be required to fill each and every zymogen granule.

6.8 THE CONDENSING VACUOLE CONUNDRUM

But our problems with the condensing vacuole are really only beginning. The distribution of condensing vacuole size is quite similar to that seen for the "mature" zymogen granule population. That is, partially, even virtually empty granules, are seen over a broad range of sizes from those equivalent to the smallest zymogen granule to those as big as the largest. Thus, one often sees large condensing vacuoles that contain very little material, and that would presumably not be enlarged further as digestive enzyme is added. The conclusion that is drawn from this observation is that condensing vacuoles do not grow from small vesicles to larger ones as they fill, but that granules of a range of fixed sizes fill. And thus according to a vesicle model, all 150,000 small vesicles must bud from the condensing vacuole after having deposited their load.* If budding at the RER is a complex process, it is no less so at the condensing vacuole. As with small vesicle fusion, the budding of vesicles from condensing vacuoles does not appear to be a regular feature of electron micrographs of these structures, and there is no other evidence that such events occur. But if we do assume such a process, what does it require? Once again:

1. A force to initiate budding.
2. And specific (chemical) membrane components to carry it out.
3. In addition, a means of leaving the transported content behind, that is, some way of not reincorporating the material just released back into the budding vesicle.
4. A means of incorporating other contents of the condensing vacuole to fill the space.
5. Since there are large differences in the surface area to volume ratio for the two vesicles (about 6:1 for the condensing vacuole and 200:1 for the small vesicles), there would be a "shortfall" in the amount of material transferred as a result of fusion, that is the volume of the larger vesicle would become larger than the space filled by the contents of the smaller vesicle (it would fill 1/30th of the newly formed space), and the opposite situation would apply due to budding, that is, too much membrane would be removed (30 times

*One is also led to this conclusion by the fact that an amount of membrane equal to that enclosing the cell is required to fill each granule.

too much relative to volume). Thus, budding and fusion would either have to be coupled to deal with these discrepancies or fluid would have to enter and exit the condensing vacuole in order to make the necessary volume adjustments.

6. If the condensing vacuoles are of relatively constant size, then to maintain size stationary, the rate of fusion of small vesicles and their subsequent budding would in any event have to be equal, that is, there would have to be a means of regulating the rates of the two processes relative to each other (as well as relative to the rate of budding and fusion at the RER).
7. Once formed such "empty" vesicles would have to migrate in a directed fashion back to the RER and be reincorporated therein, in this way replacing RER membrane lost by budding.
8. There would have to be a means of distinguishing between the transfer of empty vesicles to the RER and the transfer of filled vesicles in the other direction.
9. It would have to be unlikely for empty vesicles to fuse with condensing vacuoles and equally unlikely for filled vesicles to fuse with RER.
10. And finally the "structure" of the two vesicles would have to be different in order for this to occur, and therefore membrane components would have to be exchanged in some fashion at both the condensing vacuole and RER.

Whether or not such things are conceivable, the point is merely that such a proposal is very complex and is virtually unsupported by direct experimental evidence of any kind.

6.9 AN ALTERNATIVE FOR THE FILLING OF CONDENSING VACUOLES

At the beginning of this chapter I noted that the membrane-bound compartments within the secretory cell are thought to be separated from each other, that is, are discontinuous, and that this view, although widely held, was not universal. It has been suggested that the spaces of the RER and the Golgi apparatus are connected and therefore that material can be moved within the cisterns rather than between them. Although I do not believe that the anatomical evidence offers convincing support for this view, such a system would help us divest ourselves of the complex system of small vesicles that I have just described. In a continuous system, protein within the RER would merely diffuse, or otherwise move, within these internal spaces of the cell to the region in which condensing vacuoles are formed.

Although this idea seems far superior to the vesicle shuttle concept because of its relative simplicity, its superiority unfortunately ends at the Golgi. For even if RER cisterns are continuous with the spaces of the Golgi complex, and moreover, even if condensing vacuoles are formed from Golgi membranes, it seems that large, relatively empty condensing vacuoles exist as free-standing entities, not connected to the Golgi membrane or any other cell structures. If this is true, and if we believe that the granule membrane is impermeable to protein, then this barrier must be broached by a discontinuous vesicular process in order for granules to fill, and we are thus back where we started.

6.10 MEMBRANE FLOW

When it was first proposed that new protein is moved from the ribosome to the zymogen granule by vesicles, it was thought that transfer was accomplished as a result of the flow of membrane from compartment to compartment; RER membrane became Golgi membrane became zymogen granule membrane, and so forth. However, the result of over 25 years of research has been essentially negative in this regard, at least for the acinar cell, and the expected mixing of membrane constituents has not been demonstrated. The protein and lipid content of the various cellular membranes thus far seem to be unique and particular (Meldolesi et al., 1971). This apparent failure of a general membrane flow theory is partly responsible for the development of the "shuttle" concept that we have just discussed.

6.11 SUMMARY

There are two types of evidence that can be offered in support of the idea that vesicular events account for protein movement from its site of origin on the ribosome to zymogen granules. The first is the presence of images in cells that are consistent with such processes. As I have pointed out, there is only the most fragile and tentative visual evidence. The second line of evidence equates the intracellular movement of secretory protein and the movement of vesicles, namely, since proteins are moved from ribosomes to zymogen granules, and since all secretory proteins are sequestered and cannot cross membranes, the movement of proteins and vesicles are really one in the same. This is based on the *assumptions* that membranes are impermeable to proteins, and that secretory proteins do not exist in the cytoplasm, and not on direct evidence that proteins are moved in this fashion.

If we accept the vesicle hypothesis for intracellular protein movement, the characteristics of such a system, and what it requires of the cell, kinetically, organizationally, and thermodynamically, is quite substantial, and in some respects seemingly beyond the realm of likelihood. At this time belief in a vesicle model for intracellular transport in pancreas requires that one hold to the existence of a complex series of processes in the absence of direct evidence that they exist, based on the belief that they must.

SELECTED REFERENCES

1. Ho, J. J. L. and S. S. Rothman. Protein concentration in the pancreatic zymogen granule. *Biochem. Biophys. Acta* **755**:457–466, 1983.
2. Jamieson, J. D. The secretory process in the pancreatic exocrine cell: morphologic and biochemical aspects, in, *Secretin, Cholecytokinin, Pancreozymin and Gastrin*, J. E. Jorpes and V. Mutt, Eds. Springer Verlag, Berlin, 1973.
3. Jamieson, J. D. Membranes and secretion, in, *Cell Membranes*, G. Weissmann and R. Claiborne, Eds. HP Publishing Co., Inc., New York, pp. 143–152, 1975.
4. Jamieson, J. D. and G. E. Palade. Production of secretory proteins in animal cells, in, *International Cell Biology*, B. R. Brinkley and K. R. Porter, Eds. Rockefeller Univ. Press, New York, pp. 308–317, 1977.
5. Meldolesi, J., J. D. Jamieson, and G. E. Palade. Composition of cellular membranes in the pancreas of the guinea pig. III. Enzymatic activities. *J. Cell Biol.* **49**:150–158, 1971.
6. Palade, G. E. Functional changes in the structure of cell components, in, *Subcellular particles*, T. Hayashi, Ed. Ronald Press Co., New York, pp. 64–83, 1959.
7. Palade, G. E., P. Siekevitz, and L. G. Caro. Structure, chemistry and function of the pancreatic exocrine cell, in, *Ciba Found. Symp. on the Exocrine Pancreas. Normal and Abnormal Functions*, A.V.S. deReuck and M. P. Cameron, Eds. Churchill Ltd., London, pp. 23–49, 1962.
8. Palade, G. Intracellular aspects of the process of protein synthesis. *Science* **189**:347–358, 1975.
9. Rothman, S. S. Intracellular storage of exportable protein in functionally hypertrophied pancreas. *Am. J. Physiol.* **219**:1652–1657, 1970.
10. Rothman, S. S. Protein transport by the pancreas. Science, **190**:747–763, 1975.

CHAPTER 7

The Signal Hypotheses*

7.1 INTRODUCTION

The raison d'être of the cisternal packaging–exocytosis theory, as well as the assumption underlying the interpretation of many experiments offered as evidence for it, is the belief that biological membranes are impermeable to protein molecules as a class. Thus, it is argued, vesicular processes of one sort or another are necessary to account for the known transport of proteins across membranes. Yet, although vesicular transport models have been constructed and evidence adduced in their behalf mainly on the belief that biomembranes are impermeable to proteins, the central model, the cisternal packaging–exocytosis paradigm for secretion, itself has had to propose paradoxically that individual protein molecules cross membranes. And an even more remarkable paradox has developed. Evidence that newly forming proteins cross the membranes of the endoplasmic reticulum individually is viewed by many as the most convincing of the available evidence in support of the vesicular model for secretion. That is, the transport of individual protein molecules across membranes stands as the

*This chapter is reprinted, with minor modifications, from *Nonvesicular Transport,* John Wiley & Sons, 1984.

most imposing and substantial argument in support of the idea that proteins are moved through and out of the cell by vesicular processes because they cannot cross membranes individually. The reasoning is as follows: If proteins enter the cisternal or interior spaces of the endoplasmic reticulum individually, passing through its membrane to do so, but then having done so, are unable to cross membranes again, and are, to use the common terminology, "sequestered" therein, then how else could they be moved through and out of the cell except by vesicular processes?

This peculiar state of affairs has at its origin the fact that ribosomes, on which proteins are synthesized, are either found free in the cytoplasm of the cell or bound to membranes of the rough-surfaced endoplasmic reticulum (RER) facing the cytoplasm. Because proteins are thought to enter membrane-enclosed spaces, in particular the cisternae of the RER for secretory proteins, after, or actually during, their synthesis, the individual protein molecule must cross an intracellular membrane as it leaves the cytoplasmically oriented ribosome.

Thus, even the paradigm must embrace, however grudgingly, the concept that proteins are capable of crossing membranes individually in some fashion or another. In this chapter, I will discuss the "fashion" that has been proposed to explain this event within the context of the paradigm. In addition, I will take a step back and ask how good is the evidence that proteins are transported across the membranes of the endoplasmic reticulum in the first place.

The hypothesis that has been proposed to account for the movement of protein molecules through the membranes of the endoplasmic reticulum, even though proposing that membranes are permeable to individual proteins, views such transport as being of an extremely limited nature. Indeed, it views it in such limited terms that it permits one, while admitting to the "de facto" occurrence of protein transport through membranes, to continue to assert its "de jure" absence. This is the signal hypothesis (Milstein et al., 1972; Blobel and Dobberstein, 1975a,b). As originally conceived, it proposes that: (1) proteins can only cross one membrane, in the case of secretory proteins, that of the RER; (2) this only occurs once in the life history of each protein molecule; (3) transport only occurs in one direction across the membrane from the cytoplasmic face to a trans-compartment (for example, the cisternae of the RER) and not in the opposite direction; (4) it only occurs for the extended peptide chain, and not after the development of internal secondary and tertiary structure; (5) transport occurs directly across the membrane from the ribosome, a two-compartment system, which excludes the cytoplasm from the transport route; (6) transport only occurs during chain elongation, that is, during protein synthesis or cotranslationally, that is, as the genetic message is being translated; (7) a special aminoterminal sequence of amino acids is required on the extending peptide chain for transport to occur; and (8) this special aminoterminal sequence is cleaved during or immediately pursuant to transport to prevent its recurrence.

This series of hypotheses is called the "signal hypothesis" because the aminoterminal sequence of the transported protein is thought to act as a signal for its transport. The idea appears to be derived from the observation that in a cell-free system that synthesizes proteins, many proteins can at least in part be produced with an additional series of amino acids at the amino terminus. The originators of the signal hypothesis thought that this sequence might provide a special modus operandi for secretory proteins to cross membranes and, as it was later proposed, for the insertion of membrane proteins as well.

The idea that the aminoterminal sequence was cleaved as a result of transport seems to have been derived from the observation that the additional sequence was generally not found on secretory proteins in situ and was in a sense an artifact of the cell-free synthesizing system. Its absence in situ was proposed to be the result of the immediate cleavage of the signal sequence as it crossed the membrane, even as the peptide chain was still being formed on the ribosome. This led to another curious situation. The transport of proteins across membranes was attributed to a sequence of amino acids not generally found on the transported proteins in the cell, but seen only under special in vitro circumstances (this is not always the case; see Ho, 1985).

It is hard to imagine a more limited view of protein transport across membranes than that provided by the signal hypothesis. Transport occurs only once in the life history of a protein, in one direction only, requires a special chemical modification of the protein, occurs only when the peptide chain is elongating and before any internal structure has developed, and a covalent bond is broken to prevent its recurrence.

The signal hypothesis is also an extremely complex hypothesis, and its complexity must be viewed not only in its own right, but as adding to the overall complexity of the whole of the cisternal packaging—exocytosis concept. This may indeed reflect the complexity of the real world, but the scientific method requires that such complex hypotheses only be accepted as being descriptive of real events after the most demanding scrutiny, and most importantly, the exclusion of simpler explanations and interpretations for experimental observations. This, I hope to convince you, has not been the case. But for the moment, how complex is the signal hypothesis? Although its eight postulates listed above, might seem sufficiently complex in themselves, they only reflect the surface complexity. Underlying each postulate are necessarily other assumptions and predictions about the system, many of which remain purely hypothetical.

7.2 THE ATTACHMENT PHASE

Let us begin with the ribosomes themselves. There was an attempt over a period of many years to demonstrate that free ribosomes synthesized intracellular proteins, whereas those bound to the RER were given the task of synthesizing proteins for "export" or secretion. This attempt failed in the sense that no differences between the two groups of ribosomes could be established, and the mRNA for cellular or secretory proteins does not prefer ribosomes of one type or another as far as we can tell (but see Ho, 1985, concerning prokaryotes). Thus, it seems reasonably clear at this point that free ribosomes are capable of synthesizing proteins that are normally secreted by cells, and there is no reason to think that in the normal course of events that a meeting between a mRNA molecule for a secretory protein and a free ribosome would not occur and lead to the formation of a new secretory protein molecule that would be released into the cytoplasm. In the paradigm's terms, such an occurrence would leave the protein to languish in the cytoplasm since it would have to, but be unable to, cross membranes in order to leave. The signal hypothesis deals with this unacceptable state of affairs by the addition of a corollary hypothesis: All protein synthesis begins on free ribosomes whatever the ultimate fate of the protein may be. As the amino terminus is revealed, extending beyond the confines of the ribosome itself, if the polypeptide chain contains the aminoterminal signal sequence, then the ribosome attaches to the membrane of the RER and transport ensues. Indeed, it is this attachment that the amino terminus is supposed to signal.

This subsidiary hypothesis, that synthesis begins on free ribosomes, leads to further predictions and requirements. For example, since the signal hypothesis insists that transport only occurs for extended peptide chains, and since the formation of secondary and tertiary structure occurs spontaneously as the chain elongates, we must propose that a mechanism exists to assure the attachment of the ribosome containing the elongating chain to the membrane prior to the formation of secondary and tertiary structure. If too much time elapses, then higher-order structure would develop, protein transport would be prevented, and the new protein would nevertheless end up in the cytoplasm. If an estimated rate of formation of the peptide bond is milliseconds, then not more than a few seconds overall is likely to be available for attachment before secondary and tertiary structure develops. It is not clear whether or not the distances, number of putative binding sites, and number of ribosomes involved would permit one to account, as one must, for a high percentage of attachments on purely statistical grounds involving only the Brownian motion of the ribosome, or whether a special transport mechanism for ribosomes containing signal sequences would have to be postulated in addition. In lieu of such a modus vivendi, it has been proposed that translation is temporarily halted at the end of the signal sequence to allow the ribosome the time necessary to attach to the membrane. Moreover, the ribosome cannot inadvertently associate with other membranes in the cell, that is, inappropriate membranes, to any appreciable degree, and therefore there must be specific binding sites for the signal sequence on the membranes of the RER that are absent from other membranes. Thus, the signal hypothesis specifies the timing and location of the attachment of ribosomes to membranes.

We can summarize the various subsidiary and deductive hypotheses of the attachment phase of the signal hypothesis as follows: (1) the initiation of the synthesis of all proteins begins on free, not bound, ribosomes; (2) only ribosomes whose nascent polypeptide chains contain the signal sequence of amino acids at the amino terminus can attach to the membrane of the RER; (3) a mechanism exists to assure that these ribosomes attach to the membrane in a timely fashion and that all signal sequence-containing ribosomes do so; (4) the signal sequence is only able to associate effectively with RER membrane, and not other membranes in the cell; (5) there are specific binding sites on RER membranes for the signal sequence; (6) proteins that are not destined for "export" or insertion into membranes do not contain the signal sequence of amino acids at the amino terminal; and (7) such proteins are not synthesized on attached ribosomes.

We should add items 8 and 9 to the list. (8) There probably has to be an additional receptor site on the RER membrane that is not for the signal sequence itself, but for the ribosome proper. The reason for this can be understood if we consider the fate of the signal sequence for a moment. According to the signal hypothesis, although the signal sequence binds to the membrane of the RER that interaction is only transient, and the aminoterminal sequence moves through the membrane and as it does is cleaved by a proteolytic enzyme. Once this transport event begins, the signal sequence can no longer act to maintain the ribosome closely apposed to the membrane, and without another binding site, the ribosome might well diffuse or otherwise move away from the membrane even as protein synthesis continues. Therefore, there must be an additional receptor site on the membrane to hold the ribosome in place and of course a specific binding element on the ribosome that acts as the ligand at this site. However, since every ribosome would have to have the capability of binding to this receptor, whether or not it was synthesizing a protein with a signal sequence at that particular time, or indeed, whether or not it was synthesizing a protein at all, we have to further propose that (9) binding only occurs pursuant to the interaction between the signal sequence and the membrane. That is, in the absence of an association between the signal sequence and the membrane, the ribosome cannot bind to its membrane site; otherwise ribosomal sites on membranes would not be exclusively occupied by ribosomes engaged in synthesizing secretory and membrane proteins, and would include those synthesizing cytoplasmic and other cellular proteins, or those not engaged in protein synthesis at all.

Thus, the attachment of ribosomes to the membranes of the RER, or the "attachment phase" of the signal hypothesis, requires a considerable array of subsidiary hypotheses. What evidence is there for these ideas? We must begin with evidence for the occurrence of such a process in the first place. Unfortunately, it is at this level, a description of the process itself, that the concept lacks satisfactory experimental verification. It is not clear that a ribosome cycle, free to bound to free, occurs during the synthesis of each protein molecule, no less that such a cycle quantitatively accounts for events. It has not been demonstrated that protein synthesis cannot be initiated on attached ribosomes, as well as free ribosomes. Nor is it clear that the timeliness of such a cycle can be ensured. Thus, essential elements of the attachment phase of the hypothesis (certainly in eukaryotes) remain without adequate experimental verification, and the signal sequence is proposed to carry out a function in a process whose existence itself has not been satisfactorily and independently established. Of course, it is necessary first to establish the existence as well as the characteristics of a process before one attempts to determine *how* it occurs. Unfortunately, things have developed the other way round, and the signal sequence was proposed to account for a process whose existence had not been independently established.

Even if we accept the existence of such a cycle and its transport sequelae for the sake of discussion, there is evidence that brings the proposed role of the signal sequence into question. That is, there is falsifying evidence for its role in the signal hypothesis' own terms.

As I have said, the central postulate of the "how" portion of the signal hypothesis is that a leading amino acid sequence exists that is required for the attachment of the ribosome to the membrane of the RER and that, as a result of this attachment, protein transport across the membrane occurs. If there were proteins that did not contain this sequence of amino acids, but that nevertheless were secreted, then we could conclude that such a sequence was not required for secretion, nor for the attachment of the ribosome to the membrane, assuming that this occurs, nor for the transport of the protein across it, assuming that this occurs as well. Thus, if we assume that the process occurs as envisioned and that the general vesicular theory is correct, if such signal-free proteins were found, then the particular explanation that the aminoterminal amino acid sequence is required for ribosomal attachment and protein transport would seem to be incorrect.

Indeed, there is at least one well-established example of such a protein (there are actually numerous examples at this point, see Ho, 1985). Ovalbumin, a secreted protein which synthesized in the usual cell-free system does not contain the additional sequence of amino acids at the amino terminal, and is manufactured as seen, at least in gross structural terms, in situ and in secreted material (Palmiter et al., 1978). That is, ovalbumin does not contain the signal sequence, and yet is of course secreted. There has been an attempt to incorporate this discrepant observation into the signal hypothesis by proposing that ovalbumin does indeed contain a signal sequence, not at the amino terminus, but further along the peptide chain.

This raises the question of just what structural characteristics of a segment of a peptide chain identify it as a signal sequence if it is not its location at the amino terminus? The common structural characteristic shared by most of the numerous aminoterminal segments that have been sequenced thus far is that they contain a relatively high proportion of so-called "hydrophobic" residues, that is, this sequence tends to contain a higher percentage of amino acids with purely aliphatic side-chains than found in the polypeptide chain overall. Although this characteristic is hardly invariant or unambiguous, it does represent the best, perhaps the only, successful structural generalization made to date concerning these peptide sequences.

Returning to ovalbumin then, the internal signal sequence that has been proposed (actually, several have been proposed, see review by Ho, 1985) is a relatively hydrophobic region of amino acids in the middle of the chain; that is, this sequence is proposed to act as the signal for the attachment of the ribosome to the RER and for the subsequent transport of the polypeptide chain across the membrane. Of course, most proteins, and globular proteins in particular, contain such internal hydrophobic regions, which account at least in part for the folding of the protein, with the more hydrophobic portions tending to face inward away from the polar suspending medium. It is not clear how this particular segment in ovalbumin, or any particular signal segment for that matter, differs from any other hydrophobic region on the polypeptide chain, beyond placement at the amino terminus. If this internal region in ovalbumin is a signal, then why wouldn't internal hydrophobic regions of other proteins similarly promote the binding of the ribosome to the membrane and lead to the transport of inappropriate proteins into the cisternae of the endoplasmic reticulum? Perhaps there is something special about this sequence in ovalbumin, either in its primary or higher-order structure, that could account for such special behavior, but this has yet to be satisfactorily demonstrated. Even so, we cannot retain the topological special place of the aminoterminal sequence in the signal hypothesis (although see Ho's review regarding ovalbumin and an amino-terminal methionine).

This is not to say that this sequence is unrelated to the ability of ovalbumin molecules to cross membranes. Of course, the movement of individual protein molecules across membranes is likely to be a function of their particular structural characteristics, and, in this sense, such specific characteristics can be considered signals. However, this is quite a different matter from the signal envisioned by the signal hypothesis that leads to the attachment of the ribosome to the membrane and the formation of a special pore within the membrane to accommodate a passive extending peptide chain.

It should also be noted that secondary and tertiary structure will have developed in the elongating chain by the time that this internal signal is exposed, and therefore, even in the model's own context, transport must be envisioned as occurring after the establishment of internal structure and not solely as extended peptide chains. Thus, the postulate that transport only occurs for extended peptide chains must be incorrect, and there are many contrary examples at this point. Therefore, even in the paradigm's own terms, the transport of proteins containing internal structure must be entertained.

In regard to the absence of the signal sequence on secreted proteins, even for those protein molecules for which signal sequences have been identified, not all, nor even necessarily most, molecules recovered from the cell-free synthesizing system contain the additional series of amino acids. This discrepancy often goes unnoticed, but when it is noticed, it is usually explained by the proposal (ad hoc hypothesis) that the system is contaminated by the enzyme that cleaves the signal sequence normally and that it acts to form the postsignal protein. To evaluate the adequacy of this explanation or other similar problems, it is necessary to recover the putative signal peptide fragments and to demonstrate, not only their presence, but also that they are present in quantities that are equivalent to the number of signal-less proteins recovered. To my knowledge, such a demonstration has yet to be made, and in its absence we must consider the possibility that proteins lacking the signal sequence are synthesized in just that form.

Now let us turn the "how" of it around. Are there proteins that contain or are thought to contain signal sequences that are not secretory or membrane proteins? If so, what is their fate? Shouldn't all ribosomes that contain such signals attach to the membranes of the RER and deposit their contents therein? If this is not the case, then doesn't this mean that the signal hypothesis is incorrect, and that the signal sequence does not lead to the attachment of the ribosome to the RER? Alternatively, if there are nonsecreted, nonmembrane proteins that contain the signal sequence and are sequestered in the paradigm's terms, then the idea that the transport of proteins across the membranes of the RER is a means of separating or segregating secretory proteins from intracellular proteins must be incorrect. Examples of both situations have been reported.

Signal sequence-containing mitochondrial proteins are released as completed proteins from the ribosome into the cytoplasm of the cell, and then cross the outer mitochondrial membrane to enter the mitochondrion (Schatz and Mason, 1974; Kreil, 1981). Thus, in this case, the signal sequence does not lead to the attachment of the ribosome to the membranes of the RER. There is no obvious structural reason why the sequence of amino acids on these mitochondrial proteins should be ineffective in promoting the attachment of ribosomes to RER membrane and the transport of the peptide chain across it, whereas those of other proteins, seemingly as similar to the mitochondrial sequences as they are dissimilar from each other, should be effective. This raises the question of whether or not ribosomal attachment has anything to do with the signal sequence. Of course, as with ovalbumin, it is possible that there is some special structural characteristic of these signal sequences that has yet to be discovered and that will explain why attachment does not occur. But until such a discovery is made, we cannot assume that it is there to be discovered, and that the signal hypothesis is saved and the mitochondrial story is merely one additional exception to a general rule.

Also, as with ovalbumin, these proteins cross membranes (mitochondrial) with the protein's tertiary structure already having been established. Thus, once again, even accepting the signal hypothesis' explanation for the behavior of the signal sequence, it is clear that proteins in their full globular form cross membranes.

There are also examples of proteins that are neither secreted nor membrane proteins, but that nevertheless are thought to be sequestered; that is, they contain the signal sequence that leads to the attachment of the ribosome to the RER and their transport into its cisternal space. One example is the lysosomal enzymes; that is, enzymes that function within the membrane-bound organelle, the lysosome (Sabatini et al., 1982). The sequestration of such proteins tells us that the purpose of the transport of proteins across the membranes of the RER into its internal spaces (assuming that this is what occurs) cannot be to segregate secretory protein from cellular proteins, because this is simply not the case.

Moreover, assuming that these proteins are sequestered within the RER, that is, are transported into the cisternae of the RER without the ability of leaving across its membrane in the absence of vesicular processes, then specific vesicular processes for these particular proteins must be proposed in addition to those that lead to the en masse secretion of protein. A vesicle-sorting system similar to the small vesicle system that has been proposed for the filling of the condensing vacuole, has been proposed for "filling" the lysosome with its enzymes. Such a system requires that special vesicles be formed from the RER that would exclude nonlysosomal proteins. The formed vesicle would, of course, have to find its way to the lysosome, deposit its contents therein, and return home to the RER. In the absence of convincing evidence that such a system exists and accounts for the transport of these proteins, they cannot be assumed to be sequestered; that is, either they do not enter the cisternal spaces of the RER, or if they do, they are able to leave across its membrane without the interposition of such a complex vesicular system.

Therefore, in terms of the "how" of the matter, some proteins are secreted that do not contain signal sequences, and yet others that are not secreted contain such sequences. Among the latter group are those that are released from the ribosome into the cytoplasm to cross intracellular membranes after they are fully formed, whereas others appear to be "sequestered" in the cisternae of the RER and sorted out from secretory proteins later. The signal hypothesis, as it was originally envisioned, appears to have many exceptions to its rules. And most significantly, there is no convincing independent evidence for the existence of a ribosome attachment cycle of the sort that has been proposed, or if such a cycle exists, that the signal sequence is necessarily the responsible party. We cannot say with any certainty that ribosomes only bind to the membranes of the RER if they contain a signal peptide, nor do we know that "attached" ribosomes are necessarily engaged in protein synthesis at all times.

7.3 THE TRANSPORT PHASE

The signal sequence has been proposed to signal two events. The first is, as we have discussed, the attachment of the ribosome to the membrane. The second is the formation of a pore within the membrane of the RER through which the nascent protein travels to enter the cisternal spaces. This is the transport phase of the signal hypothesis. As was the case for ribosomal attachment, there are two questions to be asked. Do these proteins indeed cross the membranes of the endoplasmic reticulum? That is, does a transport process exist? And, if it does and they do cross the membrane, then by what means? The transport phase of the signal hypothesis is a proposal for how transport occurs, based on the assumption that it does. There seems to be evidence for the existence of the process that is being proposed and proteins may well be transported across the membranes of the RER, although, as I will discuss below, this evidence is uncertain in important respects. For the moment, we will assume that transport occurs and consider briefly the nature of the mechanism that the signal hypothesis proposes.

We can describe the subsidiary and deductive hypotheses of the transport phase of the signal hypothesis as follows: (1) the attachment of the signal sequence to a receptor leads, either by itself or in concert with other molecules, to a molecular reorganization of the membrane which produces a pore that traverses its width; (2) it is the hydrophobic nature of the signal sequence that is instrumental in forming or helping form the pore; (3) this having occurred, the signal sequence is released from its binding site and passes through the pore; (4) the pore's patency is maintained after the signal sequence has passed through, that is, its presence is not required for its maintenance; (5) the pore disassembles or otherwise closes after the carboxyl terminal of the peptide chain passes into the cisternal spaces; and (6) this leads to a decreased affinity of the ribosome for its membrane site and it is released into the cytoplasm.

In 1968 (Rothman, 1969), I suggested such a mechanism in the following terms: "Such pores could . . . be fugacious in nature opening up to let the enzyme in and then closing; . . . A fugacious pore hypothesis is more appealing (than a permanent pore) and might be achieved by the action of a third molecule or group of molecules acting on or near the ribosomal attachment to the membrane in a way that would change the membrane structure to permit the entry of the particular enzyme, but not 'other' protein molecules dissolved in the cytoplasm. After the entry of the enzyme, the third molecule, perhaps a type of permease, would dissociate from the membrane and the membrane would again be 'sealed'. If such a system were operative then only ribosomes making enzyme for export could have this 'third party', otherwise intracellular proteins would also enter the cisternae."

Thus, the signal hypothesis is a specific proposal for just such a system of fugacious pores. What the details of such a system might be is fertile soil for the imagination. How would the pore be formed? How would receptor binding be time-limited? What kind of pore would be able to permit the free transfer of the whole range of some 20 amino acids in almost any conceivable order simultaneously, accepting the polar and the nonpolar, the aliphatic and aromatic, binding neither, and yet being of such minimum width as to keep the extending peptide chain on the straight and narrow, that is, preventing "kinking" in the chain and the formation of higher-order structure within the pore? Despite these interesting questions, when all is said and done, there is not much to say about this phase of the signal hypothesis beyond describing what it proposes. It is not a hypothesis that is supported by direct experimental evidence; other than evidence for the existence of the transport process itself and the existence of the amino-terminal sequence. That is, it is not known that pores exist in these membranes, no less fugacious pores with characteristics of the type proposed by the signal hypothesis. There is no direct evidence for the particular transport mechanism proposed by the signal hypothesis as opposed to other particular potential mechanisms and explanations for protein transport across this membrane that might be considered, assuming that it does occur. Unfortunately, there is nothing more that can be said about the transport phase of the signal hypothesis as I write this. As far as I am concerned, it presently has no substantive experimental basis.

7.4 THE CLEAVAGE PHASE

This brings us to the third, final, and perhaps most problematic phase of the signal hypothesis: that of the cleavage of the signal sequence. This refers to the removal or "clipping" of the aminoterminal peptide sequence from the protein as it passes out through the cisternal face of the membrane. Cleavage is usually envisioned as occurring pari passu with passage, that is, cotranslationally as the particular peptide bond passes the active site of a specific membrane enzyme whose purpose is to cleave the bond. In this way, each and every molecule passing through the membrane would have its signal sequence "clipped" and this would explain why such sequences are not found on secretory proteins in whole cells, in secreted material, or even in cell fractions of the RER. If cleavage took place subsequent to the transfer of the protein into the cisternal spaces, rather than as it exited the pore, then at any given time it might contain some proteins with the additional sequence (but both ideas have been proposed; see Ho, 1985). Moreover, if the former were the case, then the hydrophobic amino terminus would have to be prevented from being buried within the various protein molecules, as might be expected as higher-order structure develops, or otherwise made inaccessible to the hydrolase.

In any event, wherever or however cleavage might occur, the first question is of course, does it occur? I have heard it argued that since the peptide segment is not seen in situ, then it must be cleaved. The inference being that it if had not been, it would be seen. Of course, this is merely circular reasoning. Simply because the added sequence can be demonstrated in cell-free systems, does not mean that this same translation takes place in situ. This may or may not be the case, and it requires independent evidence. The crucial evidence that must be forthcoming is for the existence of the other end product of the putative reaction, namely, the small signal-containing peptide, in situ. For some proteins, there is qualitative proof that small peptides can be recovered from a cell-free system under the right circumstances. It has not however been demonstrated that the presence of the short-chain protein (signal sequence-free protein) and the amino-terminal peptide are quantitatively related, that is, that a stoichiometric relationship exists between the two—mole for mole, in the cell-free system, no less in situ. Of course, such a relationship is to be expected if cleavage accounts for the presence of both products (the small peptide and the signal sequence-free protein). This must be demonstrated in vivo. Although the larger signal-containing proteins are usually absent in eukaryotes, there is no evidence that I am aware of for the presence of the signal sequence-containing small peptide. It is central to the validation of the signal hypothesis, not only that the small peptide remnants of the proposed cleavage be demonstrated in vivo, but that their fate be determined quantitatively. In the absence of this evidence, we cannot conclude that the larger protein is synthesized in situ, no less exclusively; nor that it is cleaved to form the observed smaller molecule, no less that the signal sequence accounts for the transfer of protein into the cisternal spaces of the RER, assuming that that occurs.

But even though we may not yet be able to say that these short aminoterminal peptides are produced in situ, or in what quantity, we nevertheless can consider the cleavage hypothesis from the point of view of the chemistry of the signal sequence itself. As I have said, at this time there is a substantial catalog of sequences, as well as presumed points of cleavage, and therefore comparisons can be made between them. Unfortunately, there does not appear to be a unique point of hydrolysis common to all of the different proteins. A variety of different bonds seem to be involved and one cannot, as of yet, generalize satisfactorily about the putative cleavage site. Cleavage appears to occur at different points in the chain in different molecules and involves different cleavage bonds. If this is correct, then either there are a variety of different cleavage hydrolases for the different bonds (even in the same cell), or there is a rather general hydrolase that can cleave them all, and both have been suggested.

Let us attempt to put such a system into place in the membrane. Since we cannot know in advance which particular protein will pass through which formed pore, each pore must have residing at its cisternal face all of the hydrolases necessary for the range of proteins that might pass through this membrane in a particular cell. If we envision the simpler idea of a rather general hydrolase capable of breaking a variety of peptide bonds, then this problem would be simplified, but we would have to propose some fashion in which a relatively nonspecific enzyme would be prevented from splitting bonds that are not at the signal sequence cleavage point willy-nilly. Although this problem might appear less severe if a variety of specific hydrolases were available, it would really remain the same: merely many enzymes rather than one being responsible. That is, our system must be designed so that hydrolysis only occurs at one site, and not at other potential loci along not only a single peptide chain but also along all other peptide chains that are able to pass through such a pore. In the absence of such restrictions, we would have a series of uncontrolled chaotic reactions producing a wide range of different peptides of different length, which apparently does not happen. One could propose that the signal sequence "signals" or denotes the appropriate site for cleavage. Perhaps other regions on the peptide chain could be made inaccessible by protein folding, if the enzyme acts lumenally rather than in the pore. But can we account for the fact that not all signal sequences are split? In any event, a very restrictive, perhaps multienzymic, system is required that would allow for the cleavage of a range of peptide bonds at the amino terminus, while preventing the cleavage of similar bonds elsewhere on the same and other proteins. These enzymes cannot merely be capable of cleaving certain kinds of bonds, but must cleave only one particular bond of a particular type. This is certainly possible and remarkably specific enzymes do exist; however, it is uncertain that they do in this case.

As I have said, the proposed cleavage must be shown to occur in situ, and be stoichiometrically established by the quantitative recovery of end product under these circumstances. In the absence of such evidence, the signal hypothesis cannot explain the absence of the signal sequence on secretory proteins in cells and in secretion, and it must be considered that its presence in cell-free systems may at least in these cases be an interesting artifact that may have some significance concerning the translation of genetic information, but that does not occur normally in situ. It cannot be assumed as it seems to have been, that the absence of the longer, signal-containing protein in situ is proof of the occurrence of cleavage in situ.

7.5 ESCAPE INTO THE CYTOPLASM

As I mentioned at the outset of this chapter, the signal hypothesis is based on the twin assumptions that transport into the cisternal spaces occurs, and that the transferred proteins are unable to leave these spaces without the interposition of a vesicle system to accomplish the task. There is also a third assumption that underlies the signal hypothesis—each and every newly manufactured secretory protein is released from the ribosome into the cisternal spaces of the RER. That is, none end up in the cytoplasm. This assumption is reflected in the signal hypothesis' concepts that transfer occurs by means of a two-compartment system, ribosome to cisternal spaces, excluding the cytoplasm from the transport route, and that the attachment of the ribosome to the membrane occurs in a timely fashion to prevent the inadvertent release of a new protein with a signal peptide into the cytoplasm.

Let us consider the basis of this assumption. Unfortunately, it is not direct experimental evidence that demonstrates that newly manufactured secretory protein is not released from the ribosome into the cytoplasm. Rather, the view that this does not occur appears to be based primarily on the belief that a cytoplasmic pool of secretory protein does not exist, and hence that release could certainly not occur into a nonexistent pool, as well as equating events such as proteolytic processing, glycosylation, protection from proteolysis, and so forth, with the cotranslational transport of the protein through the reticular membrane (see below for further discussion of these issues).

I believe that there is substantial evidence for a cytoplasmic pool of digestive enzyme in the pancreas, and this subject has been considered elsewhere in this volume (see Chapter 12). However, whatever one may think of the evidence for a cytoplasmic pool of digestive enzyme, there is certainly no experimental justification for drawing the strong negative conclusion that one can assume that such a pool does not exist and hence that it is not necessary to deal directly with the compartmental fate of new proteins released from the ribosome. This is not to say that there is no experimental evidence that bears on the question of the release of digestive enzyme from the ribosome. However, what evidence there is does not lead one to the conclusion that release into the cytoplasm does not occur. Quite to the contrary, evidence consistent with the view that release occurs into the cytoplasm can even be found in the work of proponents of the cisternal packaging model.

In a review of the subject of protein secretion in 1975, I considered one such example in data from a cell-free system in which the compartmental disposition of proteins synthesized by vesicular fragments of the RER, called microsomes, was evaluated (Rothman, 1975). In these experiments (Redman et al., 1966), important forerunners of those that later led to the proposal of the signal hypothesis, newly (labeled) manufactured protein was recovered in: (1) the ribosomes themselves (separated from the membrane of the microsomal vesicle by a detergent); (2) in composite, the internal contents of the vesicle, the membranes of the vesicle, and the proteins associated with the vesicle at its membrane's surface (elements of the microsome that were soluble in the detergent); and (3) in the medium in which the microsomes were suspended. My purpose in discussing these experiments was to note and evaluate the fact that labeled protein was recovered in the suspending medium—the in vitro equivalent of the cytoplasm. How did it get there?

What could we infer from its presence? The scientists who performed these experiments, in accordance with the belief that the protein could not end up in the cytoplasm in situ, assumed that its presence in the medium must have of necessity been artifactual, and proposed accordingly that it reflected the release of new protein from broken microsomal vesicles into the suspending medium. They did not test this ad hoc hypothesis, but thought it sufficient to note that such an explanation was possible. Certainly, it would not have been surprising if some vesicles had been broken and that it might be an artifact of this sort, although today sealed lipid vesicles are often thought to form spontaneously to minimize free energy.

In any event, it certainly was possible to imagine that some vesicles were damaged or not fully formed and that perhaps their contents were being released into the medium. However, the observed kinetics of the appearance of the new protein in the various compartments of the system raised questions about this interpretation. The appearance of labeled protein in the detergent-soluble material, assumed to be the contents of the vesicle (we will consider this assumption below), increased at a different, slower, rate than the labeled protein content of the medium (Figure 1). If broken vesicles had been the source of this material, then one would have expected the two time-courses, vesicle (detergent-soluble material) and medium content of labeled protein to increase roughly in parallel, which they did not. Moreover, labeled enzyme in the medium reached a steady-state concentration, even as that presumably entering intravesicular spaces continued to increase. This suggested that the labeled protein in the medium did not come from broken vesicles at all, or at least not in great part, and if this was not the case, then they must have come directly from the ribosomes; that is, they were released from the ribosome into the medium. Therefore, this experiment provided indirect evidence for protein release from the ribosome into the cytoplasm in situ. This interpretation was reinforced by the fact that the function describing the release of material into the medium and that for the accretion of labeled protein on the ribosomes themselves were roughly parallel; that is, they had the same form and approximately the same time course, as would be expected if the medium's labeled protein content were directly derived from the ribosomes.

Several years later, an alternative "broken vesicle" explanation was suggested (Case, 1978), in which it was proposed that a time-dependent lysis of a subpopulation of weak vesicles might account for the observed kinetics. The idea was that there were weak vesicles among the microsomes that broke open over time, and, as they did, released their contents.

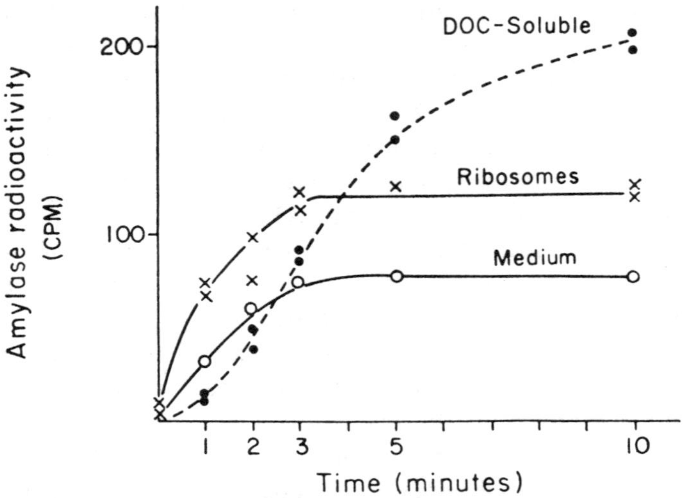

Figure 1 Radioactive amylase recovered in ribosomes, a detergent-soluble material extracted from microsomes (soluble in 0.5% sodium deoxycholate, DOC), and the suspending medium after incubating microsomes for varying periods of time in a medium containing radioactive leucine. The amylase content of the medium is substantial relative to the other fractions, and reaches a steady-state value in about the same time and in the same fashion as the labeled amylase content of the ribosomes. On the other hand, the labeled amylase content of the DOC-soluble material increases more slowly initially, but continues increasing with time after the other two compartments have approached steady-state values. These results suggest that amylase is released into the medium from the ribosome rather than as a result of leakage out of broken microsomes. See text for further discussion. (From C.M. Redman et al., 1966.)

Breakage would be greatest for a short period just after the addition of the labeled amino acid precursor to the medium, and within a few minutes only sturdy vesicles would remain. Why such a time-dependent pattern of breakage should be produced reproducibly as a result of the addition of a labeled amino acid to the medium was not explained. But even though I do not understand why such a possibility is thought credible, credible or not, it is nevertheless certainly conceivable. And that is my point here. It seems that merely being able to imagine a conceivable explanation for an event within paradigmatic borders is viewed by some as being sufficient to retain the paradigmatic view of a process. No proof that one's ad hoc hypothesis is correct, as opposed to other possible explanations, seems needed. The question that I raised in 1975 remains experimentally unanswered almost 10 years later, even though during this time the signal hypothesis, whose premises depend in important ways on the presumption that release does not occur into the cytoplasm, has been widely used and developed, at least to some degree based on the belief that an untested ad hoc hypothesis is correct.

Thus, microsomes are vesicular fragments of the rough-surfaced endoplasmic reticulum capable of synthesizing protein in vitro under the right circumstances. Simply the existence of such a system raises interesting questions about the signal hypothesis, in particular its proposal of a ribosome-binding cycle. As I pointed out above, the signal hypothesis requires that the attachment of the free signal-containing ribosome to the membrane of the RER be timely to prevent new protein from entering the cytoplasm. This either means that the ribosome has rapid access to the membranes on the purely statistical grounds of Brownian motion, or that a special process exists to ensure its timely attachment. If we can imagine such a cycle occurring in the microsome-synthesizing system, as it must if the hypothesis is correct, then we realize that there can be no special cytoplasmic transport mechanism for ribosomes to assure timely attachment, because the suspending medium is not ordered. Brownian motion is the only means available to accomplish the cycling of ribosomes, if such occurs. Although to my knowledge this system of microsomes has not been studied from this point of view, there are some interesting questions to be asked. In the acinar cell, about 1 volume of RER is suspended in 2.5 volumes of cytoplasm, based on stereological estimates. Although one cannot make accurate estimates from the literature, the microsomes in synthesis experiments probably represent less than 10% of the volume of suspending medium (and perhaps as little as 1% or even less). If the cycle concept is correct, then one should be able to replace the short, signal-free proteins recovered from the usual microsomal system, by signal-containing proteins merely by diluting microsomes in larger and larger volumes of medium, and thereby decreasing the efficiency or timeliness of ribosome binding to membranes. Moreover, this increase in signal-containing proteins should be strictly proportional to a decrease in the RNA (ribosomes) content of the microsomes and an increase in the number of free ribosomes that should also be produced by such dilution. Alternatively, if synthesis is halted while awaiting ribosome attachment, then the rate of synthesis should be reduced in proportion to such dilution in the absence of mixing. Such quantitative results need to be demonstrated if we are to prove the hypothesis of a ribosome-binding cycle and the role of the signal sequence in such an event.

7.6 PROTEIN TRANSPORT ACROSS THE MEMBRANE OF THE ENDOPLASMIC RETICULUM

It may seem somewhat odd to question the existence of the transport of individual protein molecules across the membrane of the endoplasmic reticulum given the perspective of the author, but that is just what I would like to do now. My view about the existence of this process has changed over the years from being quite certain that it occurred in 1968 ["it seems clear that newly manufactured enzyme can enter the cisternae of the endoplasmic reticulum" (Rothman, 1969)], including a consideration of how it happened, to some hesitancy in 1975 ["Do nascent peptide chains pass into the cisternal space of the endoplasmic reticulum? The (answer is) a cautious 'yes' " (Rothman, 1975)]. Today, I feel less certain. Although the evidence in favor of such a process has not actually decreased over the years, it has not in my opinion been substantially bolstered by what new evidence has been brought to bear on the subject, and many of the problems with the old evidence remain unresolved even after many years. This is not to say that I believe that such a transport process does not exist. I merely remain unconvinced and would like to consider the basis of my doubts in what follows.

There are two related types of observation that provide much of the evidence that supports the existence of such a transport process. The first is the attempt to demonstrate directly that transport occurs; and the second is an attempt to measure the contents of the endoplasmic reticulum and to demonstrate the presence of the substance of interest in this compartment. In the exocrine pancreas, there is both evidence that transport occurs and that the digestive enzymes are normally found in the cisternae of the RER, but I believe that this evidence is rather weak, particularly when one considers that it is this particular evidence that forms the experimental foundation of the signal hypothesis and the large body of work derived from it. That is, the signal hypothesis is necessarily based on the belief that, if nothing else, the existence of this transport process has been established; otherwise, it proposes a mechanism for a process whose existence is not known; and that would be an unacceptable state of affairs. In addition, the cisternal packaging theory proposes a complex system of small vesicles to carry material out of the RER, a process whose existence is almost completely unsubstantiated even after many years of experimental work (as discussed in Chapter 6), based on these observations and the conclusion drawn from them that transport occurs. It is in this rather imposing context that one must evaluate the evidence for such a transport process.

A. Does Transport Occur?

A great many experiments have been performed using microsomal vesicle preparations of the sort that I have discussed above, either to study protein transport in vitro, or to demonstrate that protein had entered the cisternal spaces in situ (by extracting microsomes from whole tissue after its exposure to radioactive amino acids). Both types of study have generally used one of two methods for evaluating whether or not a particular protein was transported into the cisternal or vesicular space. In the first, the contents of the space are extracted as part of a detergent-soluble fraction of the whole microsome; that is, this detergent-soluble fraction contains the contents of the intravesicular space. Unfortunately, however, this material also contains the vesicle membrane and its constituents, and proteins that had been absorbed or otherwise bound to the surface of the membrane. Thus, this technique does not distinguish between these three potential sources, and therefore to the degree that all three are credible possibilities, fails as a means of determining the contents of the intravesicular compartment.

The second technique involves the addition of proteolytic enzymes to a suspension of microsomes and then determining if the protein is protected from the destructive action of the enzyme. If protection is afforded, then the inference is made that the membrane is a geometrical barrier and that the protected proteins are inside of the vesicle. But of course protection is merely protection, and it can be provided in other ways—in particular, as a result of the association of the protein with the membrane of the vesicle, or as the result of protein–protein interactions in the medium. This uncertainty is particularly relevant when protection is not complete, namely, when some of the protein is damaged by the hydrolytic enzyme, as is usually the case. That is, not all or even most of the given enzyme is protected, only a fraction, and sometimes a small and unquantified fraction at that. Moreover, two proteins that have structural differences, such as signal-containing and signal-free proteins, may be differentially protected from hydrolase activity in a variety of ways as a result of these structural differences. Thus, protection from enzymic destruction is an ambiguous, as well as an indirect, criterion for determining the compartmental location of a substance. Moreover, "summed" observations using two ambiguous techniques, detergent solubility and protection, does not necessarily make the conclusion more secure.*

I would like to consider one example of this problem of compartmental analysis related to the use of a detergent-soluble fraction to determine the compartmental source of labeled protein. In the classical experiment that demonstrates the "sequestration" of protein, a tissue is pulse-labeled with radioactive amino acids, and then chased by a short incubation in a medium containing high concentrations of the same, although unlabeled, amino acid. According to the cisternal packaging concept, the "chase" in cold media should transfer essentially all of the newly synthesized (labeled) protein into the cisternal spaces from the ribosomes. Under these circumstances, a pulse followed by a brief period of chase, virtually all of the new or labeled protein in the cell should be within the cisternae of the RER. We performed this type of experiment and obtained a microsomal fraction of tissue homogenates.

Was all or most of the labeled protein associated with the microsomes within its intravesicular space as was to be expected, or was a substantial portion external to it? We tested the latter possibility simply by washing our microsomes in the usual suspension medium, resedimenting them, and then removing the ribosomes from the washed vesicles with a detergent. We found that about 10% of the labeled protein was still associated with the ribosomes despite the "chase," but, more significantly, 75% was removed by a simple wash, and thus was only "superficially" associated with the microsome, either bound at its surface, or capable of crossing the membrane of the vesicle from within to outside. Thus, only 15% of the labeled protein at the most (i.e., assuming that none was tightly associated with granule membrane) was clearly attributable to an intravesicular location (Table 1). The important point is that it is just such microsomal preparations that provide the evidence for concluding that all of the enzyme is sequestered in accordance with the cisternal packaging theory. In fact, no more than about 15% can be attributed to an intravesicular site. "Noise" accounts for almost 90% of the "signal"—not the most auspicious of experimental situations.

Scheele et al. (1978) have drawn a similar conclusion concerning the existence of substantial protein adsorption to microsomes based on studies in which they exogenously added enzyme to microsomal preparations. In accordance with paradigmatic views, they did not entertain the possibility

*It has become common recently to apply even more remote and ambiguous chemical "markers" of transport and compartmental localization such as the length (or mass) of the peptide chain ("processing"), or the glycosylation of a peptide, and so forth. In these cases, the compartmental locus is assumed, not determined.

TABLE 1. The Distribution of Labeled Protein in the Microsomal Sediment after a 3-Min [^{3}H] Leucine Pulse Followed by a 5-Min Chase

	dpm/g Tissue	*n*	% of dpm in First Microsomal Sediment
First microsomal sediment	610,803 ± 81,841[a]	6	(100)
Washed microsomes (3.0 ml resuspension volume)	196,726 ± 46,833	6	32.2
Washed microsomes (30.0 ml resuspension volume)	146,086 ± 13,596	5	23.9
(a) ribosomes	57,704 ± 5,370	5	9.4
(b) DOC-soluble fraction	88,382 ± 8,225	5	14.5

[a]Pancreatic tissue was homogenized in cold 0.3 *M* sucrose (1:10 w/v) (pH 5.5) and a microsomal fraction obtained with the following series of centrifugal spins: 1000 × *g* for 10 min for the homogenate, followed by 40,000 × *g* for 30 min for the first supernatant, followed by 105,000 × *g* for 75 min for the second supernatant. The final or microsomal pellet was the 105,000 × *g* sediment. Ribosomes were separated from the rough microsomes by suspending the sediment in 2.0 ml 0.5% Na deoxycholate (DOC) in 0.3 *M* sucrose and then sedimenting them by centrifugation at 265,000 × *g* for 5 hr. Under these conditions, approximately 95% of the RNA was sedimented. The presence of [^{3}H]leucine and [^{14}C]inulin (Table 2) was determined by liquid scintillation spectroscopy. The leucine "chase" was effective in removing free tritiated amino acid and 90% of the label contained in the microsomal fraction was TCA-precipitable. $\bar{x}$ ± SEM.

that the membranes might be permeable to the added proteins and explained the association with the microsomal fraction completely in terms of adsorption. Moreover, the conclusion they draw from their observations, in part based on their assumption that no cytoplasmic pool of digestive enzymes exists, was that although much of the enzyme content recovered in microsomal fractions was apparently external to the vesicle, this was a fortuitous artifact. This was because the material initially released from the RER during fractionation was then merely adsorbed to its surface, and the interpretation of earlier cell fractionation studies, which were important elements in generating the cisternal packaging concept in the first place, were not jeopardized by the fact that a major portion of the material was not in the vesicle, but presumably on its surface [Jamieson and Palade (1967) reported that 71 to 85% of the labeled protein of collected microsomes is released into solution in an ionic buffer at pH 8.4].

Of course, one might argue that this seems reasonable. It would not be surprising if a large percentage of the RER contents were released during its fragmentation in the homogenizer. However, we must be careful to distinguish between evidence and hypothesis. That is, even if our hypothesis that the protein was in the cisternae in the first place is correct, we cannot use the evidence derived from such experiments as proof that the protein is indeed in the vesicle. That would, of course, be fallacious reasoning. We assume our hypothesis that the material was present in the cisternal spaces in the first instance and released during cell fractionation to be correct. Then we argue that its apparent location on the surface of the vesicle is fortuitous and that our interpretation holds as if they were inside.

The amount of material that we were able to attribute to the intravesicular space at the most accounted for 1.5% of the total labeled protein recovered from the homogenate (in a microsomal fraction that contained 50% of the tissue's RNA) at a time when virtually 100% should have been within these spaces according to the cisternal packaging model. Perhaps in other hands, the number might be 2%, 5%, or even an order of magnitude higher at 10%, but there is no question that the great majority of material that is recovered is not where it should be according to our predictions. Whatever our justifications may be for its assumed natural origin within the cell, this fact cannot be avoided and it highlights the vagaries of this type of cell fractionation study (see Chapter 12 for discussion of this problem).

We must also turn the question around and ask if it is possible that the material we were able to attribute to intravesicular spaces might not be the remnants of a lost pool at all, but may have entered the vesicles during their formation when the tissue was being homogenized. Certainly, if enzyme is released from the ribosome into the cytoplasm, then some new enzyme would be likely to be trapped within the forming vesicle. We tested this possibility by homogenizing our tissue in a labeled compound that we added, the polysaccharide inulin, and by determining the percentage of the labeled inulin recovered in the portion of the microsomal fraction attributable to an intravesicular pool (the detergent-soluble phase of washed microsomes). If all of the protein had been released from the ribosomes into the cytoplasm, and if they were merely trapped inside the forming vesicle, then the percentage of the tissue's labeled protein and added inulin recovered in this fraction should be approximately the same. When this experiment was performed, the two values were indistinguishable, or about 1.8% of the inulin in the homogenization medium and 1.5% of the labeled protein of the tissue was recovered as "intravesicular" content (Table 2). This is not to say that none of the labeled protein found in the microsomal fraction was present in the cisternal spaces of the intact tissue. It merely points out that we cannot make the statement that it was, based on its mere presence within microsomal vesicles.

TABLE 2. Distribution of Inulin and Labeled Protein in DOC-Soluble Fraction of Washed Microsomes

	$[^{14}C]$Inulin, dpm/g Tissue	^{3}H-labeled Protein, dpm/g Tissue
DOC-soluble fraction of washed microsomes	113,980 ± 51,590[a] (3)[b]	88,382 ± 8,225 (5)
Homogenate	6,350,360 ± 202,750 (3)	5,930,126 ± 794,572 (5)
DOC-soluble fraction as % in homogenate	1.8%	1.5%

[a] $\bar{x}$ ± SEM.

[b] Numbers in parentheses equal number of experiments.

Thus, the hypothesis that all or even a substantial percentage of new (labeled) protein enters the cisternal spaces of the endoplasmic reticulum immediately upon its synthesis, cannot be said to be affirmed by this type of evidence. And we certainly cannot use the hypothesis that much of the material was present in the cisternal spaces in the first place and merely lost during homogenization as evidence for our thesis that the contents of the microsomal fraction reflect its initial presence. Much of what we measure in the microsome is either outside of the vesicle, or, as we shall discuss below, indicates that these vesicles are permeable to proteins contained within them. Since the hypothesis that cisternal packaging occurs is in great part based on the presence of labeled protein in microsomes recovered after labeling tissue in this manner, this hypothesis cannot be said to rest on very firm ground. Finally, the means of determining the compartmental localization of the proteins—namely, detergent-soluble fractions and protection from proteolysis—as they have been used, are ambiguous, indirect, and uncertain means of determining whether a molecule has moved from one compartment to another.

B. The Presence of Secretory Proteins in the Cisternae of the Endoplasmic Reticulum

"Transport"-oriented studies, at least as performed to date, have presented numerous interpretative difficulties, and they do not permit one to draw the conclusion unambiguously that secretory proteins are transported from ribosome to cisternal spaces, no less to conclude how such transport occurs, if it does. Perhaps the second line of evidence concerning the "contents" of the cisternal spaces of the RER is more compelling. Do they contain secretory products in situ? In considering this subject, I will once again use the digestive enzymes of the exocrine pancreas as the example and will discuss one chemical and one anatomical approach to the question.

The concept of specific activity enrichment, borrowed from protein chemistry, is often used to evaluate whether or not a substance is a normal constituent of a particular intracellular organelle or compartment. Briefly, the idea is that if a subcellular fraction derived from tissue homogenates comprised primarily of a particular organellar structure is enriched in a specific protein (usually an enzyme and usually expressed in terms of the activity of the enzyme in the fraction relative to the overall protein content of the fraction compared to that in the cell overall), then such enrichment provides supportive evidence for the presence of the protein as a natural constituent in the organelle or compartment. On the other hand, if a cell fraction is impoverished relative to the whole cell, then the converse inference is drawn; that the substance does not appear to be a natural constituent of the cellular structure. Of course, things are not this simple, and there are many problems and uncertainties associated with this technique (see Chapter 12 for further discussion).

Nevertheless, if we apply this criterion to the presence of the digestive enzymes in the subcellular fraction of the cell derived from the RER, the microsomes, we certainly cannot properly draw the conclusion from such evidence that it is likely that these substances are normally in this part of the cell. The microsomal fraction is not enriched in these proteins, but either contains them at approximately the same concentration as found in the cell overall, or is impoverished. For example, in the rat the concentration of digestive enzymes in the microsomal fraction (per milligram of microsomal protein) is between one-third and two-thirds of that in the homogenate overall depending upon the enzyme and circumstances, and no more than one-fourth of their concentration in the zymogen granule fraction of the tissue homogenate. This value *includes* the ~90% of the digestive enzyme content of such fractions that is either ribosomal or apparently external to the vesicle, and that cannot be attributed to the original contents of the RER. Thus, I think that an objective view of these data requires one to draw the conclusion that they provide no basis for the position that the digestive enzymes are normal constituents of the cisternal spaces of the RER.

Although it has not been demonstrated that these products are concentrated in this location based on cell fractionation studies, I know that in the minds of many, particularly those who are active in this field, that even though they might agree about the objective facts of data, they would feel that such a negative conclusion is not warranted. "After all," they would say, "much of the contents of the RER may have been released during homogenization." And so it may, but how is one to know that the digestive enzymes were part of these contents in the first place, or if so, to what extent? Certainly, if we found a low concentration of lactic dehydrogenase (LDH) (to choose a cytoplasmic enzyme) in the microsomal fraction, we would not conclude that it had been lost from the RER due to fragmentation during homogenization. We would not draw this conclusion, because we do not believe that LDH is a component of this part of the cell in the first place. If we are comfortable with the view that the digestive enzymes were in the cisternae of the RER in the first instance, then our comfort cannot be based on the evidence, but must either be derived from a belief in the correctness of our hypothesis regardless of the evidence, that is, based on our assumptions about the process (for example, membrane impermeability to proteins), or on other evidence, most notably that derived from the anatomical approach to the question of the contents of the RER, which I shall now discuss.

The anatomical approach involves attempts to label secretory proteins in one way or another so that they can be identified in electron microscopic sections through cells. It is commonly thought that autoradiography, in which isotopic emissions from radioactively labeled compounds in a tissue section expose a photographic emulsion immediately overlying the section, has demonstrated the presence of secretory proteins in the cisternae of the RER. Unfortunately, this technique is not capable of making spatial distinctions of the necessary resolution, and autoradiographs, at least as applied to this point, are only capable of showing that labeled enzyme is in the basolateral region of the cell. It cannot distinguish between its presence within the cisternae, on the reticular membrane, or between cisternae in the cytoplasm.

More recently, antibodies to secretory proteins, labeled with agents such as ferritin or gold that produce small particles on the surface of tissue sections where they bind to the antigen, have been used in attempts to identify proteins within the cisternae of the RER. This technique has better spatial resolution than autoradiography, but it has many potential technical problems of its own, most notably the loss or redistribution of material from the surface of the section during preparation for viewing.

Results using this technique have thus far been no more reassuring than cell fractionation data regarding the digestive enzyme contents of the cisternal spaces. Neither of two major published studies (Kraehenbühl et al., 1977; Bendayan et al., 1980) present a convincing case for their presence in these spaces, no less their presence in substantial quantities. Kraehenbühl et al. failed to find a significant number of particles over the cisternal spaces, but believing that they should be found, the authors concluded that the protein must have been lost during tissue treatment, although they provided no evidence for this view. In Bendayan et al. grains were found over the endoplasmic reticulum at between 10 and 50%, depending upon the enzyme, of the concentration seen over the zymogen granules. However, the values were for particles over the "basal-perinuclear" cytoplasm, an area that not only includes the cisternal spaces, but ribosomes, reticular membranes, and cytoplasm as well. The authors of this study present one high-magnification micrograph of the endoplasmic reticulum from which we can estimate the number of particles over the cisternal spaces, relative to those associated with ribosomes or reticular membrane, as well as in the intercisternal cytoplasm (Figure 2). By far, the great preponderance of particles appears to be associated with the ribosomes and membrane, and no more than about 25 to 33% can be relatively unambiguously located over the cisternal spaces, at least if we assume that no particles that appear over such spaces are associated with membranes sectioned "en face." If we correct the measured particle density so that only those particles clearly associated with cisternal spaces are included, then for the nine different enzymes measured, it was either the same as or less than that found over the nucleus or mitochondria.

This is not to say that the digestive enzymes are not in the cisternal spaces of the endoplasmic reticulum, nor even to say that they are not concentrated there, but rather to point out that the anatomical evidence for such a pool is as weak and uncertain as the chemical evidence. Thus, neither the "transport"-oriented studies, nor the "content"-oriented studies provide convincing evidence for the existence of a cisternal pool, no less for the accumulation of digestive enzyme at this location. At best, we can say that it may be so and that our hypothesis has not been conclusively shown to be false; that is, we cannot say that it is clear that there is no digestive enzyme stored within the cisternal spaces. In any event, in the end, at least for the pancreas, the tissue for which the cisternal packaging theory was developed, the evidence does not allow us to assert that there is a natural and substantial pool of digestive enzyme in the cisternal spaces of the endoplasmic reticulum, no less that all secretory proteins are sequestered in its spaces upon their synthesis.

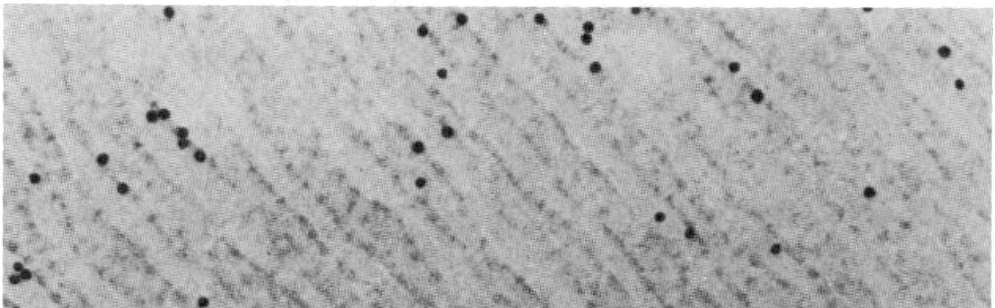

Figure 2 Immunocytochemical localization of pancreatic secretory proteins. A thin section of rat pancreatic acinar cell stained for amylase showing RER cisternae at high magnification (54,000×). Most of the gold particles (~67 to 75%) appear to be associated with RER membrane or over the intercisternal space. See text for further discussion. (From M. Bendayan, et al. 1980.)

Therefore, the central underpinning of the signal hypothesis, the existence of the transport process for which it proposes a mechanism, cannot be said to have been established with a reasonable degree of certainty in the exocrine pancreas. It is not clear beyond a substantial doubt that these secretory products are within the cisternal spaces, nor that they cross the membranes of the RER, nor that if they do, that all secretory proteins are to be found in these spaces soon after their synthesis.

7.7 THE CYTOPLASM AND THE PERMEABILITY OF THE MEMBRANES OF THE ENDOPLASMIC RETICULUM TO DIGESTIVE ENZYMES

Beyond its central assumption that the membranes of the endoplasmic reticulum are permeable to secretory proteins, the signal hypothesis assumes, as I have already noted, that this permeability is expressed across the membrane between a ribosomal compartment of protein and the cisternal spaces of the RER, and does not involve the passage of proteins through the cytoplasm en route. In this way, only the extending or elongating peptide chain crosses the membrane; that is, only it displays permeability to the membrane, and not the formed protein. In addition, the signal hypothesis assumes that the permeability of the membrane to proteins, if such it is, is unidirectional, and that once the protein is within the cisternal spaces it cannot escape. Neither of these assumptions, that is, no cytoplasmic passage and unidirectional transport, are supported by direct experimental observation.

Assuming that the reticular membrane is permeable to secretory proteins, determining their route of transport requires a kinetic compartmental analysis of that transport. Do new protein molecules pass into the cytoplasm or the medium bathing isolated microsomes prior to entering the cisternal spaces? To answer this question, one must measure the time-course of the passage of the proteins through the putative compartments. We cannot infer the route of passage by measuring the end-state distribution of material, nor can we accept as appropriate discounting a priori the presence of new protein in the medium bathing suspensions of microsomes as merely being due to an artifact of the system, without requiring that this ad hoc hypothesis be demonstrated to be correct and complete. Indeed, as I have discussed, the evidence seemingly does not permit one to draw that conclusion.

Moreover, there is evidence that cytoplasmic passage occurs during the insertion of some membrane and other intracellular proteins. That is, the formed protein with signal sequence is released from the ribosome into the cytoplasm, and then enters the membrane from this pool spontaneously as the result of a thermodynamically favorable partitioning between cytoplasmic and membrane phases (Wickner, 1979). That is, insertion does not occur, as the signal hypothesis envisions it, in the absence of intervening cytoplasmic passage and the development of higher-order structure within the protein.

This suggests that the converse event also occurs; that is, proteins leave the membrane spontaneously. This concept of the insertion and removal of membrane proteins is appealing in its simplicity and can account for differences in the turnover rate of different membrane proteins, which is otherwise difficult to account for without complex ad hoc hypotheses, namely, if one views the turnover of membrane proteins as being part of a general turnover of membrane segments rather than of individual protein molecules.

This brings us to the second assumption: that the protein, once having entered the cisternae, is trapped in that space and cannot leave of its own accord. As I have said, this is an unsubstantiated hypothesis. Just as the appearance of a newly manufactured protein in the medium bathing synthesizing microsomes cannot be properly dismissed as an artifact without the necessary experimental proof, so the release of new protein from microsomes as a function of washes, and so on, cannot be assumed to be due to desorption from its surface alone, assuming that occurs, and not transmembrane movement. This may or may not be the case, and the hypothesis that these membranes are bidirectionally permeable to their contained proteins, much as proteins may be inserted and removed from membranes spontaneously, is a hypothesis that requires experimental test and not prejudgment. To my knowledge, only one group has attempted to distinguish between the two possibilities, and this group concluded that the membranes were bidirectionally permeable to secretory proteins (Pearce et al., 1978).*

Thus, granting the permeability of reticular membrane to proteins, it has not been established that such transport does not involve cytoplasmic passage, nor that it is irreversible. These are not only unsubstantiated hypotheses, but there are experimental observations consistent with both cytoplasmic passage and reversible transport, although such observations have generally been discounted, often by untested ad hoc hypotheses that might explain them, at least to some extent, in ways that are consonant with paradigmatic precepts.

*Sjöstrand has recently made the interesting suggestion that demudded RER membranes facing the Golgi complex may be modified to provide such a permeability; Sjöstrand, Nonvesicular Transport, John Wiley and Sons, 1985.

7.8 SUMMARY

I see the following major obstacles to the acceptance of the signal hypothesis:

1. The signal hypothesis is based on the view that the existence of protein transport into the cisternae of the RER has been satisfactorily established. I do not believe that the compartmental distinctions have been clearly determined, nor have the quantitative issues of such a potential transport process been considered. Thus, the signal hypothesis is based on the occurrence of a process whose existence is still uncertain.

2. The signal hypothesis proposes a very complex mechanism for membrane permeability, given its existence, that requires many subsidiary and ad hoc hypotheses that add greatly to the already formidable complexity of the cisternal packaging–exocytosis construct. Simpler alternative explanations for observations have not been sufficiently considered.

3. Many exceptions to the postulates of the signal hypothesis have been discovered. I have discussed a small sampling here and Wickner (1979) in his review of the subject has put together a substantial list (also see Ho, 1985). I have no doubt that new exceptions are being discovered as I write this. Exceptions to the hypothesis have either been considered technical artifacts of one sort or another, or have been incorporated into the signal hypothesis by the addition of new ad hoc hypotheses introduced to explain them. For this reason, the signal hypothesis has become much more complex than the original, rather complex, series of hypotheses that I have discussed in this chapter. It is necessary to ask those who are proponents of this hypothesis, what particular piece of negative evidence would lead them to reject their general thesis about the function of the signal sequence, as opposed to introducing new and increasingly complex ideas to explain seemingly deviant observations? If they cannot point to such tests, then the signal hypothesis is a hypothesis that is not capable of falsification, and therefore is not a proper scientific hypothesis in my view.

4. The signal sequence is not usually present on secreted proteins in the cell or its endoplasmic reticulum in eukaryotes. Thus, it is a hypothesis that is constructed in part on the belief that something exists that cannot be demonstrated in situ. Indeed, the absence of the signal sequence in situ, the inability to find proteins that contain it, is a prediction of the hypothesis. How can one prove such a hypothesis false? This is a very disconcerting aspect of the signal hypothesis.

5. The existence of the leader sequence alone is not evidence of its biological function, and it has not been demonstrated that the signal sequence functions to *produce* either ribosomal attachment or protein transport, no less in the way envisioned by the signal hypothesis. These are hypotheses that remain essentially unsubstantiated by experimental evidence. This is not to say that the presence or absence of this sequence of amino acids does not effect the transport of proteins across membranes, but rather that the specific hypotheses that state that signal sequences function to cause ribosomal attachment and the formation of a pore through which an elongating peptide chain passes are unsubstantiated. For example, there is evidence that particular leader sequences can effect the transport across or insertion into membranes of proteins in prokaryotes. However, their "mode of action" is another matter.

It is one thing to say that a particular segment of a protein has an effect on its ability to associate with membranes or cross them, and quite another to say that the segment acts specifically as a signal for a particular sequence of events. From what we know, I do not believe that we can say that the signal sequence signals anything in a real sense, or that the specific events that it has been proposed to signal in the signal hypothesis occur. Moreover, based on the available evidence, we cannot even say that the presence of the signal sequence, however it may function in one particular situation or another, always facilitates or permits the transport of proteins into or across membranes, even though it has such an effect in certain cases; that is, we do not have sufficient information to conclude that this is its natural and general role that is applicable to all proteins. And we certainly cannot draw the negative conclusion, nor is it true, that as a class, proteins that lack the leader sequence are unable to enter or cross membranes individually.

6. Although the existence of leader sequences on new proteins in some circumstances and for some proteins is clear, it is not at all clear how the diverse variety of such peptide chains, and their diverse putative cleavage bonds, as well as their seemingly quixotic presence or absence on various proteins and in various situations, can be explained coherently in terms of the various sequences being required for the attachment of ribosomes to membranes and the transport of proteins through or into them.

7. It has not been established experimentally that new secretory or membrane proteins only leave the ribosome to enter the membrane or cisternal spaces, rather than the cytoplasm of the cell. Indeed, there is substantive evidence that at least some do just that.

8. The proposition that new proteins are sequestered within the

cisternal spaces, unable to leave of their own accord, has required the proposal of a complex, experimentally unsubstantiated, and I believe organizationally and thermodynamically difficult, proposal that small intracellular vesicles carry different products around the cell.

As I said at the outset, the signal hypothesis is a proposal for the movement of individual protein molecules across membranes that is so narrow in its view of such events, it retains the notion that there is no such thing as a natural permeability of membranes to proteins. However, it is becoming increasingly clear that this general view is incorrect, and that membranes are permeable to many different proteins, and that such permeability, is often, if not always, expressed by the mature formed protein "posttranslationally." Thus, the idea that the particular modus operandi that we have been discussing here represents a general means of overcoming the essential impermeability of membranes to these macromolecules, no longer seems attractive. It has been suggested (Sabatini et al., 1982) that, instead, the opposite might be the case; that is, that such a mechanism may represent a latter-day evolutionary development and specialization that acts to *restrict* the natural permeability of membranes to proteins in eukaryotes. I cannot think of compelling reasons why such a complex, nonselective (different proteins are moved together en masse), and, in the end permeable system would have any particular selective advantage. Perhaps I am wrong, and it will be shown to have just such a value; but whatever purpose or lack thereof we might imbue such a process with, the central point at issue is whether or not the evidence in hand is sufficient to draw the conclusion that such a mechanism, or one much like it, is at work in cells. I believe that the signal hypothesis, as well as the overall cisternal packaging–exocytosis construct, although supported by a variety of experimental evidence, remain essentially unproven hypotheses: they are not processes whose existence has been established and that account in full for the events that they concern.

SELECTED REFERENCES

1. Bendayan, M., J. Roth, A. Perrelet, and L. Orci, *J. Histochem. and Cytochem.*, **28**:149, 1980.
2. Blobel, G. and B. Dobberstein, *J. Cell Biol.*, **67**:835, 1975a.
3. Blobel, G. and B. Dobberstein, *J. Cell Biol.*, **67**:852, 1975b.
4. Case, R. M. *Biol. Rev.*, **53**:211, 1978.

5. Ho, J. J. L. in S. S. Rothman and J. J. L. Ho, Eds. *Nonvesicular Transport.* John Wiley & Sons, New York, 1985, p. 63.
6. Jamieson, J. D. and G. E. Palade. *J. Cell Biol.,* **34**:597, 1967.
7. Kraehenbühl, J. P., L. Racine, and J. D. Jamieson. *J. Cell Biol.,* **72**:406, 1977.
8. Kreil, G. *Ann. Rev. Biochem.,* **50**:317, 1981.
9. Milstein, C., C. G. Brownlee, T. M. Harrison, and M. B. Mathews. *Nature New Biol.,* **239**:117, 1972.
10. Palmiter, R. D., J. Gagnon, and K. A. Walsh. *Proc. Natl. Acad. Sci. USA,* **75**:94, 1978.
11. Pearce, P. D., B. R. May, and W. H. Elliot. *Biochem. J.,* **176**:611, 1978.
12. Redman, C. M., P. Siekevitz, and G. E. Palade. *J. Biol. Chem.,* **241**:1150, 1966.
13. Rothman, S. S. *Science,* **190**:747, 1975.
14. Rothman, S. S. in S. Y. Botehlo, F. P. Brooks, W. B. Shelley, Eds., *The Exocrine Glands.* Univ. of Pennsylvania Press, Philadelphia, 1969, p. 169.
15. Sabatini, D. D., G. Kreibich, T. Morimoto, and M. Adesnik. *J. Cell Biol.,* **92**:1, 1982.
16. Schatz, G. and T. L. Mason. *Ann. Rev. Biochem.,* **43**:51, 1974.
17. Scheele, G., G. E. Palade, and A. Tartakoff. *J. Cell Biol.,* **78**:110, 1978.
18. Wickner, W. *Ann. Rev. Biochem.,* **48**:23, 1979.
19. Wickner, W. *Science,* **210**:861, 1980.

PART 3

Testing the Paradigm

The truth is, that man cannot frame sufficiently abstract hypotheses; imagination is always intruding upon logic, and causing men to make pictures of occurrences which are essentially incapable of being visualized.

Bertrand Russell

CHAPTER 8

The Equilibrium Hypothesis and Tests of the Paradigm

The vesicle or cisternal packaging–exocytosis model was perhaps the first paradigm to be spawned by the relatively new discipline of cell biology. As a paradigm it is more than just a popular model for a particular process. It is a complex, although incomplete, system of knowledge and belief, which is only in part a specific theory that makes explicit predictions about a process. It is associated with certain kinds of experimental and technical approaches to the study of cells that includes electron microscopy, cell fractionation, and electron microscopic autoradiography. Furthermore, and perhaps most important for our discussion, it makes certain assumptions about the nature of cells, some of which we have already considered. The cisternal packaging–exocytosis model is itself a model for other models; and as such it embodies a gestalt of the cell, how it functions, and how one goes about learning its secrets. It views the cell as an extremely complicated factory crammed full of machines of various sorts that account for the movement of the cell's chemical products hither and yon. For this reason we should not only ask whether it provides us with a relatively satisfactory explanation for the secretion of protein by the pancreas and other protein secreting cells, but whether or not it is an adequate *general model* for the range of cell functions and their study that it touches upon. This is the broader question, and no doubt the more important one.

My work and that of my colleagues on secretion and protein transport has gradually led us to question the cisternal packaging–exocytosis paradigm and to formulate an alternative general view. This alternative, the equilibrium theory, has two primary sources. The first is a series of negative tests of the exocytosis model in the mammalian pancreas. A discussion of some of these will be the major focus of the chapters that follow. These negative tests obliged us to develop, if we could, an alternative model that could more successfully explain the behavior of this system.

The other source of the equilibrium theory was the belief that another paradigm, a diffusive model for the transport of molecules across membranes, had deeper theoretical roots and needed to be experimentally tested for goodness of fit to the observed kinetics and thermodynamics of protein transport. A diffusion-based model has precedence because it embodies a more fundamental conceptualization of transport events, and for this reason has been the major experimental and theoretical tool guiding the study of biological transport for more than a century. It is founded on two central beliefs. The first is that simple physical explanations for the movement of molecules must be favored over the more complex alternative of chemical or reaction-mediated movement, unless and until the evidence requires us to consider such processes. The second is that observed molecular movements should be assumed to result from the intrinsic kinetic energy of the moving molecule itself, and not from an external energy source of some kind, unless and until the evidence requires that we consider such complex processes.

When a diffusion-based hypothesis is applied to a process such as the transport of protein by the pancreatic acinar cell, it becomes apparent that in addition to its more general character, such an hypothesis is superior in at least one important additional respect. Although unambiguous experimental proof may be a chimera, the equilibrium hypothesis presents us with a crystal-clear predictive situation when compared with the murky case of the cisternal packaging–exocytosis model. This is so because all diffusion-based models stem from a single, well-established mathematical relationship, the Fick diffusion and similar equations, which correlate the observed flux of a molecule to the variable relevant to that flux, namely, concentration gradient, as well as to the phenomenological constraints of the system such as surface area and diffusivity. These established relationships make predictions clear and permit us to test the hypothesis in a fashion that allows for the falsification of its premises, leads to quantitative predictions for various specific models derived from it, and permits an orderly development of hypotheses. Of course, a superior hypothesis in these terms has no greater hold on truth or reality, but it does offer a rational path to follow. It offers the opportunity of starting with the simplest possible case, and being drawn to the more complex only by gained knowledge of the system. As I see it, even if we were to end up rejecting each simpler case, being drawn time after time necessarily to the next more complicated scenario, inexorably to the paradigmatic view of things in the end, it would still be the proper road to travel as we attempt to discern the real situation.

The equilibrium hypothesis has yet another value. It can be viewed as antithetical to the cisternal packaging–exocytosis model. This permits us to ask whether or not a particular observation, consistent with the cisternal packaging–exocytosis model, is not also perfectly consistent with its logical antithesis. Without the foil of an antithetical statement we might mistakenly think that a particular experimental inference lent considerable weight to our theory, whereas such a foil might disclose that our inference was very weak indeed since the observation fit not only our model but an opposing conjecture as well. Similarly, the existence of an antithetical model offers us the possibility of designing experiments to distinguish between the two views. Thus, there is particular strength in being able to test one model against another.

Perhaps this can be appreciated better by considering the way in which the equilibrium model helps us formulate tests of the cisternal packaging–exocytosis model; using it as a comparative case. In this way, it requires clarification of the paradigm. For example, a diffusion-based model predicts that: (1) the movement of molecules is bidirectional across planar membranes, and is therefore experimentally reversible; and (2) the rate of transport can be altered by altering the chemical activity gradient for the transported substrate. These, of course, are merely two aspects of equilibrium behavior characteristic of diffusive processes. The exocytosis model can be distinguished from a diffusive construct in terms of these two predictions. It views transport as "vectorial" or unidirectional; that is, as a process incapable of reversal after membrane fusion. Moreover, it proposes that the rate of transport is not a function of the extant concentration gradients for the transported molecules, but is instead a function of an external force or event which leads to movement (namely, membrane fusion-fission).

In addition to these two rather general predictions, there is a third that can help us distinguish between a diffusion-based and exocytosis explanation for transport, at least for the particular cell that we will be considering. In the acinar cell much of the digestive enzyme is stored in secretion or zymogen granules, and therefore secretion, as envisioned in an equilibrium construct would occur, at least in part, as the result of the movement of protein from zymogen granule to cytosol, and then out of the cell. That is, secretion can involve three compartments, at least when there is a substantial amount of enzyme in a zymogen granule storage pool. The exocytosis model, on the other hand, is clearly different in this regard and proposes that the transfer of matter only involves two compartments, zymogen granule and extracellular space (the duct system in the case of the pancreas).

Thus, as our starting point, we can define the equilibrium hypothesis simply as the application of the diffusion equation to a three-compartment system. More complex constructs may have to be considered, but only as the evidence requires.* As I have already discussed, the cisternal packaging–exocytosis model presents us with a much more complex picture. Unlike a simple diffusion hypothesis, it offers an incomplete structural framework that is often rather difficult to put to the test. Predictions are not always, or perhaps not even usually, clear or explicit, even if they are implicitly so. This vagueness protects the paradigm from ready falsification. Although most of the predictions that I will consider below have been put forth explicitly in one place or another, some are based necessarily on what I view the implicit paradigmatic belief to be.

To the degree that my view does not accurately reflect that generally held in one specific area or another, I can only apologize, and say that it has seemed that way to me. In any event, setting predictions for the paradigm can still be a useful tool to help us organize these ideas and define what the paradigm does indeed predict about events. If predictions about one aspect or another of the system's behavior cannot be made in the paradigm's terms, then this reflects upon its weakness as a model.

In this short discussion I have touched on several potential experimental tests, and in what follows these and other tests of the paradigm will be

*The equilibrium hypothesis is described here in its simplest form, as a three-compartment system capable of diffusive equilibrium. A more complete model would include other intracellular and extracellular compartments such as the ribosomes, the basolateral extracellular space, and even the intestinal lumen. In addition, although there is evidence for diffusive equilibria across the apical membrane of the acinar cell, it is not clear whether or not the interposing membranes are capable of maintaining steady-state concentration differences or disequilibria across the surface, and if so, which membranes and in what fashion. With these caveats in mind, we gain little beyond complexity, by considering our model here in other than its simplest form initially, even though we know that in the end we must consider these issues.

considered. Each will be discussed separately, in its own experimental and theoretical context. In general, I will present some background for each; the prediction or predictions of the paradigm; the results of experimental work or theoretical analysis; and then, since all of these cases provide negative or falsifying evidence, I will discuss the riposte or explanation given these observations by supporters of the paradigm; my comment on these explanations; and finally the results will be discussed in terms of the equilibrium model.

8.1 THE PARADIGM AND CONJECTURE

The vesicle model is more than a complicated grouping of hypotheses, and seems to involve deeply embedded beliefs, not unlike other widely held scientific theories. Perhaps this aspect of the paradigm can be appreciated by considering the role of conjecture in maintaining a paradigmatic perspective in the face of what appears to be negative tests of its validity. As I have pointed out, such negative tests are often countered by ad hoc hypotheses; that is, hypotheses that are proposed to account for discrepant observations within the borders of the paradigm. They modify the paradigm, almost invariably by making the model more complex. It is usually and perhaps uniformly true that one can explain negative tests of a paradigm by the addition of such new, ad hoc elements, and in this way protect the larger model from falsification. Of course, at some point such modifications can no longer be integrated and the model remain intact. It is interesting that these conjectures often are not actually tested or thought necessary to test. The mere fact that one can imagine a way to explain a discordant observation in paradigmatic terms may be thought sufficient in itself.

CHAPTER 9

The Disappearing Zymogen Granule

A decrease in the number of zymogen granules in the acinar cell during augmented protein secretion was first noted by Rudolf Heidenhain in the 19th century, and has been reported many times since. To the degree that such a decrease is a prediction of the exocytosis model, it is borne out by the evidence. Indeed, I think that this, perhaps more than any other observation, is responsible for the widely held belief that exocytosis occurs and accounts for secretion, because such a process could account for the disappearance of granules during active secretion. In this chapter, I will consider this, as well as other potential explanations, for their disappearance.

9.1 WHEN GRANULE NUMBER DOES NOT CHANGE

Before we discuss the disappearance of secretion granules, we should consider the possibility that granule number may not decrease during augmented secretion. How would such an eventuality affect our consideration of the exocytosis hypothesis? In the first place, it is often unclear when a reduction in number has or has not occurred. In Heidenhain's original studies on dog pancreas, as in most subsequent light microscopy studies, the term granule "depletion" is used to indicate that the area of the cell occupied by zymogen granules, the granular zone, has become smaller. This does not necessarily mean that the number of granules in the cell has decreased. Of course, the same number of granules might merely occupy a smaller volume of cytoplasm, and this could occur for a variety of reasons, one of which (a decrease in their size) I will consider below. This distinction is often unappreciated, and decreases in the number of granules in the cell has become the common explanation for decreases in the size of the granular zone whether or not such an inference is actually justified by the data. Most reports concerning changes in granule number have not been quantitative, and, as a result, often reflect little more than the impressions of the microscopist, however carefully made. Quantitative measurements from which we can more objectively estimate granule number in various functional states are needed and necessary.

Recently, Ermak made one such estimate. He found that although active secretion (induced by feeding) may lead to a reduction in granule number,

TABLE 1. Mean Diameter, Volume, and Number of Zymogen Granules in Acinar Cells of Fasted and Fed Rats

Experimental Group	Mean diameter[a] (μm)	Calculated volume (μm^3)	Mean number[b]
Fasted	0.85 ± 0.15 $n = 916$	0.32	45 ± 3 $n = 54$
Fed	$0.65 \pm 0.15^{\delta}$ $n = 1660$	0.14	$29 \pm 2^{\delta\delta}$ $n = 119$
Sample 1[c]	$0.70 \pm 0.20^{\delta}$ $n = 256$	0.18	43 ± 4 $n = 12$
Sample 10	$0.60 \pm 0.15^{\delta}$ $n = 140$	0.11	$15 \pm 3^{\delta\delta}$ $n = 15$

[a]Arithmetic mean to nearest 0.05 μm $\pm$ SD, n = number of granules.

[b]Per section through a single cell, arithmetic mean $\pm$ SE, n = number of cells.

[c]Sample 1 and 10 refer to two different tissue samples taken from fed rats.

$^{\delta}$Value significantly different from that in fasted rats (two-tailed t test, $p < 0.001$)

$^{\delta\delta}$Value significantly different from that in fasted rats (two-tailed t test, $p < 0.001$)

Source: From Ermak and Rothman, 1981.

the number of granules in some individual cells does not decrease (Table 1, sample 1; Figure 1, quadrant II) (Ermak and Rothman, 1981). That is, active secretion does not *require* a reduction in the number of granules in cells.

At first glance, this appears to contradict the paradigm's prediction that granules should disappear under these circumstances. There are, however, two obvious ways in which we can account for the absence of changes in granule number during active secretion within the context of an exocytosis mechanism. The first is that if the number of granules in a particular cell or group of cells does not decrease during augmented secretion, then these cells may be refractory; that is, they may not be secreting at all, or may only be secreting at relatively low rates. In order to test this hypothesis properly it

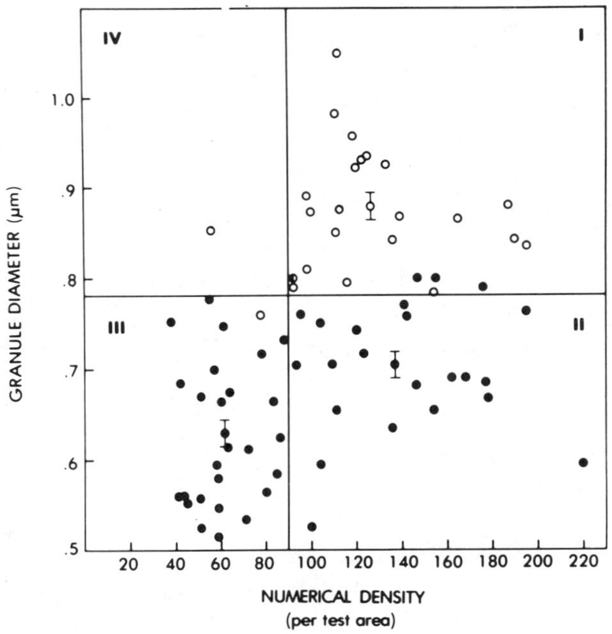

Figure 1 Scattergram of the mean diameter of granules versus numerical density (the number of granules per test area) (test area approximately 275 μm^2) from fasted (○) and fed (●) rats. Symbols with SE bars are mean values for granule diameter ± SE for either fasted or fed points above or below 90 granules per test area. The graph is subdivided into quadrants by lines at $\bar{N} = 90$ granules and $\bar{d} = 0.78$ μm. Over 90% of the samples from fasted rats fit into quadrant I (large number and diameter). Samples from fed rats fit mainly into two quadrants (II, large number but small diameter; and III, both small number and diameter). Quadrant IV (small number but large diameter) did not contain samples from fed rats. (Reprinted, with permission, from Ermak and Rothman, 1981.)

is necessary to compare rates of protein secretion for different individual cells, relative to changes in the number of granules in the same cells, and such measurements are not yet practicable. Ermak has however reported that after a period of augmented secretion, changes in the *size* of granules are found in cells in which granule number is unchanged (Figure 1; compare quadrants I and II). This is a "cytologic" sign of activity, and although in itself it does not prove that such cells are actively secreting protein, it does indicate that they appear, at least in this fashion, responsive to the applied stimulus. That protein secretion is indeed occurring in such cells can be inferred from the extent and nature of the observed changes in granule size, and this will be discussed below.

The second explanation is in a sense the opposite of the first; that is, the rate of new granule formation in some cells is so active that granule number remains constant despite a large increase in the rate of secretion. That is to say, cells that do not show changes in granule number *are more active.* However, for a number of reasons it does not seem likely, at least in the pancreas, that the rate of granule formation can keep pace with active secretion. From cases in which large decreases in granule number are observed during augmented secretion, we can estimate the rate of granule formation under these circumstances for the tissue overall, at least as we consider these questions through the eyes of the paradigm. For example, in one study, the granule content of the cell was reduced by some 95% over a 3-hr period of augmented secretion (Figures 2 and 3). For this to have occurred, the rate of granule formation would have been only 5% of the rate of loss in the paradigmatic view of the process.

Of course, the point is that their rate of formation may be higher in the particular cells or tissues in which depletion is not seen, but it should be kept in mind that the rate of granule formation in such cells, given an equivalent rate of protein secretion, would have to be much greater than in cells in which substantial depletion is observed; more than 20 times as great for this particular example. Moreover, under conditions in which protein secretion is maximally increased (by 10–100 times), protein synthesis is hardly altered in relative terms (a brief increase that is maximally about two- to threefold for most reports). To the degree that protein synthesis reflects the rate of granule formation, it does not appear able to keep pace with active secretion.

In any event, potentially one can account for the absence of changes in granule number during active secretion within the context of the exocytosis model by means of these ad hoc hypotheses, namely, increased granule

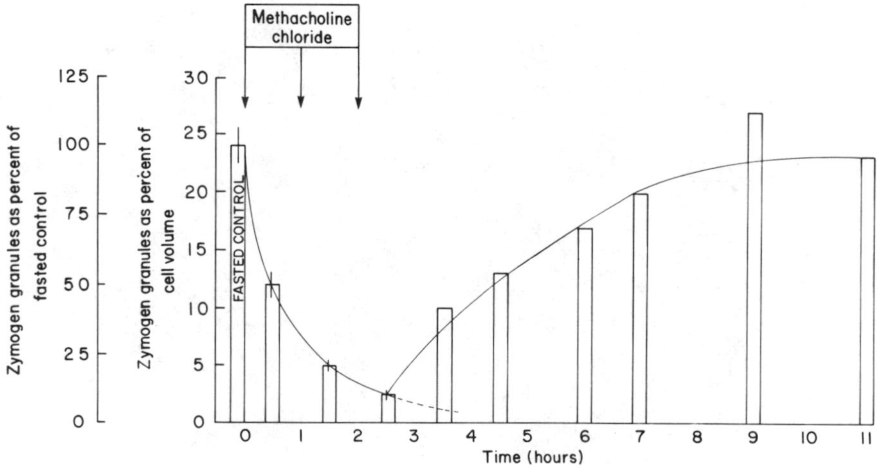

Figure 2 The effect of a cholinergic agonist (methacholine chloride) on the percentage of cell volume occupied by zymogen granules as determined by a random point-count method in light micrographs (see Figure 3). Bars placed between hours are for samples taken at every 0.5 hr. Error bars show SE of measurement for cells with varying volume density of zymogen granules. (Reprinted, with permission, from Rothman, 1975.)

formation or decreased secretion. However, it should be remembered that these explanations are simply conjectures, that is, unsupported by evidence. We will consider the issue of granule formation relative to the rate of secretion further as we now discuss the question of granule turnover in the pancreas in more general terms.

9.2 GRANULE TURNOVER

It is believed that granules are formed continuously and that their formation accounts for the increase in the number of granules in the cell during the period of recovery following active secretion (Figure 2), and furthermore, as I have said, a decrease in their number during active secretion is thought to occur because the rate of granule formation is less than the rate at which granules are lost or disappear.

Sometimes these ideas about granule turnover or cycling are tied to the vesicle concept of secretion, as if the existence of the former implied the latter. This confusion extends two attempts to measure the rate of granule turnover and formation. The usual fashion, in which "measurements of

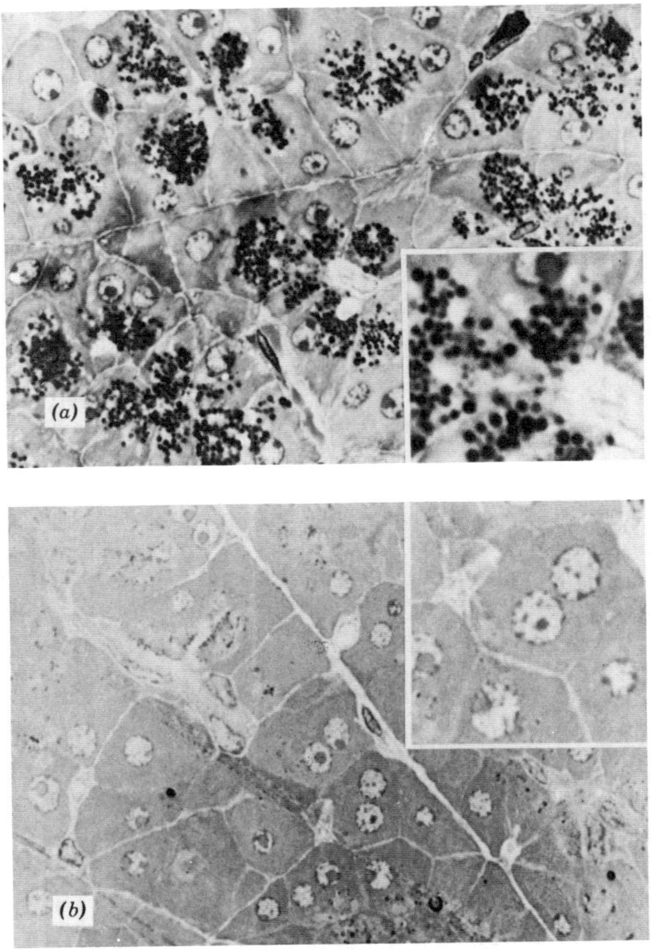

Figure 3 Light micrographs of rat pancreatic tissue after an overnight fast (A), and after three injections of a cholinergic agonist subsequent to the fast (B) (see Figure 2 for protocol). Tissue was removed 0.5 hr after third injection in treated condition. Inserts are areas of tissue magnified to show an acinus. (Reprinted, with permission, from Rothman, 1975.)

turnover" are made, either in the pancreas or in other secretory tissues, is by labeling the product that is to be secreted with a radioactive element, and by then estimating the rate of appearance of labeled product in granules (formation) or the rate of appearance of labeled product in secreted fluid (turnover). Of course, this experiment has to do with the movement and turnover of secretory product, and not granules, and the assumption that the appearance of product in granules or its secretion is analogous to measuring the rate of formation of granules or granule turnover or cycling, respectively, is just as assumption; and an assumption that ties granule turnover arbitrarily to the vesicular concept of secretion. That is, it assumes a priori that the turnover of granules and product are one in the same and that the secreted product cannot and does not move from place to place independently of the formation, movement, and loss of granules. This assumption cannot be made. Indeed, it is known, not only for the pancreatic zymogen granule (as I shall discuss), but for many secretion granules, that the secreted contents enter and leave the granule across its membrane, as well as enter and leave the cell across the plasma membrane; and therefore the two processes are often known to be independent.

It is an important test of the exocytosis hypothesis to determine whether or not the rate of appearance of product in granules and the rate of turnover of product are indeed the same as the rate of formation and turnover (or cycling) of granules. If they are the same, then this would be support for the exocytosis concept. Nevertheless, we have no right to assume the equality of granule and product turnover a priori. To measure the rate of granule formation and turnover we must have a specific granule marker that is *independent of the secreted product,* perhaps a membrane marker (N.B.: Meldolesi has reported that the rate of turnover of granule membrane protein is lower than that of secretory protein).

Although it is true that without a granule marker we cannot properly measure the rate of granule turnover, we can nevertheless ask a simple question that can help us evaluate the situation and set limits in the absence of this measurement. How much of the product present in the cell at the outset of augmented secretion is secreted during an active period? Is it several times that originally present in the tissue or is it less? If it is the former, then at least we know that the product, if not the granules, has turned over numerous times during the secretory period, and this would at least give us cause to consider the role of granule turnover under these circumstances. On the other hand, if the amount that is secreted is only a fraction of the product present in the granule pool or the cell at the outset of augmented secretion, then substantial turnover of product would not be necessary to account for secretion, and in its absence, in paradigmatic terms, neither could substantial granule turnover or cycling have occurred.

If the latter were indeed the case, then it would considerably simplify our attempts to understand the nature of changes in granule number, or the absence of such changes, and their relationship to protein secretion. What are the facts? In the fasted rat pancreas, the amount of secretory protein in the cell is about 70 mg/g of tissue (Ho and Rothman, 1983). Depending upon our assumptions, the granule pool would contain between 30 and 60 mg of protein/g of tissue. Three hours of maximal stimulation with a cholinergic agonist in the rat causes secretion of about 20 mg of protein/g of tissue (Rothman, 1975). Therefore, only a fraction, between one-third and two-thirds, of the original granule pool of secretory protein is released from the cell even during a prolonged period of active secretion. Thus, the enzyme pool does not turn over numerous times in order to account for active enzyme secretion. The initial store is sufficient.

In the context of the exocytosis model, if this secretion is not accompanied by a change in the tissue content of secretory protein, then new granules presumably would have been formed at a rate equal to their rate of loss. On the other hand, if tissue content decreased in direct proportion to the secretion of protein, then no new granules would have been formed (assuming unchanged intragranular protein concentration). When such measurements were made for the above experimental situation, the observed decreases in tissue content ranged between one-third and one-half [Ho and Rothman, 1983; Miyasaka and Rothman, 1982], suggesting that the latter view, that is, the rate of new granule formation is low relative to their loss during active secretion, is closer to the truth.

In this case, we would also expect to find a reduction in the volume of the cell occupied by granules (granule volume density) of one-third to two-thirds under the same circumstances. When quantitative estimates were made, the reduction was even larger, about 95% (Figures 2 and 3), reinforcing the view that the rate of new granule formation is low relative to their loss under these circumstances. As I discussed above, we can also use this decline in the granule content of cells itself to estimate the rate of granule formation. When estimated in this way, a value is obtained of about 1.5% of the initial pool being formed per hour. We can also estimate the rate of granule formation from the rate at which granules reappear during recovery from stimulation. In this particular case, full recovery is observed in 6–7 hr (Figure 2) or at an average rate of about 14–16% per hour.

Whether the estimate is 1 or 15%, the rate of new granule formation cannot be very large relative to the size of the granule pool found initially in the pancreas of the fasted rat, and changes that are observed in the granule pool over a short period can be considered to involve, with some small error, the same granules that were present at the outset, even during the most active secretion. It should be kept in mind that these estimates assume the exocytosis model for secretion, and in other theoretical contexts the true rate of granule turnover might be quite unrelated to these observations.

9.3 THE RELATIONSHIP BETWEEN THE DISAPPEARANCE OF GRANULES AND THE SECRETION OF PROTEIN

As I have suggested, if there are decreases in the granule volume of a cell when the rate of digestive enzyme secretion is increased, then we would expect this decrease to be correlated to the amount of protein secreted. The paradigm is quite specific in this regard.

It is generally believed that the concentration of digestive enzyme protein is either essentially the same in each zymogen granule, or varies randomly from zymogen granule to zymogen granule. As a result, the amount of enzyme stored in a granule, or more generally in the granule pool of the cell, can be said to be roughly proportional to its volume, or to the fraction of the cell volume that it occupies, its volume density. Therefore, if granule sizes are normally distributed, as they more or less are, then the product of the number of granules in a cell and the average volume of the individual granule is a measure of the amount of enzyme stored in the granule pool of that cell.

In this case, changes in granule volume density should be inversely proportional in a linear fashion to changes in protein secretion, as long as the formation of new granules is slow in comparison to the rate at which the contents of formed granules are secreted. That is, if secretion is accomplished by the direct transfer of mass from one compartment (intracellular granules) to another (the duct lumen), then changes in the content of the two compartments should be reciprocal and proportional; as more and more enzyme is secreted, fewer and fewer granules should remain in the cell, and volume density should decrease accordingly. Eventually the rate of secretion should decrease to zero or slightly above zero (to a rate at which granule formation equals the rate at which protein is secreted) as the cell becomes depleted of its granule contents. If we plot the decrease in granule number against the amount of protein secreted, a linear function should be formed that intercepts the axes at approximately the origin (no loss of granules, no secretion of enzyme*), and in addition, our function should reach a limit at 100% of the granules lost (no more can be secreted than can be accounted for by 100% of the available granules).

Therefore, in terms of the exocytosis model, during the augmented secretion of digestive enzyme: (1) decreases in granule volume density should be proportional to decreases in granule number; (2) decreases in

*Actually this is a low rate of secretion, at which granule formation and release are equal.

granule number and volume density should be inversely proportional to increases in the rate of protein secretion; and (3) augmented secretion should cease when the cell is depleted of its granules. Is this the case?

9.4 DOES A DECREASE IN GRANULE NUMBER ACCOUNT FOR DECREASES IN VOLUME DENSITY DURING ACTIVE SECRETION?

As I have already mentioned, decreases in the size of granules are seen during active secretion, even in the absence of changes in number. In this case, *all* of the decrease in granule volume density is attributable to changes in the *size* and not the number of granules. Changes in granule size during augmented secretion were first noted by Heidenhain's contemporaries, Kühne and Lea, viewing living rabbit pancreas in the light microscope, and were confirmed by Heidenhain in fixed dog pancreas. While the changes in the number of granules in cells during active secretion that were reported by Heidenhain formed the foundation of the vesicular theory of secretion, the roughly contemporaneous observation that granules change size has lived a life of complete obscurity until quite recently. The literature contains numerous, and sometimes rather striking cases, where subsequent to the application of a secretory stimulant, the size of granules in cells decrease. Often the fact is not noted, and perhaps not even noticed. Or if noticed, assumed to be due to the presence of newly formed granules. Recently, Ermak undertook a reexamination of Kühne and Lea's observations of size change in the pancreas under the electron microscope. He measured changes in the size and number of granules in cells by feeding rats whose acinar cells were filled with zymogen granules after an overnight fast. He found that large changes occurred in the size of granules and that the average size decreased by over 50% after 90 min of feeding (Ermak and Rothman, 1981) (Figures 4, 5 and 6; Table 1). These decreases were seen in cells that appeared to have fewer granules, but also in those in which no change in granule number was seen (Figure 1 and Table 1). Moreover, changes in granule number were not seen without accompanying changes in granule size (Figure 1; note the absence of closed circles in quadrant IV). This is important because it suggests that changes in size and number are related. Beyond that, it suggests that a decrease in granule number may *depend upon* a decrease in granule size; that is, it may be that if and only if granule size decreases, does number decrease.

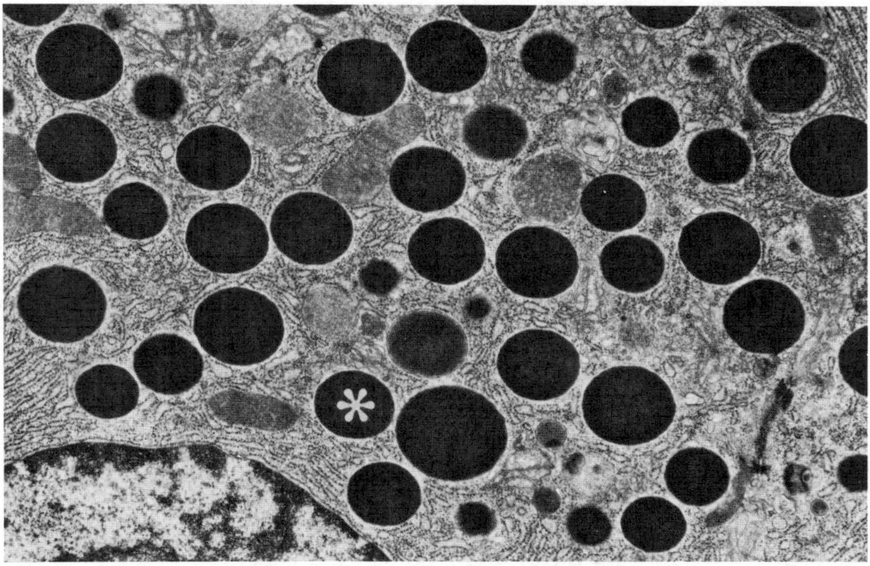

Figure 4 Zymogen granules from the pancreas of a fasted rat. Granules range in diameter from 0.7 to 1.3 μm and average 0.9 μm (*). Magnification, 12,000×. (Reprinted, with permission, from Ermak and Rothman, 1981.)

This is also consistent with the fact that smaller granules are sampled less frequently in electron micrographs, and therefore lead to an underestimate of true granule number relative to larger granules. As I have already mentioned, it is often overlooked that substantial apparent decreases in granule number may merely be illusory, due to differences in sampling probability when granule size is reduced. Therefore, granule disappearance, real or illusory, may occur *because* size decreases.

Thus, the prediction of the paradigm that changes in volume density are the result of changes in granule number is clearly incorrect, and changes in size can account for a large percentage, and under some circumstances all, of an observed change in volume density during active secretion; and most significantly, decreases in number may even be attributable to changes in size.

Decreases in granule size can be explained within the paradigm in two ways: (1) decreases in size reflect the formation of new, smaller granules, and (2) the decrease in size does not indicate the loss of protein from the granule but is attributable to an "osmotic" event. In the former case, for the feeding experiment just described, we would have to propose that during

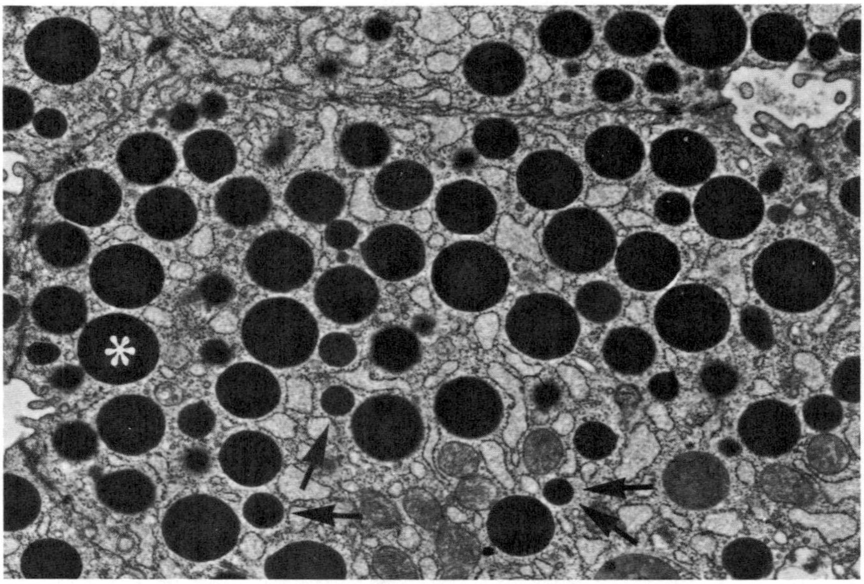

Figure 5 Zymogen granules from the pancreas of a fed rat. Granules are slightly smaller in size (average diameter 0.7 μm), ranging from 0.3 (double arrows) to 0.9 μm (*), but not fewer in number. Small granules with distinct outlines (arrows) are several micrometers smaller than the smallest granules from fasted rats (Figure 4). Compare the distribution of granules sizes in Figure 4 with that in Figure 5 (same magnification, 12,000×) relative to the two granules with asterisks which are the same size. In Figure 5 the profile containing the asterisk is the largest in this section, whereas in the section from the fasted rat (Figure 4) granules of this size are about average. (Reprinted, with permission, from Ermak and Rothman, 1981.)

the 90-min period of feeding, the granule pool had turned over numerous times. If the pool had been replaced just once in 90 min, then some 50% of the granules originally present would still be in the cell and of course be of the original size, thereby producing a bimodal distribution of granule sizes. This is not seen. The distribution of granule sizes after feeding is essentially the same unimodal log-normal distribution observed in the fasted animal, but displaced to lower mean granule diameters (Figure 6).

A more complex notion can also be considered, that is, old granules could be secreted first, new ones following in line. This hypothesis would only require that the granule contents of the cell turn over once during the experimental period in order to produce a new normal distribution of granule sizes. However, there is no indication in micrographs of these tissues that large granules are lined up at the apical surface of the cell, while

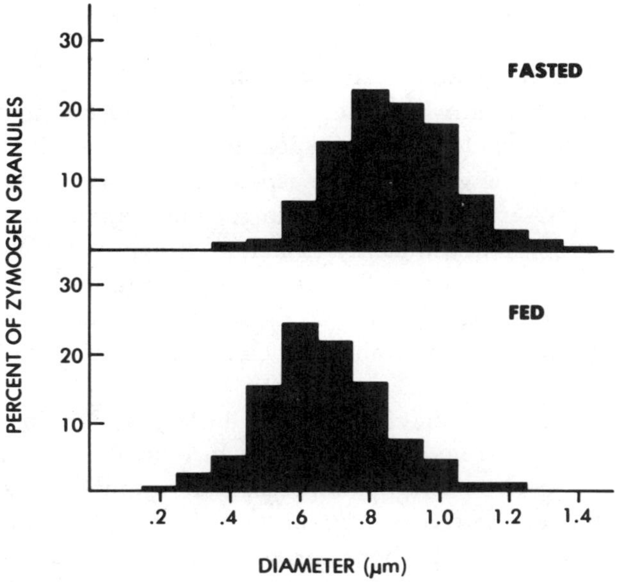

Figure 6 Distribution of the diameter of zymogen granules in the pancreas of fasted and fed rats. Data expressed as percentage of total population (see Table 1). In fed animals, granules 0.1 μm in diameter were also seen, although not shown in this figure, constituting less than 0.1% of the population. (Reprinted, with permission, from Ermak and Rothman, 1981.)

smaller ones are placed more basally. The granules appear to be more or less randomly distributed in terms of their size.

In any event, from our previous discussion, it seems clear that only a small fraction of the granule pool could have been replaced during the 90-min period of feeding, and therefore the decrease in the size of granules, some 50% on the average, was an occurrence that primarily involved the same granules that were present at the outset of the study; that is, individual granules decreased in size during active secretion.

We can estimate the amount of protein secreted into the duct system during this 90-min feeding period maximally at about 15 mg/g of tissue. If we estimate the amount of enzyme that would have been lost from granules, if release from them were proportional to the changes in granule *volume* [on the basis of estimates of intragranular protein concentration (see Ho and Rothman, 1983)], then the observed change in granule volume (approximately 50%) would account for between 15 and 30 mg of protein, or an amount of protein similar to that secreted. Although this is consistent with the hypothesis that protein is being released from the granules under these circumstances, an osmotic explanation that does not involve protein transport out of the granule is also possible.

An osmotic effect in which the protein content of the granule would remain unchanged could be of two types. In the first, the osmolarity of the medium in which the granules are suspended, the cytoplasm, would be increased. The observed 50% decrease in granule volume, would require a doubling of cytoplasmic osmolarity, namely, from approximately 300 to 600 mOsm/L. There is no evidence that such a large change in osmolarity takes place during augmented secretion on the basis of measurements of tissue and secretory fluid osmolarity, and this possibility can I think reasonably be eliminated.

The other type of osmotic effect involves the loss of solute, other than the digestive enzymes, from the granules, which in turn would cause a dependent flow of water in its wake, thereby leading to a decrease in granule size. For such an effect to account for substantial changes in granule size, the solute being moved must be the major solute of the intragranular medium, that is, the solute that accounts in great part for its osmolarity. In addition, there would have to be a concentration gradient for this solute during unstimulated secretion, because its dissipation would be required to provide the force for the efflux of water. A twofold concentration gradient could account for a 50% reduction in granule volume. Potassium salts would seem to be the most likely candidate based on their high concentration in tissue, however, the membrane of the zymogen granule does not contain a Na^+, K^+-ATPase thought necessary for the maintenance of a concentration gradient for K^+.

Whatever gradient-producing mechanism we might consider, an osmotic effect seems unlikely for another reason. In the feeding experiment just described, the change in granule volume was 50%, but only *on the average.* Since volume is a cubic function of radius, a given change in the radius of a large granule would produce a smaller relative decrease in volume than an equivalent change in the radius of a small granule. Since roughly equivalent changes in the radius of granules of various sizes is what occurs (as evidenced by a shift in a normal distribution of granule diameters) (Figure 6), the same percentage change in volume does not apply to each granule. Indeed, because we are dealing with a cubic function, the range of differences can be immense. An osmotic effect requires that the changes in volume be proportional; that is, that the volume of large, medium, and small granules all change by the same proportionate amount—about 50% in the study cited. We can appreciate the very great deviation from this prediction by considering the case of the small granule. If we consider conservatively that the smallest granules seen after stimulation were derived from the smallest present before stimulation, then very large changes in the volume of these granules, in relation to the average change, must have taken place. For example, a change from a 0.4-μm diameter granule to one of 0.1 μm, the approximate lower limits in both fasted and fed groups, would result in a decrease in granule volume of more than 99%. Far greater than the average or mean value of 50%!

This example also shows why the size changes must include the loss of protein from the granule. Estimates of the contents of the zymogen granule indicate that about 60% of its mass is water and about 40% solids, in great part the digestive enzymes themselves (Ho and Rothman, 1983). Thus, an osmotic effect could maximally reduce the volume of the granule by 60%, if it were totally depleted of intragranular water. Thus any change in volume greater than 60% would require the loss of solids (protein) contained in the granule. For the smallest granule case considered above, virtually all its protein contents would have been released (over 99%).

Thus, the prediction of the paradigm that changes in the volume of the cell occupied by zymogen granules, that is, changes in their volume density, are due to and are therefore strictly proportional to changes in granule number is clearly incorrect. Granule size also changes with changing functional state, and not only must be factored into the equation as a cause of changing volume density, but almost wholly accounts for such changes in some situations. Certainly, the "depletion" of granules and reductions in the volume of the cell that they occupy cannot be assumed to solely reflect decreases in their number.

During the prenatal period when granules first form in the acinar cell, over 90% of the increase in their volume density is attributable to changes in granule *volume,* and not number (Ermak and Rothman, 1983). Similarly, immediately after birth when active secretion first starts, granule volume decreases by about an order of magnitude during the first days and weeks of life (Ermak and Rothman, 1980, in submission 1985). Thus, the size of granules varies with secretory state, and a "granule size" secretory cycle exists that is analogous to the classical "granule number" secretory cycle in which granule number decreases with feeding and increases after digestion is complete. That is, it also appears to be true that granule size decreases during active secretion and increases during periods of inactivity and increased storage.

9.5 ARE DECREASES IN GRANULE VOLUME DENSITY DURING AUGMENTED SECRETION INVERSELY PROPORTIONAL TO THE AMOUNT OF ENZYME SECRETED?

The paradigm predicts that the loss of granules should be reflected in a proportional increase in the amount of protein secreted by the gland, that is, it predicts that these two variables are inversely proportional to each other in a linear fashion. Some years ago we tested this prediction in rats, measuring both the volume density and the amount of enzyme released during highly augmented secretion. When changes in the two variables were plotted against each other, the relationship was found to be nonlinear (Figure 7). The ratio of granules lost to protein secreted was greatest at the onset of augmented secretion when the gland was filled with granules, and it decreased as the number of granules in the tissue diminished. Thus, the relationship between the change in volume density (number of granules lost in the paradigm's terms) and the amount of protein secreted was variable. It was not fixed or constant in accordance with the exocytosis model. Less protein was secreted for a given amount of "granule loss" when the cell was filled with granules, than when granule volume density was low.

To explain this nonlinear relationship in terms of exocytosis, we must propose that the rate of granule formation increases during augmented secretion, becoming increasingly greater as fewer and fewer granules remain. In this way it would only appear as if fewer granules accounted for a given amount of protein secretion, whereas in fact the rate of cycling of granules had been increased instead. Such an increase in the rate of granule formation would have to be linked to the number of granules present in the cell at a given time, the rate increasing as the number of granules per unit volume decreased. That is, the cell would have to know how many granules it contained and adjust the rate of granule formation accordingly.* A simple stimulus-induced increase in the rate of granule formation would not suffice; this would merely change the slope of the linear function in an exocytosis context, increasing the amount of enzyme lost per granule, but would not produce a nonlinear function.

Looking at this question in another way, if the ratio of granules lost to protein secreted seen at the onset of active secretion represents the true,

*One could develop an analogous model for variations in intragranular protein concentration. The same issues apply.

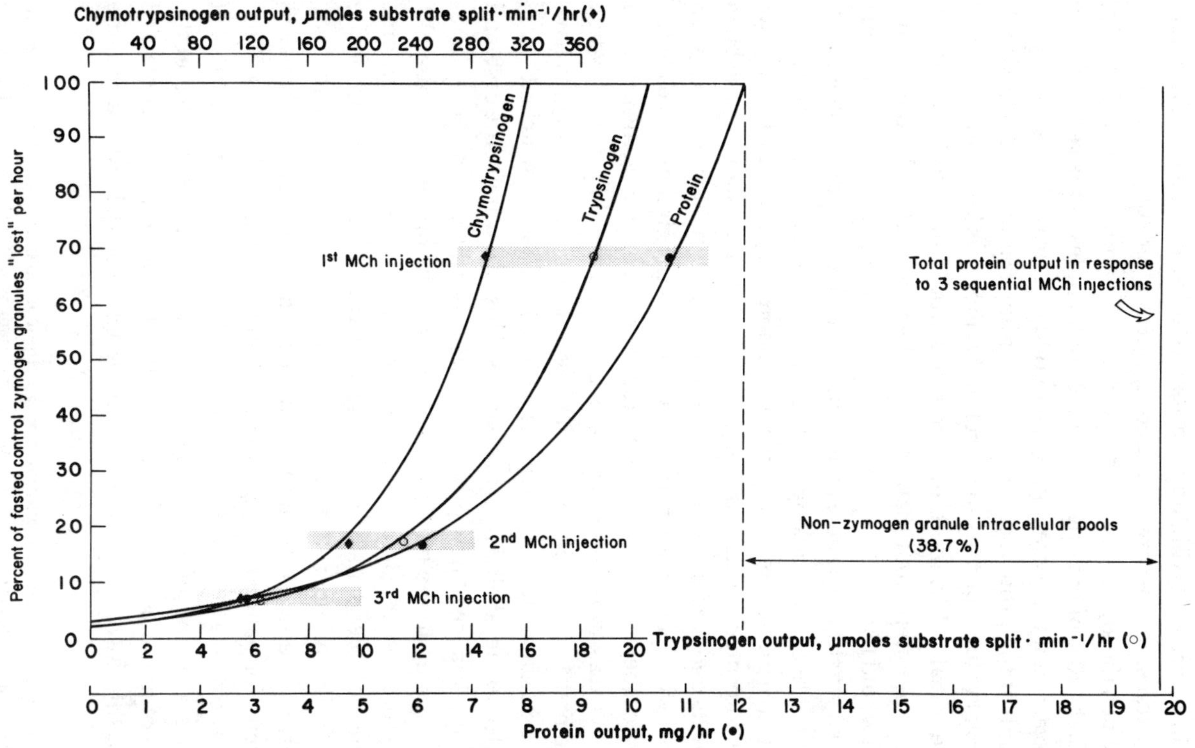

Figure 7 Relationship between degranulation and chymotrypsinogen, trypsinogen, and protein outputs. A breakdown of intracellular sources is shown for protein secretion; 100% of granule contents of fasted rat pancreas only accounts for about 60% of total protein secretion. The remainder must be ascribed to nonzymogen granule intracellular sources. The relationship between the amount of enzyme secreted and the loss of granules that occurred pursuant to sequential injections of a cholinergic agonist (methacholine, MCh) is nonlinear ($y = ae^{bx}$) for all three parameters. (Reprinted, with

fixed ratio in accordance with the paradigm (or 0.15 mg protein/1% granule volume for the study we are considering), then after some 3 hr of active secretion when the amount of protein secreted was approximately 3 mg/hr, approximately 20% of the original number of granules in the cell would be required to account for secretion over a period of 1 hr (N.B., less would be required only if the protein concentration within granules had increased). Since the total volume density in the cell at this time was less, approximately 17% of the initial value, more new granules would have to have formed during the subsequent hour than were initially in the cell at the outset, that is, the pool would have to turn over completely within the hour. If this were the case, then we should be able to determine it by labeling the proteins with a radioactive tracer and looking for the appearance of labeled protein in secreted fluid. If granule release occurred randomly by exocytosis, then half of the labeled protein made at the beginning of the hour should have been secreted by its end, and 25% should be secreted within 30 min. When such experiments were performed for this particular set of circumstances in the rat, almost no labeled protein was collected in secretion during the first hour (Figure 8). If we assume ordered release, in which case old granules would be released first,* then by the end of the first hour we would expect to have reached the maximal attainable specific radioactivity as we "switched" from the unlabeled to the labeled pool of granules. This also was not the case, and the rate of appearance of new protein was actually diminished in the stimulated, as compared to the unstimulated state, not the opposite (Figure 8).†

Furthermore, over the total period of study there was a decrease of some 95% in the volume density of zymogen granules, or in terms of the paradigm, in the number of granules. Therefore, whatever changes may have occurred, the rate of granule formation and loss were nevertheless very different, formation having occurred at only 5% of the rate of loss. That is, as I said above, we can consider changes that occur in the granule pool of the cell over a period of a few hours primarily to reflect changes in the population of granules that were in the cell at the outset. In the context of exocytosis, this low rate of granule formation would hardly affect the

*As I said above regarding granules of different size, there is no indication that granules are lined up in the cell in an ordered sequence in which first made would be first released. In fact, to the contrary, it is clear that new protein tends to be secreted first while the older pool remains relatively intact (Rothman and Isenman, 1974).

†This observation is explained by the equilibrium model in terms of the release of older, stored enzyme into the cytosol during active secretion, thereby competitively inhibiting the release of new (labeled) protein which is present at a much lower concentration (see Chapter 12).

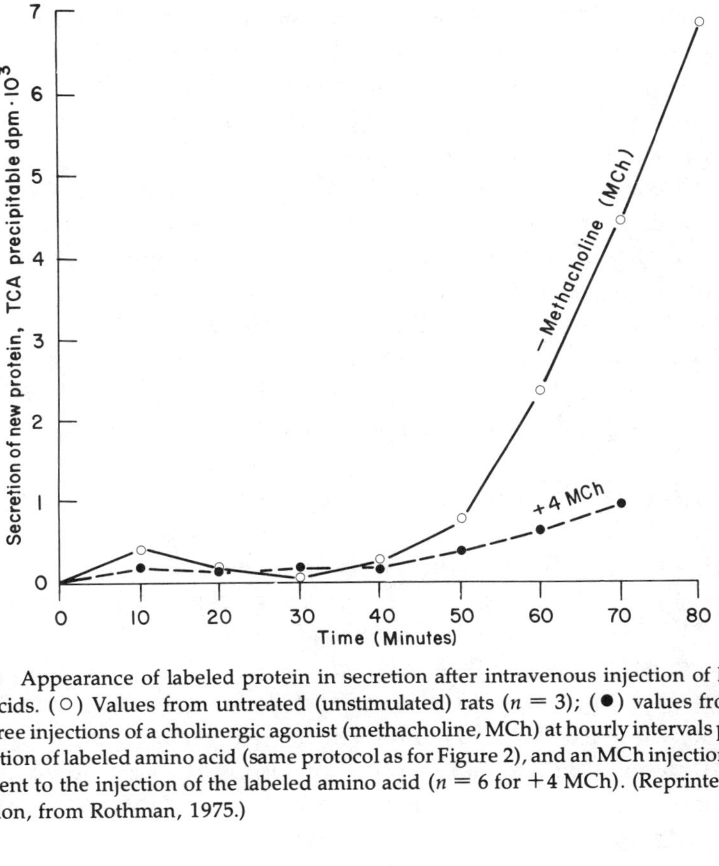

Figure 8 Appearance of labeled protein in secretion after intravenous injection of labeled amino acids. (○) Values from untreated (unstimulated) rats ($n = 3$); (●) values from rats given three injections of a cholinergic agonist (methacholine, MCh) at hourly intervals prior to the injection of labeled amino acid (same protocol as for Figure 2), and an MCh injection 0.5 hr subsequent to the injection of the labeled amino acid ($n = 6$ for +4 MCh). (Reprinted, with permission, from Rothman, 1975.)

observed relationship between the amount of protein secreted and the number of granules lost. Therefore, the nonlinear relationship between granules lost and protein secreted seen during the course of this study appears to reflect the kinetic pattern of secretion itself, not a balance between the formation of granules and their "release," and as such is not consistent with the transfer of material by a two-compartment exocytosis-type system.

This view is reinforced by the following. Tissue content of amylase decreases after this course of stimulation by between one-third and one-half of that initially present (Ho and Rothman, 1983; Miyasaka and Rothman, 1982). If we assume that this decrease applies to all secretory protein, then from an initial pool of some 70 mg/g tissue, we would lose between 23–30 mg/g of tissue, or more or less the same amount that was secreted (approximately 20 mg). That is, as I discussed above, we have every reason to think that the material secreted was predominantly derived from a preexisting pool that was being depleted as a result of secretion, which, if it was being replenished, was being replenished at a relatively insubstantial rate compared with rate of secretion.

In addition to the question of the nonlinear nature of the function, if protein secretion were wholly accounted for by the direct transfer of material between two compartments, as in exocytosis, then more protein could not be secreted than could be accounted for by the loss of all the granules. In this case, if we extrapolate the function in Figure 7 to 100% granule loss, then this value should correspond to the maximum amount of protein that could be secreted by the gland under the particular set of circumstances, whatever the shape or slope of the function and whatever changes in turnover might have occurred. However, the value for 100% granules lost was about 12 mg of protein or only about 60% of the amount secreted during the experiment. The rest (approximately 40%) must be attributed to nongranule sources within the cell.

Thus, a two-compartment model appears unable to explain the observed relationship between protein secretion and granule loss. The curved form of the relationship between these two variables seems to require the interposition of a third, and presumably cytoplasmic, compartment in the transport sequence that leads to the release of digestive enzyme from the cell, as the equilibrium model proposes. In this case the granule would act as a "protein capacitor," releasing its contents into the cytoplasm of the cell on demand. As such, the loss of material from the granule and its appearance in ductal fluid need not be equal in the short run. Such a three-compartment model accounts in a simple way for the nonlinear character of the function that relates protein secretion to granule disappearance, as well as for the 40% of protein secretion that cannot be attributed to granules in a paradigmatic context.*

*This model can also explain why estimates of the rate of granule formation during depletion (the falling curve in Figure 2) are about 1.5% of the initial pool per hour, whereas estimates based on the recovery of granule content after depletion (the rising curve in Figure 2) are 10 times this value. Since we are merely dealing with the "discharging" and "recharging" of a capacitor in this model, these estimates of the rate of granule formation do not apply and the disappearance and reappearance of granules reflect this capacitive function and not granule turnover.

9.6 DOES SECRETION CEASE WHEN THE GRANULES ARE NO LONGER PRESENT?

It was first noted many years ago by Grossman and his colleagues (Lin and Grossman, 1956) that the secretion of digestive enzyme by acinar cells was maintained at highly augmented rates, even though by light microscopy standards, granules no longer seemed to be present within them. Somewhat more recently, Jamieson and Palade (1971) also reported digestive enzyme secretion in the absence of zymogen granules. My colleagues and I have also studied this question. As I have discussed, in the fasted rat after 3 hr of highly augmented protein secretion, the volume density of zymogen granules was decreased by about 95%. Therefore, these cells were essentially devoid of granules at this time (at least as estimated under the light microscope) (Figure 3). However, digestive enzyme secretion nevertheless continued at a highly augmented rate (approximately one-third of the maximum or peak rate and a 10-fold increase over time-paired unstimulated controls) (Table 2). Thus, digestive enzyme secretion does not cease when the cell is emptied of its zymogen granules. Secretion can continue and at a highly augmented, although somewhat diminished, rate. Once again the prediction of the paradigm is not borne out by the evidence and a cytoplasmic origin for at least some secreted material seems indicated.

TABLE 2. Enzyme Output from Degranulated Cells in Response to Methacholine Chloride

		Enzyme Output (U/h)	
	Protein (mg/h)	Trypsinogen (μmoles substrate split $\cdot \text{min}^{-1}$/hr)	Chymotrypsinogen (μmoles substrate split $\cdot \text{min}^{-1}$/hr)
Control[δ]	0.23 ± 0.06	0.89 ± 0.18	3.45 ± 0.67
Methacholine	2.83 ± 0.32	6.51 ± 1.04	89.20 ± 13.75

Values are means ± SE. n = Six animals in each group. The degranulated cells had approximately 7% of the control number of granules. Secretion was measured for the hour following the third methacholine chloride injection (see Figure 2) from the cannulated pancreatic duct of anesthetized rats.

[δ]Time-paired control group at 3 hr, unstimulated.

Source: From Rothman, 1975)

Supporters of the paradigm have attempted to explain the secretion of enzyme in the apparent absence of zymogen granules by proposing that under these circumstances protein is secreted in much smaller granules (0.1 μm or less in diameter) that cannot be seen under the light microscope. This argument is based on the fact that small granules are seen in electron micrographs of depleted cells. In accordance with the paradigm they are viewed as new, not old smaller, granules, because granules of this size were not present initially. In any event, the number of such granules required to account for transport by exocytosis is substantially greater than for the larger zymogen granule. For example, the volume of a 1-μm diameter zymogen granule is about 0.5 μ^3, whereas that of a 0.1-μm diameter granule is 1000-fold less. Therefore, 1000 times the number of granules would be required to move an equal amount of material (at the same concentration), and there is no indication that there are a sufficient number of small granules in "depleted" cells to accomplish such a task. In their absence, the rate of turnover of small granules would have to be prodigious in order to account for secretion. Of course, if the same number of small, as normal-sized, granules were present in the cell, then for a given amount of secretion, they would have to be formed and release their contents at 1000 times the rate. In addition, the amount of membrane that would have to be inserted into the plasma membrane would also be increased because the surface/volume ratio of spheres increases as diameter decreases. In this case (from 1 to 0.1 μm diameter), the surface-to-volume ratio would increase by 10-fold; that is, 10 times as much membrane would be required to move an equivalent amount of material. Thus, if we propose the exocytosis of smaller granules, then we must account for the fact that such transport would be much less efficient and would require magnitudinally greater numbers of granules, fusion events, and membrane to move an equivalent amount of material. Of course, if we propose even smaller granules, as has been suggested recently (the exocytosis of the small smooth-surfaced vesicle), the geometrical problems become even more severe and border on the ridiculous.

9.7 SUMMARY

The observation that the number of granules in secretory cells decreases during active secretion is the foundation of the exocytosis model. Indeed, it is this observation that appears to have originally led to its proposal.

Exocytosis is an example of one of two possible types of explanation for the disappearance of granules—granule removal in toto. The other, of course, is their dissolution within the cell. Even though evidence has been on hand for more than a century suggesting the latter possibility, the idea of granule dissolution within the cell has been virtually ignored; and conversely it has been widely held that their disappearance is attributable to their removal from the cell, even though evidence for this point of view in the pancreas has in my view been no more compelling. Three things seem clear at this juncture: (1) granules can change size during active secretion and this change involves the loss of protein from formed granules, (2) the relationship between the rate of protein secretion and granule disappearance is not consonant with the traditional exocytosis construct, and (3) active secretion can be maintained in the absence of zymogen granules. Some of these facts are agreed upon by supporters of the paradigm, who have offered conjectures and ad hoc hypotheses to account for them, most significantly changes in the rate of turnover of the granule pool. However, the evidence on hand does not indicate that even the most active secretion is accompanied by the necessary increase in granule turnover, nor do the kinetics appear explicable solely in a paradigmatic context in any event.

SELECTED REFERENCES

1. Ermak, T. H. and S. S. Rothman. Large decrease in zymogen granule size in the postnatal rat pancreas. *J. Ultrastruct. Res.* **70**:242–256, 1980.
2. Ermak, T. H. and S. S. Rothman. Zymogen granules of pancreas decrease in size in response to feeding. *Cell Tiss. Res.* **214**:51–66, 1981.
3. Ermak, T. H. and S. S. Rothman. Increase in zymogen granule volume accounts for increase in volume density during prenatal development of pancreas. *Anat. Rec.* **207**:487–507, 1983.
4. Ermak, T. H. and S. S. Rothman. Zymogen granule size in pancreas of nursing rats. In submission.
5. Ho, J. J. L. and S. S. Rothman. Protein concentration in the pancreatic zymogen granule. *Biochem. Biophys. Acta* **755**:457–466, 1983.
6. Jamieson, J. D. and G. E. Palade. Synthesis, intracellular transport, and discharge of secretory proteins in stimulated pancreatic exocrine cells. *J. Cell Biol.* **50**:135–158, 1971.
7. Kühne, W. and A. S. Lea. Über die Absonderung des Pankreas, Verhandl *Naturhist. Med. Ver Heidelberg.* **1**:445–450, 1876.
8. Kühne, W. and A. S. Lea. Beobachtungen über die Absonderung des Pankreas. *Untersuchungen Physiol. Inst. Heidelberg.* **2**:448–487, 1882.

9. Lin, T. M. and M. I. Grossman. Dose response relationship of pancreatic enzyme stimulants: pancreozymin and methacholine. *Am. J. Physiol.* **186**:52–56, 1956.

10. Miyasaka, K. and S. S. Rothman. Redistribution of amylase activity accompanying its secretion by the pancreas. *Proc. Natl. Acad. Sci.* USA **73**:5438–5442, 1982.

11. Rothman, S. S. Enzyme secretion in the absence of zymogen granules. *Am. J. Physiol.* **228**:1828–1834, 1975.

12. Rothman, S. S. and L. D. Isenman. The secretion of digestive enzyme derived from two parallel intracellular pools. *Am. J. Physiol.* **226**:1082–1087, 1974.

CHAPTER 10

Vectorial Transport

A vector is a quantity that possesses the attributes of both magnitude and direction. In the transport of substances across planar surfaces two vector quantities exist, one describing movement in each direction. When we talk of cellular transport, we generally refer to the vector for the process (or processes) that carries material into a cell as the influx (J_i) and the one that carries material out as the efflux (J_e). The net vector quantity, or net flux (J_{net}), is the algebraic sum of the two vectors whose orientations differ by 180°: $J_{net} = J_i - J_e$. Thus, we assign two vectors to the transport of substances across surfaces; no more than two because only two are possible, that is, they describe the system's potentialities fully—no less than two, because transport processes are generally reversible and require both vectors.

Saying that a transport process is reversible does not mean, however, that things move equally in quantity or facility in both directions. Indeed this often is not the case. Moreover, biological transport processes tend to be at least operationally irreversible in that time passes and entropy grows. The point is merely that such processes are generally *capable* of reversible or equilibrium behavior, even if in the normal course of events a particular substance is moved preferentially and irreversibly in one direction or another. This concept of transport as a reversible process that involves two vector quantities is different from that proposed for exocytosis. Although exocytosis is often called a "vectorial" transport process, it is actually a proposal for a single vector process for the movement of substances across surfaces. That is, a single vector, efflux, accounts fully for the movement of the substance, and therefore the net vector quantity or net flux is equal to, indeed identical with, efflux: $J_{net} = J_e$ because $J_i = 0$.

In concrete terms the exocytosis model proposes that concentrated material in the secretion granule enters the dilute extracellular environment through a hole, produced as a result of membrane fusion, either by diffusion or as a result of fluid flow. This is the efflux vector and the only vector, and fusion and hole formation are its underlying mechanisms. Therefore, in the exocytosis model the efflux (and net flux) of material, that is, the rate of its secretion, is determined by the frequency with which granule and cell membranes fuse and holes form, other things being equal. The greater the frequency of fusion events, the greater the rate at which the substance is secreted. Thus, in this model J_e or J_{net} is strictly proportional to the rate of fusion.

Can we devise tests of this "univector" hypothesis that can judge its goodness of fit to a particular secretion process? In a series of experiments in pancreas, Isenman, Ho, and I applied one simple test. If it is true that we are dealing with a transport process that has only an efflux vector, then increasing the concentration of the secreted substance or substances in the extracellular medium would not alter the rate of secretion or the net flux of material out of the cell, because it would not affect J_e, the rate of which is set by intracellular events (fusion), nor could it enter the cell because an influx process does not exist ($J_i = 0$). As a result, J_{net} or the rate of secretion would be unchanged and independent of the concentration of material outside of the cell. This, of course, is different from the behavior that we would expect if both influx and efflux vectors, and as a result the potential for equilibrium behavior, did exist; in this case, increasing the concentration of a secretion product in the extracellular medium would increase its movement into the cell. If there had been a net flux of material out of the cell, namely, secretion, before we increased the extracellular concentration, then it would be reduced because $J_{net} = J_e - J_i$ and J_i would have been increased.

Indeed, in a reversible system if we increased the concentration of the transported substrate in the extracellular medium sufficiently, then we should be able to reduce the net flux out of the cell to zero (by increasing J_i to equal J_e so that $J_{net} = 0$). In a system in which there is only a single vector, this could not happen, and, as I have said, secretion should continue at essentially the same unchanging rate whatever the external concentration of the transported substrate.

Therefore, we have a relatively clear predictive difference between the exocytosis model and an equilibrium model for secretion, at least for exocytosis as it was originally proposed and has generally been viewed: a transport process for which only an efflux vector exists. In any event, if the secretion process is equilibrium dependent or at least equilibrium capable, then we should be able to reduce the rate of secretion merely by letting secreted material accumulate in the external medium; in this way influx or backflux would increase over time. If only an efflux vector existed, then such an accumulation of product should have no discernible effect on the rate of secretion, which in this case should continue unabated.

10.1 DUCT BLOCKAGE

In an exocrine gland such as the pancreas, the simplest way to test this idea is merely to block the secretory duct and to prevent the outflow of material contained in the duct system. In this way, the removal of secreted product from the immediate environment of the secretory cell can be prevented. If exocytosis is the operating mechanism, then under these circumstances we would expect to see secretion continue at a constant rate into the duct system, and as a result we would expect the concentration of product in the duct lumen to increase in a roughly linear fashion during the period of blockage as secreted protein continues to accumulate in the blocked duct.

If, on the other hand, an equilibrium-based process is involved, then preventing the removal of material from the site of its secretion and letting it accumulate in the immediate vicinity of the secretory cell should reduce the rate of secretion (as J_i increases) and eventually prevent it altogether (when $J_i = J_e$).

This experiment was performed using the rabbit pancreas in organ culture, an experimental system in which the whole pancreas is placed in a chamber, bathed by a physiological salt solution, and secretion collected directly from the cannulated pancreatic duct (Figures 1 and 2) (Rothman, 1964; Rothman and Brooks, 1965). Blockage is achieved merely by closing off the indwelling catheter. When the catheter is opened, the accumulated material can then leave the confines of the duct (be excreted and collected). A period of blockage was chosen that would lead to a roughly magnitudinal increase in the concentration of protein in the duct system if secretion continued at a constant rate during blockage (Isenman and Rothman, 1979).

What was observed? After a substantial period of duct blockage, the amount of protein secreted during the collection period immediately after

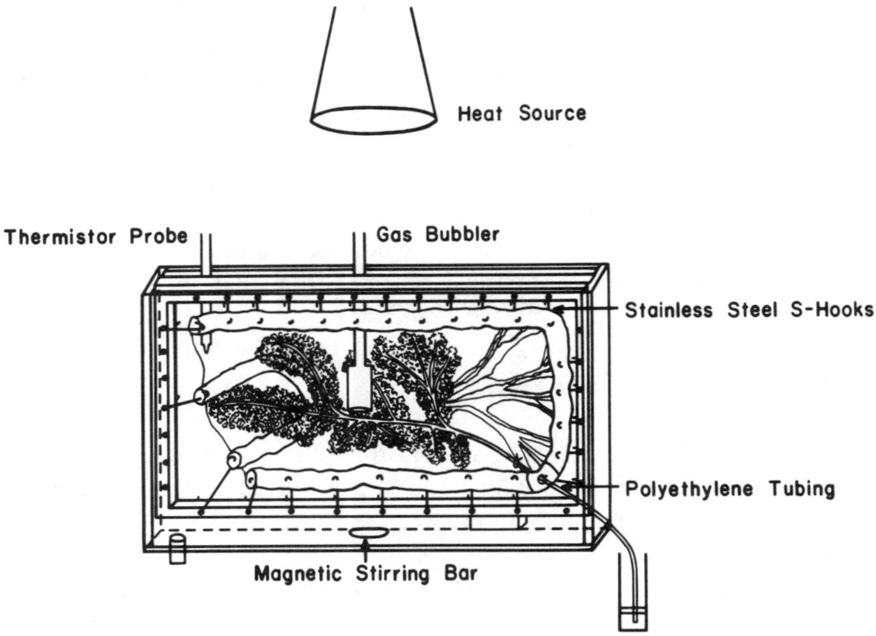

Figure 1 Schematic diagram of rabbit pancreas and duodenum suspended in chamber. The pancreas is diffusely organized and is nearly as thin as the surrounding mesentery (averaging about 0.2 mm thick). The pancreas is entirely located within the mesentery of the first loop of duodenum distal to the pylorus (the mesomentum). The gland has a single major pancreatic duct, separated from the common bile duct, that can be cannulated either intraduodenally via the intestinal orifice or extraduodenally with polyethylene or other similar tubing. The duodenum is attached to a Plexiglas frame either by stainless steel hooks or silk suture, and is in this way used as a scaffolding to suspend the gland. Secretions are collected by passing the indwelling catheter through a hole in the chamber wall. In more recent versions of the device, temperature is controlled by means of an external water jacket, rather than an infrared light source, through which fluid of varying temperature can be circulated, and gas dispersion filters are built into the chamber rather than added free standing. Chamber volume is normally about 400 ml, and can be varied by means of shims and other volume-altering devices. The gland is mounted vertically as shown and secretion collected dependently as is the case in situ in the rabbit. The gland is usually bathed in a balanced salt solution, gassed with 95% O_2 and 5% CO_2 at 30°C containing 5.5 m*M* glucose. An amino acid mixture is added when one wishes to have protein synthesis occur actively. The preparation secretes fluid and electrolytes at rates comparable to that found in situ and maintains an effective osmotic gradient by means of active ion transport, as well as a four- to sixfold bicarbonate gradient, across the epithelial surface. Protein secretion is augmented by all of the common stimulants by up to 10–100 times, a responsiveness that is unmatched by other in vitro preparations of this tissue. (Reprinted, with permission, from Rothman, 1964.)

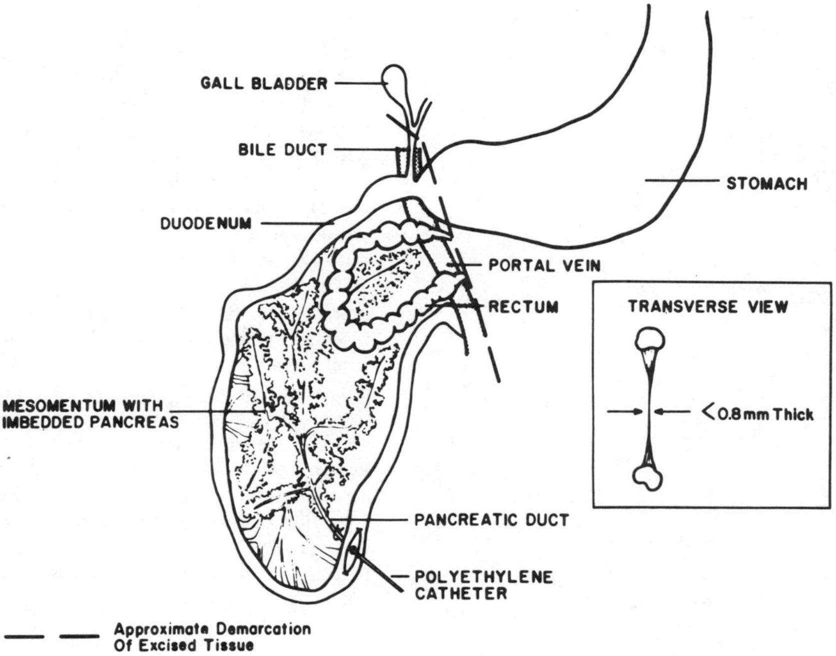

Figure 2 Schematic diagram of rabbit pancreas in situ interposed between ascending and descending limbs of the mesomental loop of duodenum. A polyethylene catheter is shown in place in the pancreatic duct. The dashed line indicates the approximate line of demarcation for excision. The tissue is maximally 0.8 mm thick, and averages about 0.2 mm thick overall. Due to the lobular structure of the gland, the effective thickness is substantially less. (Reprinted, with permission, from Rothman and Brooks, 1965.)

flow had been reestablished was essentially the same as seen prior to duct blockage; that is, no substantial accumulation of material had occurred during a period in which a tenfold increase in intraductal concentration should have developed if a univector event accounted for secretion. Blockage had led to an essentially complete cessation of secretion (Figure 3).

Thus, the results fit an equilibrium and not a "univector" exocytosis view. We explored these relationships further by applying a small end pressure to the contents of the duct system in order to reduce the flow of material out of the duct in a graded fashion. Flow (Q) is determined by the hydrostatic pressure gradient between the site of cellular secretion (P_g) and the end of the collecting duct (P_e); that is, $Q = f(P_g - P_e)$ where P_g is the generating

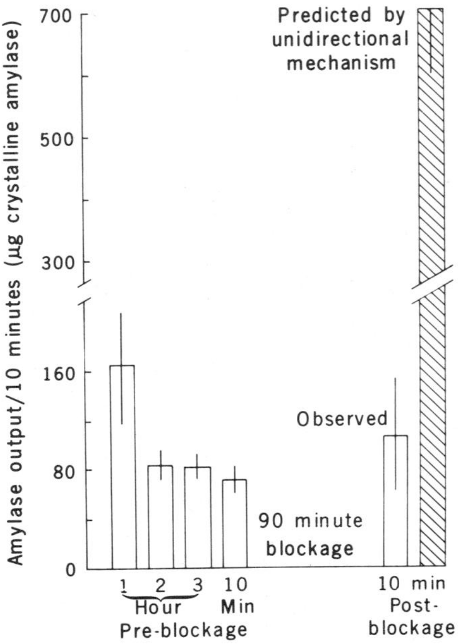

Figure 3 The effect of ductal blockage on amylase secretion (open bars). Data are presented as mean amylase output per 10 min ± SE ($n = 4$). Samples were collected at hourly intervals for 3 hr and for a 10-min period immediately before blockade. Post-blockage data are given for the first 10-min period and compared with the value predicted by a unidirectional transfer mechanism (hatched bar). Secretion during the first period after blockage (10-min post-blockage) was close to that observed for a similar time period just prior to blockage, indicating that secretion did not accumulate to any degree in the duct system during blockade, as would have been expected if a vectorial or unidirectional secretory process accounted for enzyme secretion. Longer collection periods after blockage (up to 45 min) gave the same result. (Reprinted, with permission, from Isenman and Rothman, 1979.)

pressure (due to the osmotic flow of water out of the cell) and P_e is the pressure at the end where the material flows out of the duct system. Therefore, if we elevate P_e, then we decrease the gradient and decrease Q as well. As we decrease flow, we also reduce the rate at which secreted material leaves the immediate extracellular environment of the secretory cell. The greater the reduction in flow, the less readily material enters the large conduits of the duct system. When such studies were performed, we found that as flow was reduced, so was protein secretion and in a directly proportional fashion (Figures 4 and 5). Specifically, the concentration of protein in the duct remained essentially unchanged, and protein "output," or the amount secreted, decreased in proportion to the reduction in flow. When the process was reversed and flow was gradually increased, protein secretion increased in proportion as well.

These observations could not be ascribed to changes in the rate of transit of already secreted material through the duct system, and a true reduction in the rate of release of protein from the cell appeared to be involved.* The fact

*The time required to pass through the physiological ductal dead space in this gland is of the order of a minute and the observations shown in Figures 4 and 5 are for collections made between 10–20 min after a particular pressure was established. See below for further discussion of ductal transit time, as well as Isenman and Rothman, 1979, and Ho and Rothman, 1982.

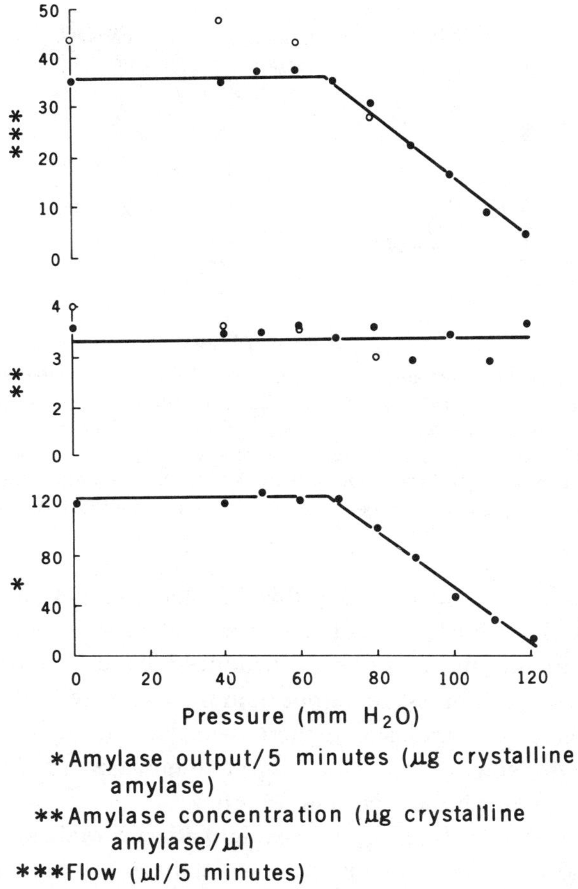

Figure 4 Fluid secretion (flow), amylase concentration, and amylase output as a function of backpressure applied to the column of fluid in the pancreatic duct for a single experiment (●). As backpressure was increased, both flow and output decreased in tandem leaving concentration essentially unchanged. Pressure is given in millimeters of water above the level of the pancreatic papilla of the organ mounted vertically in a Plexiglas chamber (see Figure 1). (○) Values for flow and concentration when the pressure was reduced back toward zero. (Reprinted, with permission, from Isenman and Rothman, 1979.)

that inhibition was completely reversible indicated that there was no permanent damage to the duct system produced by blockage, and, as I have said, the equilibrium and not the exocytosis view of things seemed to be supported by the data.

These observations might nevertheless be reconciled with the exocytosis model in two different ways. Exocytosis itself might have the characteristics of an equilibrating process. There is no a priori reason to exclude such a

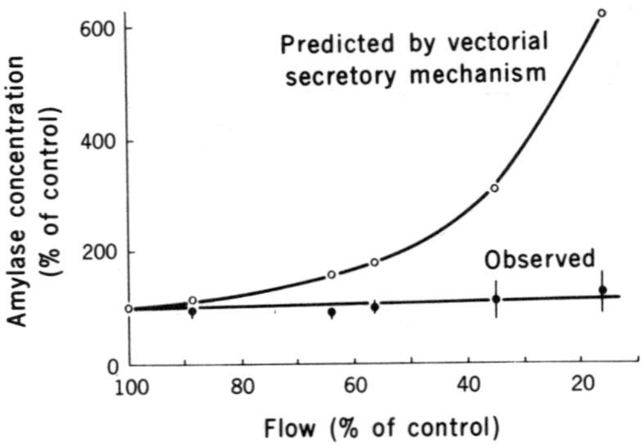

Figure 5 The concentration of amylase in secretion as flow is progressively decreased by the application of a graded backpressure. The amylase concentration observed in secretion (•) remained essentially constant over the wide range of flows studied. Data represent the mean ± SE for four experiments. (○) Increase in concentration predicted if enzyme secretion continued at a constant rate independent of fluid flow, as in a unidirectional (univector) secretory process. (Reprinted, with permission, from Isenman and Rothman, 1979.)

possibility, and we will consider this idea at the end of the chapter. Alternatively, the decrease in protein secretion could have been the result of an increase in intraductal pressure produced by duct blockage or the application of a back-pressure, rather than the decrease in flow. Since the duct system is somewhat elastic, an increase in the volume contained within it is possible (indeed, occurs), and this might increase intraductal pressure which in turn might inhibit protein secretion.

An increase in intraductal pressure might affect protein secretion either directly or indirectly. It could directly prevent the exocytosis of granules. In order to explain the observations in this way: (1) exocytosis would have to be very sensitive to low pressures (the end-pressure required to abolish protein secretion totally is only about 10 mmHg); (2) increases in intraductal pressure would have to produce proportional reductions in the rate of exocytosis (because protein secretion decreases in a linear fashion with increasing back pressure); and most significantly, (3) two seemingly independent processes, exocytosis and fluid secretion, would have to be altered to the same degree, over the same time course, by identical pressure increments, by chance (because protein secretion and fluid flow changed to the same proportional extent at the same applied pressure over the same period of time).

An increase in intraductal pressure might also indirectly reduce protein secretion, or at least appear to reduce it. If duct blockage and back-pressure led to the development of a sufficient transmural pressure gradient to open paracellular shunts through which the bulk movement of material out of the duct system could occur, then secretion by the acinar cells might actually continue at the same "univector" rate despite duct blockage or the application of an end-pressure, but the bulk of secreted material would either wholly (blockage) or in proportion to the applied pressure (back-pressure) be diverted from the confines of the duct into interstitial spaces. Such a shunt would have to be responsive to low pressures, open gradually with increasing pressure, and close reversibly as pressure was reduced in order to explain the observations. There is no indication that such a system exists (see Rothman et al., 1980, for a discussion of this idea, as well as Chapter 15).

10.2 METABOLICALLY PRODUCED REDUCTIONS IN FLOW

Whatever one may think of these pressure-dependent possibilities, the issue is best resolved by reducing flow in ways that avoid the potential for increasing intraductal pressure. This can be done by reducing flow metabolically, that is, by reducing P_g, rather than increasing P_e. In this case, decreases in flow are produced by decreases in intraductal pressure. If the same effect were observed under such circumstances, then flow, and not pressure, could be assumed to be the variable directly responsible for the reduction in protein secretion. We used two different techniques to perform this test (Ho and Rothman, 1982). In the first, the concentration of sodium in the medium bathing the gland was lowered. Water secretion by the pancreas is a sodium-dependent process, and a sufficient reduction in the external sodium concentration reduces the transepithelial osmotic force (by reducing the transport of sodium and associated anions) and hence the transepithelial movement of water. The second method, the addition of the Na^+, K^+-ATPase inhibitor ouabain, inhibits fluid flow for the same osmotic reason, although in a somewhat different manner.

Would a reduction in fluid flow produced in this way also reduce protein secretion? The answer was not only affirmative, but both low-Na^+ media and ouabain treatment produced essentially the same effect as each other and that seen with mechano-physical methods, namely, reductions in protein output proportional to reductions in fluid flow, thereby leaving the concentration of protein in the duct essentially unchanged (Figure 6). The

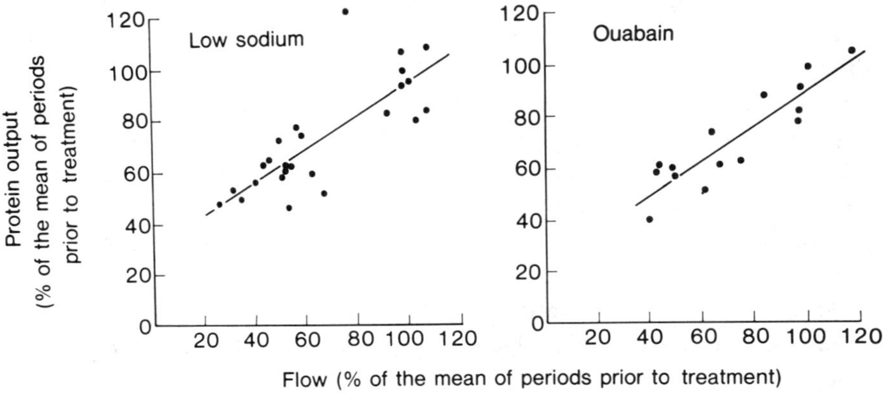

Figure 6 Regression of protein output against flow in unstimulated state. Both protein output and flow are expressed as percent of mean of three 15-min collection periods before either replacement of medium by one containing 50 m*M* sodium ($n = 5$) or addition of 10^{-5} *M* ouabain ($n = 3$). Mean flow rate for this 45-min control period was 27.9 ± 2.2 (SE) μl/5 min ($n = 8$), and mean protein output for the same period was 83.6 ± 24.4 μg/5 min ($n = 8$). Data shown are for 15-min collection period before and four 5-min collection periods after treatment for each individual experiment. Low-sodium: $y = 0.647x + 30.4, r = 0.787, n = 25, p < 0.01$; Ouabain: $y = 0.654x + 23.9, r = 0.903, n = 15, p < 0.01$. (Reprinted, with permission, from Ho and Rothman, 1982.)

same result was obtained when the concentration of protein in the duct system was elevated initially by stimulating protein secretion with the gastrointestinal hormone, cholecystokinin-pancreozymin. That is, reductions in flow were accompanied by roughly proportional reductions in protein secretion, even though in this case, the concentration of protein in the duct system was about an order of magnitude higher than in unstimulated tissues (Figure 7). Thus, even when exocytosis is presumably most active, during augmented secretion, one can reduce protein secretion in a proportionate manner merely by reducing fluid flow.

Even so, these observations, at least if considered alone, might still be reconciled with the "univector" exocytosis model. For example, the reduction in protein secretion may not have been due to a true reduction in cellular secretion, but may merely have reflected a transient decrease in the *excretion* of already secreted material from the duct system. That is, the effect could merely be due to a reduction in the rate of passage of secreted protein through a ductal dead space, as noted above. If this were the case, then once the dead space of the conduit had been cleared, there would be a compensatory increase in protein output, followed by recovery to normal rates over time, reflecting the continued secretion of protein at the cellular

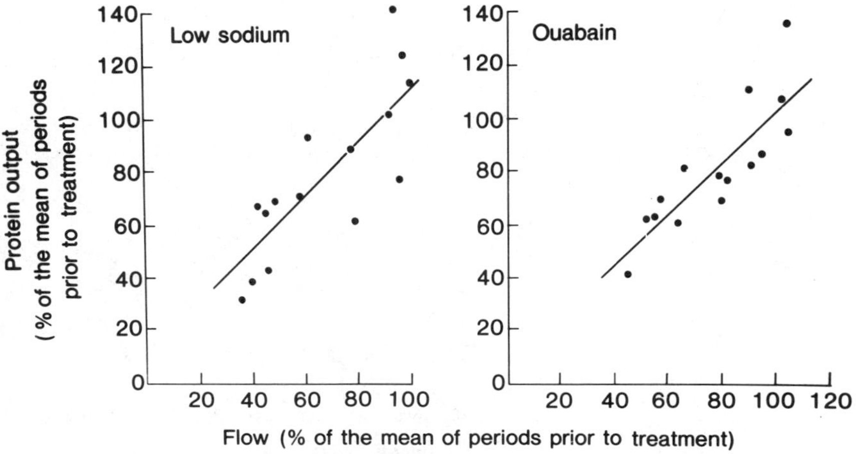

Figure 7 Regression of protein output against flow after stimulation with cholecystokinin-pancreozymin. Both protein output and flow are expressed as percent of mean of four 5-min collections before bath replacement or addition of ouabain (low-sodium, $n = 3$; ouabain, $n = 3$). Mean flow flow rate for this 20-min period was 27.8 ± 1.9 (SE) μl/5 min ($n = 6$), and mean protein output for the same period of time was 1004 ± 156 (SE) μg/5 min ($n = 6$). Data shown are from 5-min collection period before and four 5-min collection periods after treatment for each individual experiment. Low-sodium: $y = 0.999x + 10.9$, $r = 0.788$, $n = 15$, $p < 0.01$; Ouabain: $y = 0.978x + 6.8$, $r = 0.837$, $n = 15$, $p < 0.01$. (Reprinted, with permission, from Ho and Rothman, 1982.)

level. This was not observed. In the low-Na^+ situation, the initial reduction in protein secretion was maintained for many hours during which time the functional, as well as the anatomical dead space, of the duct system had been cleared many times over (Figure 8) (Ho and Rothman, 1982).

Another possibility is that altering ion gradients might affect the activation of a putative exocytosis mechanism. A low-Na^+ medium of course reduces the sodium gradient into the cell, and this might alter the frequency with which exocytosis occurs in some fashion or another. Alternatively, ouabain, even in the short run, might increase intracellular sodium locally which might also in some way reduce the rate of exocytosis. We chose these two particular metabolic interventions in part to help deal with these possibilities. Ouabain leaves the Na^+ gradient into the cell intact (as opposed to a low-Na^+ medium), and conversely, a low-Na^+ medium is likely to reduce, not increase, intracellular Na^+ (unlike ouabain). Nevertheless, both treatments produced almost identical effects, reinforcing the idea that flow was indeed the relevant independent variable.

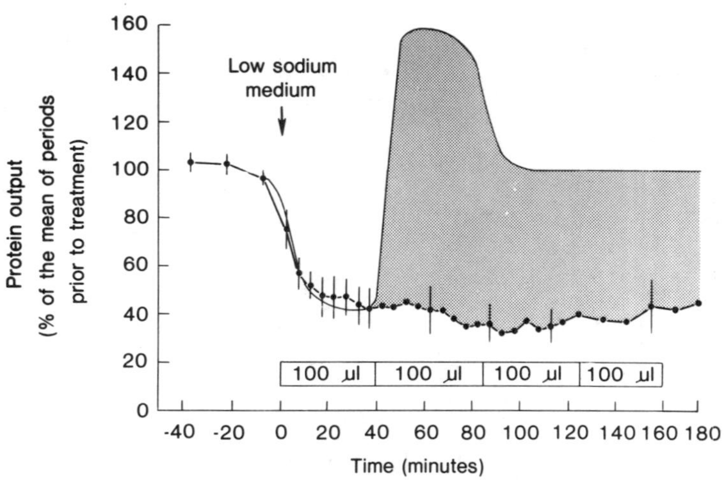

Figure 8 Effect of low-Na^+ medium on protein output in unstimulated state. At time 0 (arrow), normal Krebs-Henseleit medium was replaced by a medium in which Na^+ concentration had been reduced to 50 m*M*. After medium was changed, secretion was collected at 5-min intervals for 2 hr and then at 10-min intervals for an additional hour. Data are expressd as percent of mean of three 15-min collections immediately before change in medium. Mean protein output for this control period was 113 ± 33 (SE) µg/5 min ($n = 5$). Points, means; error bars, SE. One hundred-microliter bars represent average time required to clear 100 µl from duct system, a volume of fluid approximately equivalent to the estimated anatomic plus cannula dead space. Shaded area represents protein output predicted if protein secretion by acinar cell had been unaltered by reduced flow (assuming an initial transient decrease in output due to slowed passage through the anatomic dead space). The inhibition in protein output is due to a true decrease in the rate of cellular secretion and not merely to a delay in the passage of secreted fluid through a ductal dead space. (Reprinted, with permission, from Ho and Rothman, 1982.)

Whatever questions one may be able to raise, it must be borne in mind that not only did two different metabolic methods produce the same effect, but they were essentially identical to that seen when flow was reduced by mechano-physical methods. Moreover, the same effect was seen in both the unstimulated state and during highly augmented secretion produced by a secretagogue. In all cases, reductions in flow were accompanied by proportional reductions in protein secretion. Although different methods reduced flow in different ways for different reasons, they all affected protein secretion in a remarkably similar fashion. Seemingly, the sole common link between the various treatments was the reduction in flow itself.

We tried a final means of reducing flow metabolically. In this case we merely reduced the temperature of the medium bathing the gland (Rothman, 1980). Fluid flow is dependent, as I have said, on ion transport, and ion

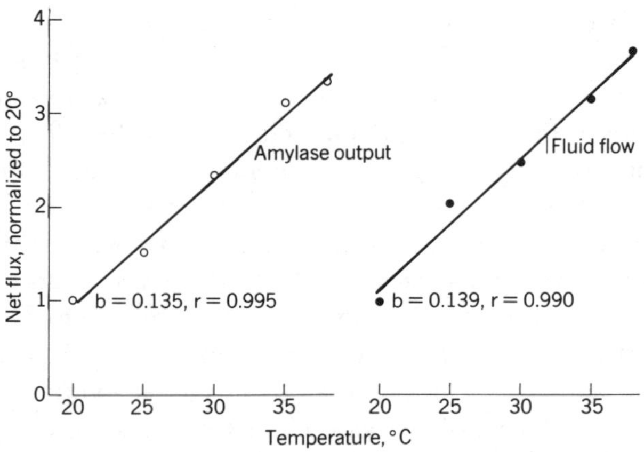

Figure 9 Temperature dependence of fluid and enzyme secretion by whole rabbit pancreas in vitro. Both measurements were made with the same glands and results are presented as mean values for six separate preparations. The temperature dependence of both processes was virtually identical ($b = 0.139$ for fluid flow and $b = 0.135$ for protein secretion). (Reprinted, with permission, from Rothman, 1980.)

transport is a temperature-sensitive process. Although it would certainly not be surprising to find that a reduction in temperature reduced both flow and protein secretion even if they were totally independent processes, one would not expect to find that their temperature sensitivity was almost identical. This is what was observed, that is, over a temperature range of 20–37°C both processes exhibited the same temperature dependence (Figure 9), once again suggesting a connection between the two.

10.3 OTHER WAYS TO EXAMINE THE EQUILIBRIUM BEHAVIOR OF THE SYSTEM

With whatever attendant uncertainty, it seemed reasonable to draw the conclusion that flow was the independent variable that altered protein secretion in these studies. If so, its effect could have been produced in two ways. We have discussed the first, namely, altering the rate of removal of product from the site of secretion. It is also possible that the reduction in ion transport itself might have reduced protein secretion if the two processes were linked in some fashion. The fact that the same proportional effect was observed in both stimulated and unstimulated states, that is, a given reduction in ion transport (and flow) produced very different absolute inhibitory effects on stimulated as opposed to unstimulated secretion, argues against this possibility. However, this question, as well as some of the others that I raised above, could best be dealt with if we were able to alter protein secretion in a volume-dependent fashion that did not require chemically or physically altering the secretion of fluid and electrolyte (and that might also bypass some of the problems related to the passage of secreted material through a ductal conduit). That is, one might still argue that although these results are certainly consistent with an equilibrium explanation for the process and do not appear consistent with the exocytosis model, each test involved an artificial circumstance (end-pressure, duct blockage, the addition of a chemical, removal of sodium, or altering temperature), and it is possible that each in some other unknown fashion could have reduced the rate of protein secretion in a proportional manner to the reduction that they produced in fluid flow.

Partly to deal with this possibility, Ho and I used a system in which one can test the equilibrium behavior of secretion without introducing such "artificial" variables (Ho and Rothman, 1983). It has been known for some time that if one places pieces of pancreatic tissue in a bathing medium and follows the release of an enzyme (usually amylase) into that medium, that the rate of secretion falls off exponentially with time and that after a few hours approaches zero. This decrease has usually been attributed to metabolic fatigue, indicative of the time-dependent failure of the tissue in vitro. We observed the same phenomenon with pieces of rabbit pancreas. Over a period of about 3 hr, the rate of secretion fell off by some 80% relative to that seen initially [from 51.3 ± 7.5 (SE) $\mu g \cdot min^{-1} \cdot g^{-1}$ tissue wet weight for the first 15 min to 8.7 ± 5.1 for the last ($n = 10$)] (Figure 10). However, when the same tissue was studied in the whole organ preparation discussed above and secretion collected directly from the duct system, such a falloff was not observed, and indeed the rate of amylase secretion was maintained relatively constant over the same 3-hr period (Figure 11) (Isenman and Rothman, 1977). Why should the difference between these two very similar situations exist? It occurred to us that the reason for the falloff might be related to the equilibrium nature of secretion and not tissue fatigue. In the organ preparation, secreted material was removed (by fluid flow) from the site of its secretion into a collecting vial, whereas when pieces of tissue were incubated in media, the secretory product accumulated in the bathing medium. Therefore, it was possible that in the second case, the accumulation of product led to the development of a backflux or influx (J_i) of already secreted material into the tissue, in this way reducing the net efflux (J_{net}) over time.

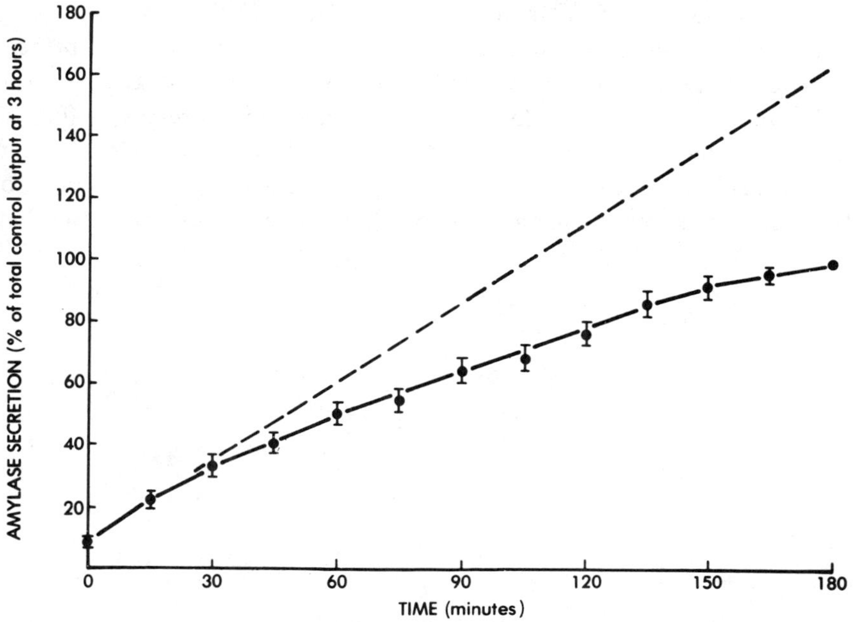

Figure 10 Secretion of amylase by strips of rabbit pancreas. Amylase secretion is normalized relative to total amylase output into bath at end of 3 hr for each individual experiment. Points are means; error bars are the SE; $n = 10$. Dashed line is an extrapolation of the initial rate as determined from the first 15-min collection period. (Reprinted, with permission, from Ho and Rothman, 1983.)

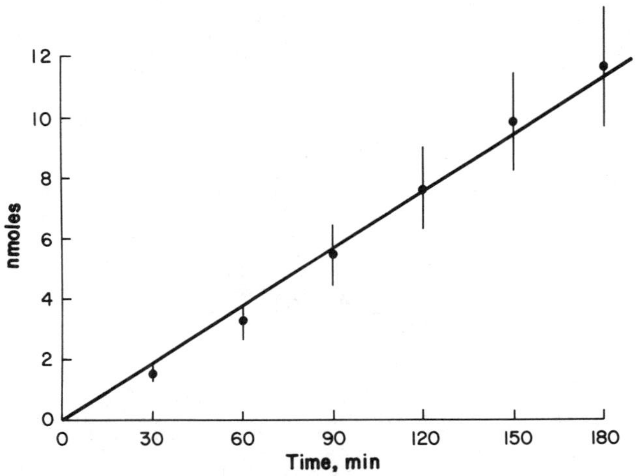

Figure 11 Cumulative amylase output into ductal secretion. Secretion was collected by means of a catheter in the pancreatic duct of the organ in vitro as shown in Figure 1. The points represent the means of nine experiments and the vertical bars the SE. (Reprinted, with permission, from Isenman and Rothman, 1977.)

We first determined whether or not the pieces of tissue were actually viable after a 3-hr incubation period, by adding a stimulant of protein secretion at this time (when the rate of amylase release was negligible). The addition of a cholinergic agonist led to a prompt and full response (Figure 12), and protein secretion was increased to about the same extent as when the stimulant was added immediately after the tissue was incubated in vitro. Thus, the tissue was viable and able to respond to an appropriate stimulus.

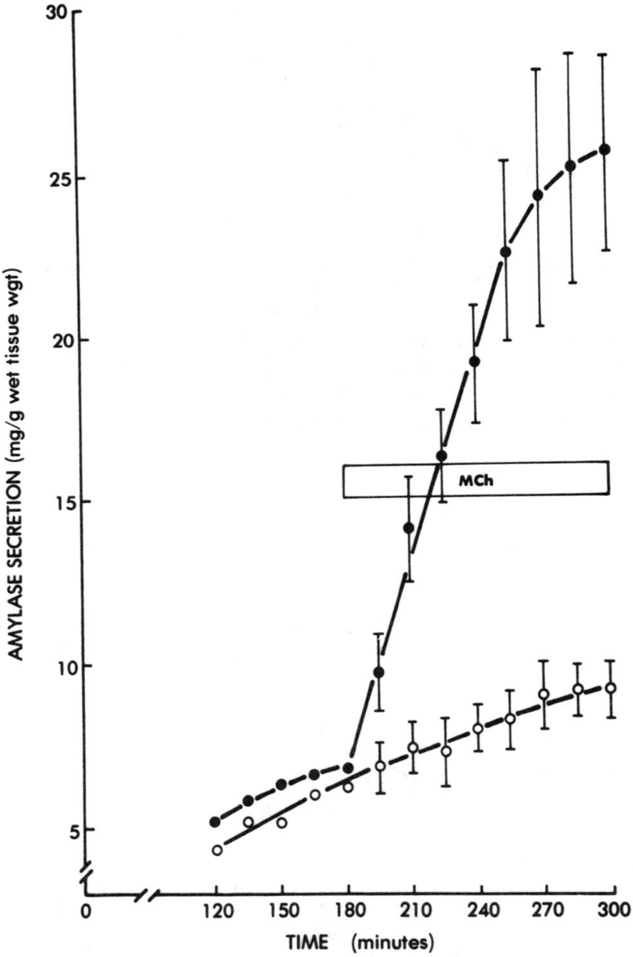

Figure 12 Effect of addition of cholinergic drug acetyl-β-methylcholine chloride (MCh) to tissue strips after 3-hr incubation in vitro. MCh (1 mg/100 ml bath volume) was added to medium at end of the third hour (bar). +MCh (●) and −MCh (○). $n = 5$ for each group. Points are means; error bars are the SE. (Reprinted, with permission, from Ho and Rothman, 1983.)

Next we attempted to determine whether we could prevent the falloff in secretion. If it was true that the reason for the decrease in amylase secretion over time was its accumulation in the bathing medium, then we should be able to prevent the falloff merely by replacing the medium periodically, thereby preventing the development of a substantial J_i and maintaining secretion at or close to the initial rate of efflux (J_e in the absence of J_i). When this was done, secretion was indeed maintained at a high rate over the 3-hr period of study, and the falloff was no longer observed (Figure 13).

In order to determine whether the cause of the falloff was the release of a large molecule from the tissue, rather than the result of the release of a small,

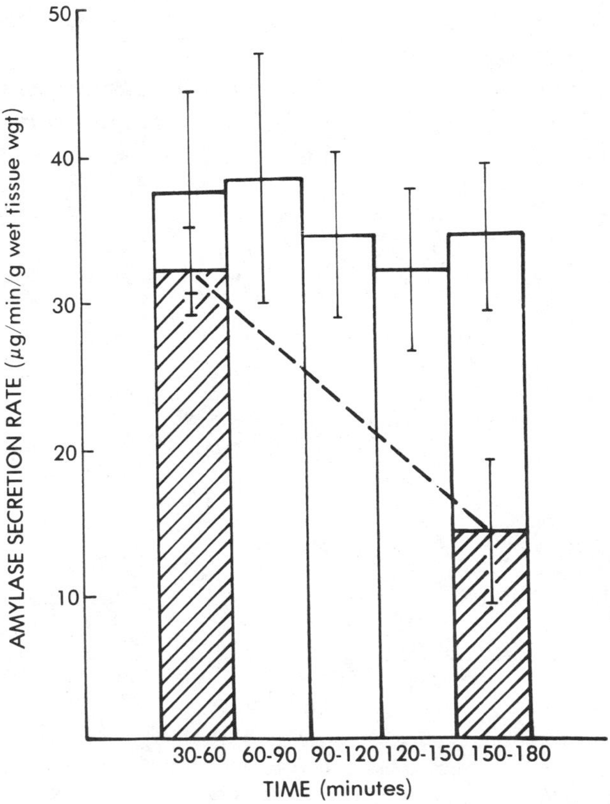

Figure 13 Effect of periodic replacement of bathing medium on amylase secretion by pancreas strips. Medium bathing tissue strips was replaced by an equal volume of fresh medium every 30 min (open bars). In paired controls medium was not replaced (hatched bars) and secretion rate decreased greatly (also see Figures 10, 12, and 14). Medium was sampled just prior to its replacement; $n = 6$ for both groups. (Reprinted, with permission, from Ho and Rothman, 1983.)

dialyzable substance or the rapid utilization of some component of the incubation medium, all of which were dialyzable, we placed the tissue pieces and media in a dialysis sac, which was then bathed by a large external medium. We changed the external medium periodically, but left the medium within the sac unchanged over the 3-hr course of study. When this was done, the falloff in amylase secretion was still seen, as if the medium had not been changed (Figure 14). Thus, it seemed clear that the cause of the falloff was the release of a nondialyzable compound from the tissue.

In order to determine whether the released compound was indeed amylase, we added the amylase substrate glycogen to the medium. We reasoned that if amylase accumulation in the medium had been the cause of the falloff (due to the development of a back-flux), then perhaps an amylase substrate would bind amylase and compete with the influx process, and we might be able to competitively inhibit the development of a back-flux. In this case, we should be able to maintain the rate of amylase release at a higher rate than seen in the absence of the substrate. Not only was this observed, but glycogen acted to prevent any falloff in amylase secretion at all; that is,

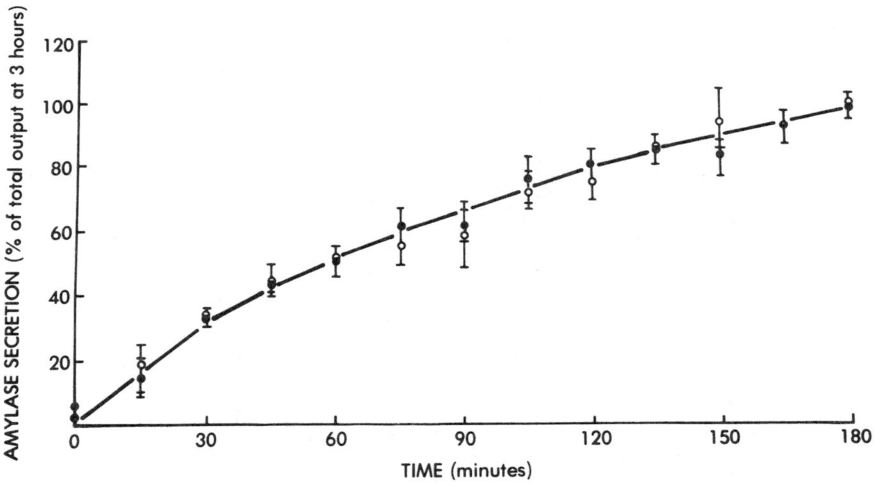

Figure 14 Effect of periodic replacement of an external bathing medium on amylase secretion by tissue in a dialysis bag. Pancreas tissue strips were placed in a dialysis bag (mol. wt. cutoff 6000–8000) containing 100 ml of bathing medium (internal bath), which was then placed in 200 ml of the same medium (external bath). In half of the experiments, the external bath was replaced every half hour (●), whereas in the other half, neither bath was changed over the 3-hr experimental course (○). Data are expressed as percent of bath concentration at end of 3 hr for experiments in which there was no bath change (○). Points are the means; error bars are the SE; $n = 3$ for each group. (Reprinted, with permission, from Ho and Rothman, 1983.)

the initial rate was maintained over the 3-hr course of study (Figure 15). Thus, glycogen was apparently able to prevent the development of a substantial backflux and maintained the rate of amylase release at J_e.

In addition to providing further evidence for the equilibrium nature of the secretion process, including evidence that the released enzyme itself was responsible for the behavior, these experiments demonstrated in addition that it was possible to increase digestive enzyme secretion using equilibrium-based methods, as well as inhibit it, as had previously been shown with the whole gland.

We followed this line of investigation one step further. We attempted to inhibit the secretory response of strips of pancreatic tissue to a cholinergic

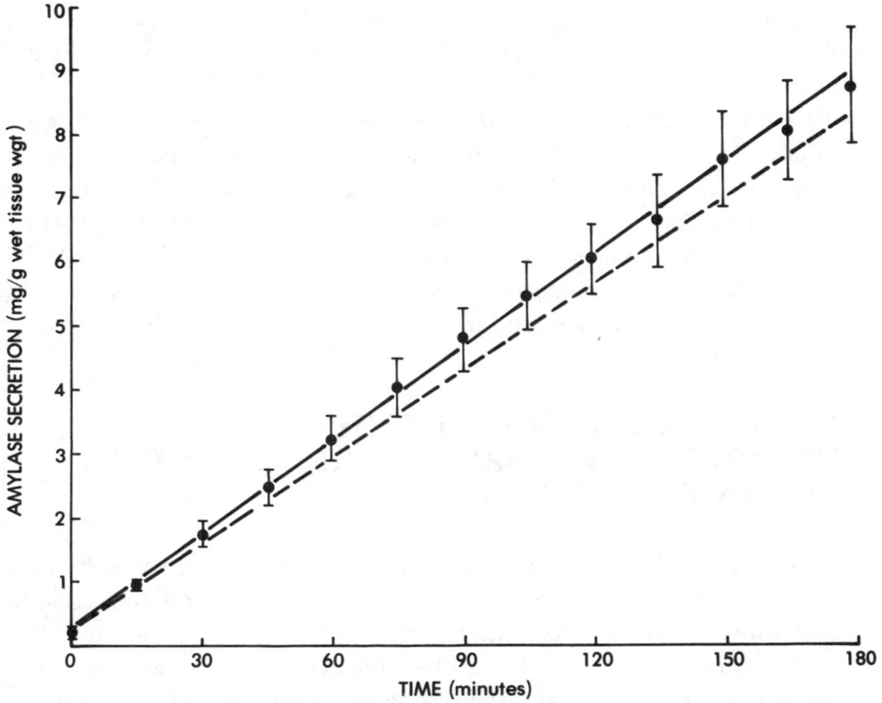

Figure 15 Effect of addition of 0.5% glycogen to medium on amylase secretion. Points are means; error bars are the SE; $n = 3$ for both + glycogen and − glycogen groups. Solid line is for + glycogen data and was determined by linear regression of the means of individual time points ($y = 0.047x + 0.37$, $r = 0.999$, $p < 0.01$). (●) Mean values for each period for the + glycogen group; (----) extrapolation of the initial rate of amylase release from − glycogen group as determined from the first 15-min collection period. (Reprinted, with permission, from Ho and Rothman, 1983.)

agonist by limiting the volume of medium in which secretion was allowed to accumulate. To do this, it was necessary to reduce the volume of the medium to approximate the volume of the tissue itself, so that the sizes of the compartments were roughly comparable. Towards this end, we placed pieces of tissue and a small amount of medium in dialysis sacs. These sacs were then incubated in a much larger external bathing medium. In this way, the tissue had access to the large volume of external medium and all of its dialyzable contents, but nondialyzable substances were of course retained within the sac. Using such an experimental system, we were able to inhibit the response to a cholinergic stimulus by about 90%; that is, merely by reducing the nondialyzable volume of the medium, we were able to abolish almost completely the response to a powerful secretagogue (Figure 16). Moreover, we were able to establish a relationship between the volume of the internal incubation medium and the magnitude of the response, indicating its volume dependence; the larger the volume, the greater the response.

The fact that inhibition was indeed due to the volume restriction for nondialyzable compounds was confirmed in the following manner. Tissue was incubated for several hours in a small sac volume that inhibited the response to a cholinergic drug by about 90%. The dialysis sac was then perforated with numerous holes through which nondialyzable material could pass readily. When this was done, amylase secretion escaped from the inhibition and rapidly increased to highly augmented rates (Figure 16).

10.4 THE PERMEABILITY OF THE BASOLATERAL MEMBRANE OF THE ACINAR CELL TO DIGESTIVE ENZYMES

Another means of examining the equilibrium nature of digestive enzyme transport that can avoid some of the problems of dealing with a ducted gland is to look for movement across other membranes that might be permeable to these proteins. For a number of reasons it seemed likely that the basolateral membrane of the acinar cell might be, and Isenman and I attempted to demonstrate such a permeability (for another example see Chapter 11 for a discussion of zymogen granule permeability). If it existed, studying transport across this surface might be particularly useful, because there were no preconceived notions about such a process, nor was there evidence of vesicular transport across this membrane (although it has since been

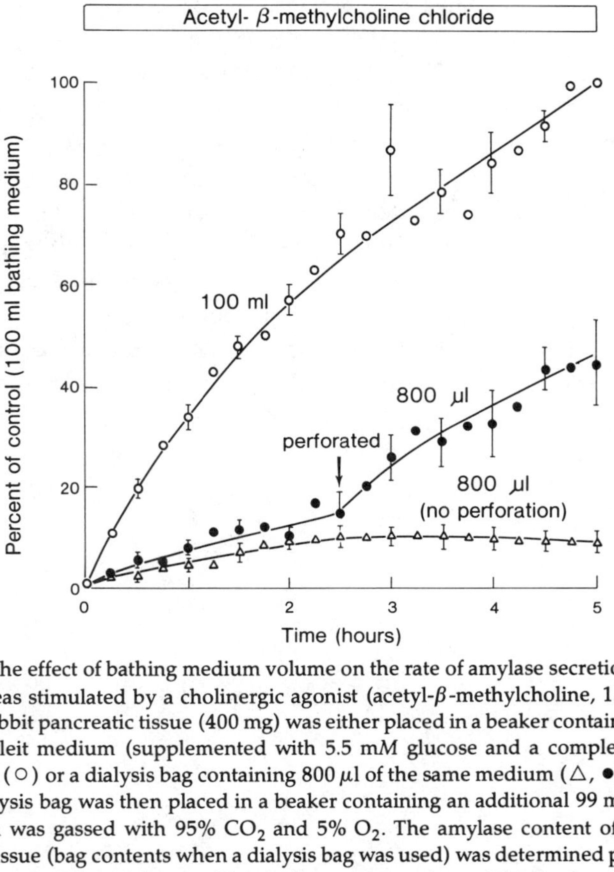

Figure 16 The effect of bathing medium volume on the rate of amylase secretion by strips of rabbit pancreas stimulated by a cholinergic agonist (acetyl-β-methylcholine, 1.0 mg/100 ml medium). Rabbit pancreatic tissue (400 mg) was either placed in a beaker containing 100 ml of Krebs-Henseleit medium (supplemented with 5.5 m*M* glucose and a complete mixture of amino acids) (○) or a dialysis bag containing 800 µl of the same medium (△, ●). In the latter case, the dialysis bag was then placed in a beaker containing an additional 99 ml of medium. The medium was gassed with 95% CO_2 and 5% O_2. The amylase content of the medium bathing the tissue (bag contents when a dialysis bag was used) was determined periodically as noted. In one set of dialysis bag studies, the bags were perforated 2.5 hr after the start of the experiment (arrow). Amylase output is expressed as percent of control; that is, output at 5 hr for experiments where tissue was placed directly in 100 ml of medium (○). The points are means and the error bars the SE. Six experiments of each kind are shown.

proposed). Exocytotic figures had not been reported, nor was there any indication that granules were normally oriented in an endocrine direction in the cell, although they were at times found in the basolateral region.

We were able to demonstrate amylase release across the basolateral surface into interstitial space, and our studies suggested, as later work confirmed, that there is a natural endocrine secretion of at least certain digestive enzymes that is responsive to stimulants and is distinct from ductal secretion [Isenman and Rothman, 1977; Miyasaka and Rothman, 1981; Grendell and Rothman, 1982].

Amylase transport had the characteristics of an equilibrating process. Isenman followed the appearance of amylase in the medium bathing the rabbit pancreas in culture (Figures 1 and 2). Using this preparation, ductal and basolateral secretions can be followed and collected separately. She found much the same kinetic pattern of release as was seen several years later in the tissue strip experiments that I have just discussed. Release was greatest initially and the rate decreased over time in a roughly exponential fashion, establishing a steady-state or stationary concentration in the bathing medium in several hours (Figure 17). As with tissue strips, the addition of a cholinergic agonist after equilibration, elicited a prompt and substantial secretory response. Release was also found to be temperature sensitive, with a "Q_{10}" for transport of 2.3, suggesting a reaction-dependent

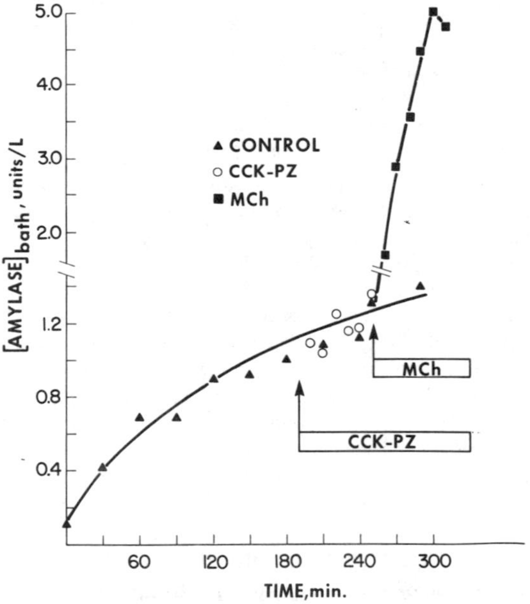

Figure 17 Effect of pancreozymin (cholecystokinin) (CCK–PZ) and acetyl-β-methylcholine chloride (MCh) on the concentration of amylase in the bathing medium. Data for each experiment are normalized to the bath amylase concentration at 180 min. (▲) Control data; $n = 8$ for all control points up to 190 min, $n = 4$ thereafter; (○) CCK–PZ; $n = 4$ for all points except at 250 min, where $n = 3$; (■) MCh; $n = 3$ for points at 260, 270, and 280 and 2 at 290, 300, and 310 min. CCK values were not distinguishable from control data (higher doses of CCK have since been shown to increase basolateral secretion from this preparation; Miyasaka and Rothman, 1981), but when MCh was added, the concentration of amylase in the bath increased greatly. (Reprinted, with permission, from Isenman and Rothman, 1977.)

process (Figure 18). Furthermore, the steady-state concentration of amylase in the medium was very similar to normal levels found in rabbit plasma. Isenman was able to demonstrate the equilibrating nature of the process by periodically changing the medium bathing the tissue and reestablishing the initial rate of release. That is, the rate of release was not a function of some event endogenous to or within the tissue that occurred over time but rather was determined by the concentration of amylase in the bathing medium. If the medium concentration was reduced by replacing it, then the initial rate of release was reestablished (Figure 18). Thus, digestive enzyme appears to be transported across the basolateral surface of the acinar cell in an equilibrium-dependent fashion in much the same way as across its apical membrane, and, as we shall discuss in the next chapter, across the membrane of the zymogen granule.

10.5 THE NATURE OF THE EQUILIBRIUM

I believe that the experiments that I have considered in this chapter, when taken together, provide compelling evidence for the equilibrium nature of the secretory process although its details, such as the number and size of the involved pools, the maintenance of steady-state concentration gradients, the

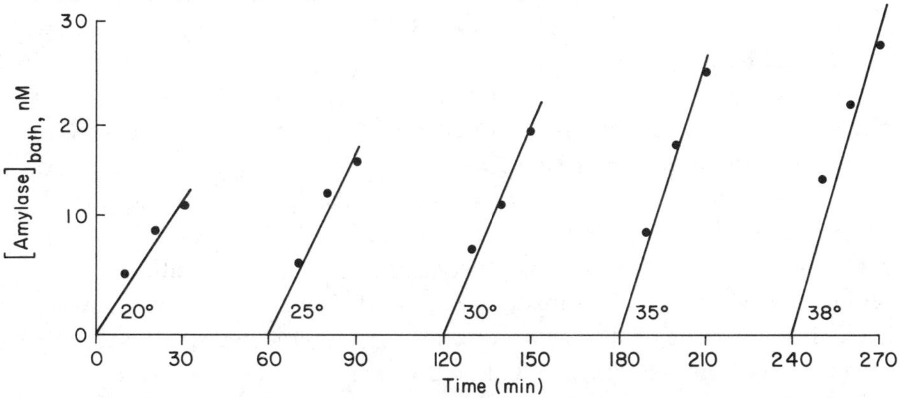

Figure 18 Reestablishment of initial rate of basolateral amylase release with repetitive bath changes. For 5 hr the bath was changed at hourly intervals and the amylase concentration of the bathing medium was monitored for 30 min. In addition, the temperature of the bath was also varied with each bath change as indicated. For all data points, $n = 6$. (Reprinted, with permission, from Isenman and Rothman, 1977.)

effect of different stimulants on the distribution, and so forth, remain to be worked out.

This having been said, however, the existence of equilibrium-based transport does not necessarily mean that individual protein molecules pass through the relevant membranes of the acinar cell accounting for this behavior, although the observations are certainly consistent with this point of view. It is possible to imagine that vesicular processes are not truly "(uni)vectorial," as has been thought, but are themselves capable of equilibrium behavior. This can be considered in two ways.

It is commonly proposed today that although exocytosis is a "univector" process, it is accompanied by another, separate vesicular process, also "univectorial" but directed oppositely, that exists to recover the membranes inserted into the plasma membrane by exocytosis. This is the idea of endocytosis in which invaginations in the cell membrane form intracellular vesicles. Such vesicles could carry secreted material back into the cell, ergo, the appearance of equilibrium behavior.

We could also propose that exocytosis itself is reversible and that the amount of material released by a given fusion event is not an all-or-none phenomenon but can be graded, being responsive to the concentration of secreted material in the extracellular environment. This must not be confused with varying the rate at which material leaves the vesicle environment after fusion, which is not a steady-state effect and would not alter the steady-state rate of secretion, but only the rate at which a new steady-state is achieved. To alter the steady-state rate of secretion, partially emptied granules would have to reseal after having released only a portion of their contents.

The evidence from the experiments that we have discussed in this chapter poses three problems for either of these hypotheses (endocytosis or resealing) insofar as their being able to account for the decreased rate of protein secretion seen during flow reductions, in which circumstance exocytosis itself, determined by intracellular events, presumably would continue at a constant rate. First, increases in reuptake, by either means, require a driving force. The only obvious possibility is the released material itself. For example, if the concentration of enzyme were increased in the immediate vicinity of the cell, then this might act to increase the rate of endocytosis or resealing. But in the studies discussed above, the concentration of protein in the secreted fluid was not significantly increased as output was reduced. Indeed, it was possible to reduce secretion by some 90% without any observable increase in concentration. Under such circumstances we would have to propose a 10-fold increase in endocytosis or resealing, in the apparent absence of a force to produce such a change.

The second problem is also a curious one. If either mechanism accounted for our observations, then how would we be able to vary the rate of protein secretion under normal circumstances? As more material is secreted, an equivalent increase in reuptake would be expected. Also, at least in the unstimulated state, there would be a considerable dilution of the secreted material released by exocytosis into the watery environment of the duct lumen (or bathing medium for tissue strips) from concentrated stores in the secretion granule. In such a circumstance, reuptake would seemingly have to be a considerably more active process, perhaps magnitudinally so in terms of membrane utilization, than the original exocytosis event.

Moreover, the reduction in enzyme secretion by tissue strips over time occurs in a constant volume of medium; that is, in this case the rate of protein secretion is time dependent, not volume dependent, and likewise so would a putative reuptake process have to be viewed as time dependent, and not volume dependent. That is, the process would have to be volume dependent in one situation and time dependent in another, a seemingly unlikely possibility although one that is explained simply in terms of a diffusive equilibrium.

10.6 SUMMARY

Whether or not complex vesicle reuptake systems exist and account for these results, or whether other vesicular explanations for them are possible, cannot be answered at present. In any event, perhaps the best way in which to consider the hypothesis that the bidirectional transport of individual protein molecules across membranes accounts for this equilibrium behavior further is to attempt to provide direct evidence for the existence of such processes. Can these proteins be shown to cross cellular and granule membranes?

SELECTED REFERENCES

1. Grendell, J. H. and S. S. Rothman. The effect of changes in circulating amylase levels on amylase output in bile. *Am. J. Physiol.* **243:**G54–G59, 1982.
2. Ho, J. J. L. and S. S. Rothman. The nature of the flow dependence of protein secretion by the exocrine pancreas. *Am. J. Physiol.* **242:**G32–G39, 1982.

3. Ho, J. J. L. and S. S. Rothman. Origin of the decline in pancreatic amylase secretion with time in vitro. *Am. J. Physiol.* **245:**C21–C27, 1983.

4. Isenman, L. D. and S. S. Rothman. Transport of α-amylase across the basolateral membrane of the pancreatic acinar cell. *Proc. Natl. Acad. Sci.* USA **74:**4068–4072, 1977.

5. Isenman, L. D. and S. S. Rothman. Diffusion-like processes account for digestive enzyme secretion by the pancreas. *Science* **204:**1212–1215, 1979.

6. Miyasaka, K. and S. S. Rothman. The endocrine secretion of α-amylase by the pancreas. *Am. J. Physiol.* **241:**G170–G175, 1981.

7. Rothman, S. S. Exocrine secretion from the isolated rabbit pancreas. *Nature* **204:**84–85, 1964.

8. Rothman, S. S. The passage of protein through membranes—old assumptions and new perspectives. *Am. J. Physiol.* **238:**G391–G402, 1980.

9. Rothman, S. S. and F. P. Brooks. Electrolyte secretion from rabbit pancreas in vitro. *Am. J. Physiol.* **208:**1171–1176, 1965.

10. Rothman, S. S., L. D. Isenman, and J. J. L. Ho. Protein secretion by the pancreas. *Science* **209:**1554, 1980.

CHAPTER 11

The Membrane Transport of Protein

In 1970, Liebow and I began to explore directly the possibility that proteins move individually through both the plasma membrane of the acinar cell and the membrane of the zymogen granule. Our reasons for undertaking these studies were at the time threefold:

1. Several lines of evidence pointed to the existence of a cytoplasmic pool of digestive enzyme from which secretion appeared to be derived (see Chapter 12 for a discussion of the cytoplasmic pool). The existence of such a pool suggested that both the granule and cell membrane might be permeable to these proteins.

2. Isolated zymogen granules released their protein contents in environments in which one would have expected them to be stable; that is, in the same environment in which they presumably exist within the cell, an isotonic medium at neutral pH (Hokin, 1955; Rothman, Burwen and Liebow, 1974). Although it was not clear that release under these circumstances was due to membrane transport, as opposed to granule lysis, beyond the observation that different proteins were not released in parallel, the fact that granules released their contents in what was presumably their "natural"

environment raised the possibility that they might be permeable to proteins; more so in the ionic, neutral, high-temperature (release is temperature sensitive, see below) medium of the cell than in nonionic media, at slightly acid pH, and low temperature in which environment they are generally extracted from tissue homogenates as "stable" granules.

3. The digestive enzymes were contained within the granules in a bound state, and were not, as had been thought, merely suspended in an aqueous medium held in place by a limiting lipid membrane. They were bound in high-affinity, enzyme specific arrays, indicating significant intragranular organization (Rothman, 1971; Rothman, 1972). This internal organization would appear to be without purpose if secretion were merely the result of the mass transport of mixed bags of protein by exocytosis. On the other hand, such organization might account, at least in part, for the selective release of different enzymes from granules that had been observed, as well as for the ability of granules to retain their protein contents if the membranes were indeed permeable to them.

In any event, we began a series of experiments to determine whether or not digestive enzymes could be released from isolated zymogen granules. Our approach was simple. If enzyme was released from the granule in a concentration-dependent fashion across its membrane, then if we reduced the concentration of enzyme in the medium in which the granules were suspended, more enzyme would be released. We chose to study the potential for release in what in a sense was the most unfavorable situation; that is, in nonionic media at slightly acidic pH, conditions under which granules were known to be most "stable." Release would be expected to be least and slowest under these circumstances; but it also might be easier to follow over time, and furthermore there would be fewer concerns about the nature of the starting material, since granules were thought to be "stable" or "normal" in this environment.

When isolated granules were diluted in larger and larger suspension volumes, thereby progressively reducing the external concentration of enzyme, release into the medium was observed (Figure 1) (Liebow and Rothman, 1972; Liebow and Rothman, 1976). Amylase, chymotrypsinogen, and protein in general were all redistributed, as a result of dilution, in a concentration-dependent fashion, although each to a different extent. Release varied in a continuous, as opposed to a discontinuous manner (as might be expected if lysis were the cause of release) as a function of suspension volume. The larger the volume, the greater the amount released,

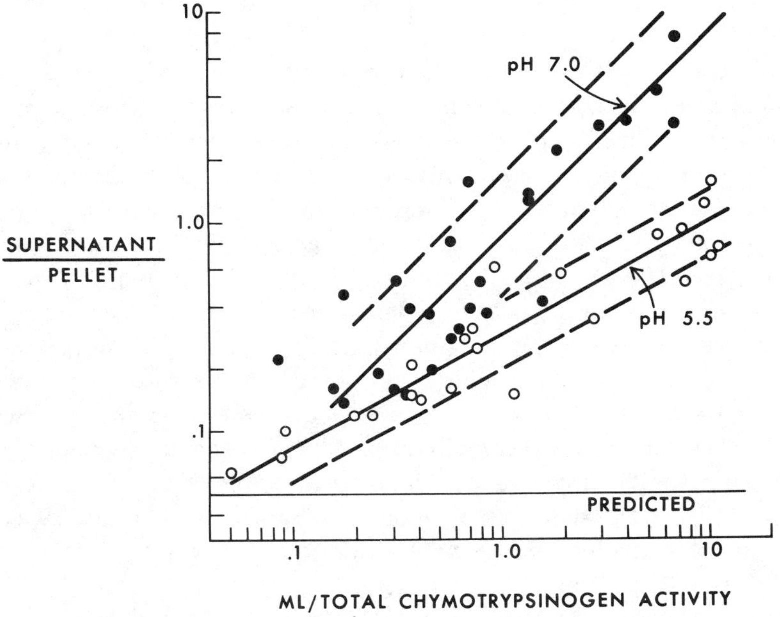

Figure 1 Release as a function of the relative dilution of granule suspensions. The ratio of log activity of nonsedimenting (supernatant) to sedimenting (pellet) chymotrypsinogen versus log volume per total chymotrypsinogen activity (dilution). (●) pH 7.0; (○) pH 5.5. Linear regression lines are drawn through the points (pH 7.0 has a correlation coefficient of 0.93, pH 5.5 has one of 0.92). The line labeled "predicted" indicates the absence of dilution-dependent release as would be observed if granule contents were independent of the enzyme concentration in the surrounding medium. The dashed lines indicate the standard error of the estimate. (Reprinted, with permission, from Liebow and Rothman, 1972.)

and the ratio of enzyme contained in the medium to that in the granule could be varied continuously by over an order of magnitude in this fashion.

At the more "labile" pH of 7.0, the distribution of protein between granule and medium was altered to an even greater extent (by about two orders of magnitude) merely as a function of dilution; from relatively concentrated suspensions in which almost 90% of the protein was retained within the granule, to those in which 90% was recovered in the suspending medium (Figure 1). It is important to point out here that even our "concentrated" suspensions were rather dilute in terms of the solubility of the digestive enzyme proteins themselves. They ranged downward more than two orders of magnitude from a total protein concentration of approximately 1 mg/ml, at which fractional concentrations virtually all of the enzymes are soluble in water, as is the whole mixture of proteins in secreted fluid.

This 100-fold change in distribution occurred over an approximately 100-fold range of suspension volume. That is, the ratio of released to bound enzyme varied in an almost linear fashion with suspension volume, from wholly bound to fully released. This meant that virtually all of the enzyme in the population of granules was in equilibrium with enzyme in the medium. Release could not merely be attributed to an equilibrium between some small percentage of the total granule content and the medium, such as would be the case for the release or desorption of protein from the granule surface or the release of the contents of a subpopulation of broken granules. If either were the case, then only small changes in the ratio of released to bound enzyme would occur for a given dilution since most of the material would be excluded from the equilibrium-dependent events (for example, a change of 10 or 20% in the ratio rather than a 100-fold change for a 100-fold dilution). Thus, the whole population of granules and virtually all of its contents were involved in the equilibrium.

11.1 THE TIME COURSE OF RELEASE

Weak binding interactions, such as in protein adsorption, usually respond rapidly to changes in an equilibrium state; that is, no substantial period of time is required to establish a new state, and association or dissociation is usually complete within a small fraction of a second. Delays in attaining a new equilibrium state are usually attributable to "access" or the need for substances to move to or from the site at which the interaction takes place. Thus, if a new equilibrium state produced by diluting a suspension of zymogen granules were merely a function of weak interactions (ionic or hydrogen bonding) without a significant intervening permeability barrier, then it would in all likelihood be rapidly attained without a macroscopically observable time course.

With this in mind, suspensions of granules were diluted and the approach to a new equilibrium state followed over time. Reequilibration was not instantaneous, but required a considerable period, suggesting the presence of a significant permeability barrier (Figure 2). For example, at a dilution that altered the distribution of chymotrypsinogen from about 90% in the granules initially, to about 50% outside, reequilibration took more than 4 hr at 37°C (Figure 2).

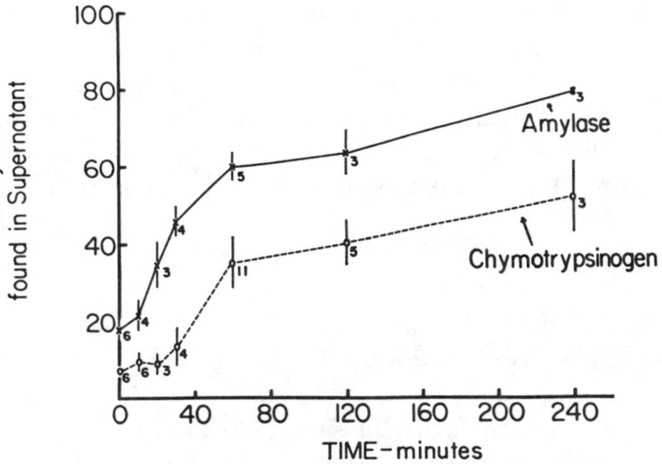

Figure 2 The time course of enzyme redistribution after dilution. The percentage amylase (-x-) and chymotrypsinogen (--- o ---) activity found in the medium (supernatant) of a pH 5.5 suspension of granules at various times after dilution (time 0) of granule suspensions from ~1 ml/mg protein to ~10 ml/mg protein. Vertical bars represent the standard error of the mean and the number of experiments is noted by their side. No loss in enzyme activity was observed in samples incubated up to 120 min. (Reprinted, with permission, from Liebow and Rothman, 1976.)

Furthermore, release occurred at different rates and to different extents for different enzymes (Figure 2). Amylase was more "labile," than chymotrypsinogen; that is, it was released at a more rapid rate and to a greater extent for a given dilution. For the example just noted, $t_{1/2}$ was 30 min for amylase as compared with 60 min for chymotrypsinogen, and 80% of the total was released, or a medium to granule ratio of 4:1 was established for amylase as compared with 50% release or a 1:1 ratio for chymotrypsinogen. This difference argued against the possibility that a slow, time-dependent granule lysis accounted for enzyme release, because in this case, both the rate and extent of release would have been expected to be essentially the same for the different enzymes.

We also considered whether a membrane barrier was responsible for the delay in equilibration in the following manner. It is possible to solubilize the contents of zymogen granules and to then reaggregate them in non-ionic media at slightly acid pH in a form that displays similar binding characteristics to enzyme contained within the native granule (Rothman, 1971, 1972). However, this reaggregated material lacks an enclosing membrane barrier (Ermak and Rothman, 1978). The question was whether or not a delay in reequilibration would still be seen after dilution in the absence of a membrane barrier? The answer was no. In this case, reequilibration occurred within the time required to separate the aggregate from its suspending medium (Figure 3). No time dependence for release could be demonstrated, although the state function, that is, the equilibrium distribution, itself was much the same as seen with the native granule (Liebow, 1985).

11.2 TIME VERSUS CONCENTRATION-DEPENDENT RELEASE

As I have said, it was possible that the slow time course of enzyme release was due to the slow lysis of labile granules rather than the release of enzyme from intact granules. Although the different release characteristics for different enzymes, as well as the evidence just discussed, argued against this possibility, we attempted to distinguish between such time versus con-

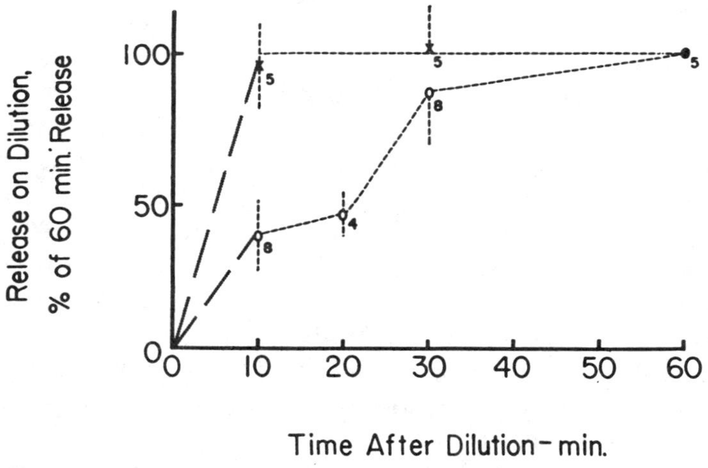

Figure 3 The time course of enzyme redistribution after dilution for native granules and "reaggregated" granule material. The approach with time to the distribution found at 60 min for chymotrypsin activity between supernatant and pellet for whole granules and "reaggregated" granule material (with disrupted granules membranes) after an eight-fold dilution. Error bars represent standard error of the mean and the numbers represent the number of experiments. (x) Reaggregated material; (o) native granules. The reaggregated material was at the new equilibrium state within the 10-min period required to separate granules from medium, whereas native granules continued to release enzyme with time. (Reprinted, with permission, from Liebow and Rothman, 1976.)

centration-dependent release in another fashion as well. Granules were suspended in a filtration chamber through which a flow of fluid could be maintained and the released enzyme, although not the granules, filtered. If release was due to time-dependent granule lability, then we would expect to see much the same phenomenon that was seen when granules were merely incubated in medium in which the released contents accumulated. That is, the removal of released material should have no effect on the rate of release (rate of lysis), and the same time course and pattern seen for simple incubation would have been expected to be reproduced. This was not the case. The filtration of released enzyme led to the continuous release of enzyme from granules (Figure 4). That is, the rate of release did not fall off over time as in the simple incubation study, but continued at a relatively

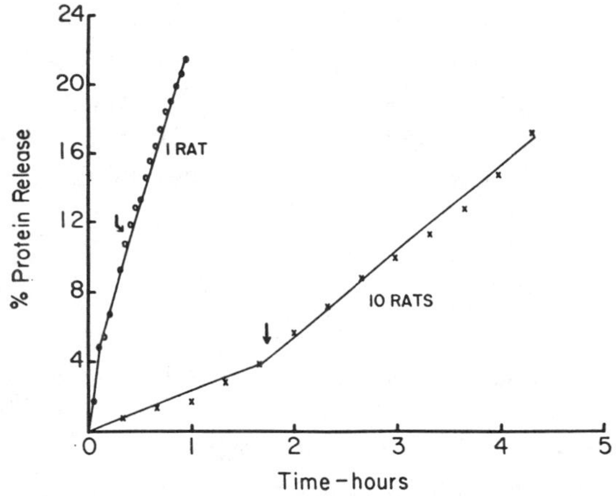

Figure 4 Protein release from granules in response to continuous washing. Granules were placed in a mixing chamber and washed continuously with 0.3 *M* sucrose (pH 5.5). Granules were retained in the chamber by a 100,000-molecular-weight filter through which the solvent and soluble secretory proteins passed. Protein collected in the effluent was considered "released." Two preparations are shown: one with a relatively high concentration of granules (collected from 10 rats) maintained at 25°C (x) and one with a relatively low concentration of granules (from one rat) at 4°C (○). The release function shown is equated for the temperature difference using the estimated Q_{10} of chymotrypsinogen release (see Figure 5). Flow rates were increased by 20% at the arrows by increasing filtration pressure from 8 to 24 lb/in^2. The system with a low effluent protein concentration (low concentration of granules, one rat) showed no increase in the rate of enzyme released in response to increased flow rates. The system with the high effluent protein concentration showed an increased rate of enzyme released in response to increased flow. This suggests that backflux of enzyme into granules is a significant factor in net enzyme release from the concentrated granule preparation (at high temperature), but not from the low concentration of granules (at low temperature). (Reprinted, with permission, from Liebow and Rothman, 1976.)

constant rate. Thus, external concentration and not time appeared to be determining the rate of release.

That release was dependent upon the rate of removal of already released enzyme, and hence the concentration of enzyme in the medium, was further evaluated by varying the rate of removal of released material either by increasing filtration rate or decreasing the absolute amount released from the granules for a given rate of filtration (by decreasing the number of granules in the suspending medium). In the first case, it should be possible to increase the amount of material released from the granules by increasing filtration rate, if release is not already occurring at a maximal rate. This would be most likely when relatively large numbers of granules are added to the filtration chamber, so that the amount released from the granules would be relatively great compared with its rate of removal by filtration. This would provide the opportunity for the flux of material back into the granules (that is, backflux or influx). If release were concentration-dependent, then under these circumstances we should be able to increase it merely by increasing the rate of filtration, and thereby reducing backflux (J_{net} is increased by decreasing J_i since $J_{net} = J_e - J_i$). When this was done, an increase in the release of enzyme from the granules was seen (Figure 4; "10 rats").

If this interpretation of the cause of increased release were correct, then we should also be able to set the system up so that release would occur at a greater rate, as a percentage of the amount of material in the granules, merely by suspending fewer granules in the chamber. In this case, for a given filtration rate, the concentration of released enzyme in the medium would be lower; that is, less material would be released (from a smaller number of granules) but into the same volume (a constant filtration rate). If release were due to some type of lytic event, independent of external concentration, then we would expect the same amount of material to be released, in proportionate terms, regardless of the number of granules in suspension. On the other hand, if release were a concentration-dependent phenomenon, then a lower concentration in the medium would reduce the backflux into the granules and hence the net flux out would be greater (that is, we would see an increase in the amount released per unit time per granule). When this was done and the number of granules suspended was reduced by about an order of magnitude ("one rat"), the rate of release per granule increased greatly (that is, J_{net} was increased because J_i was reduced due to the reduction in the enzyme concentration in the external medium). When filtration rate was increased in this case, no further increase in enzyme release occurred, which supported the interpretation that under these circumstances release was already occurring at the maximal rate, namely, at the initial rate of efflux, so that $J_{net} = J_e$ because J_i approaches zero and therefore J_{net} could not be further increased by varying filtration rate.

Thus, it was demonstrated in a variety of ways that the release of enzyme from isolated zymogen granules was the result of their concentration-dependent release from the granules into the medium, and that this flux of protein was being expressed across the granule membrane.

11.3 THE TEMPERATURE DEPENDENCE OF ENZYME RELEASE

The release of enzyme from granules was found to be highly temperature sensitive. For example, when a suspension of granules was diluted and release followed over a 20-min period, either at 20°C or at 30°C, the temperature dependence of release, namely, its apparent Q_{10}, was about 5.0 for protein overall and for chymotrypsinogen, and about 2 for the more "labile" amylase (Figure 5). That is, the rate of release increased by two to five times over a 10° temperature range.

These observations offered further support for the view that a membrane transport process was involved in the release of enzyme from granules. Such a substantial temperature dependence would not be expected if simple diffusion in an aqueous medium out of broken granules or away from their surface, or dissociation from weak bonding at either site, was responsible for release. In these situations, Q_{10} values of little more than 1.0 would have been expected. The observed values were as high or higher than those usually reported for reaction-dependent transport or for enzymatically catalyzed reactions in general ($Q_{10} = 2$–3).

The view that the temperature sensitivity of release was due to a membrane transport process, as opposed to a temperature-sensitive granule lysis, was supported by the fact that the temperature dependence of release was substantially different for amylase and chymotrypsinogen. A differential temperature sensitivity for different proteins would not be expected if either lysis, simple dissociation, or a combination of the two were the cause of release.

One additional characteristic of the temperature dependence of release also points to a transport, and not a lytic, process. Although the rate of release is highly temperature sensitive, the steady-state distribution of these

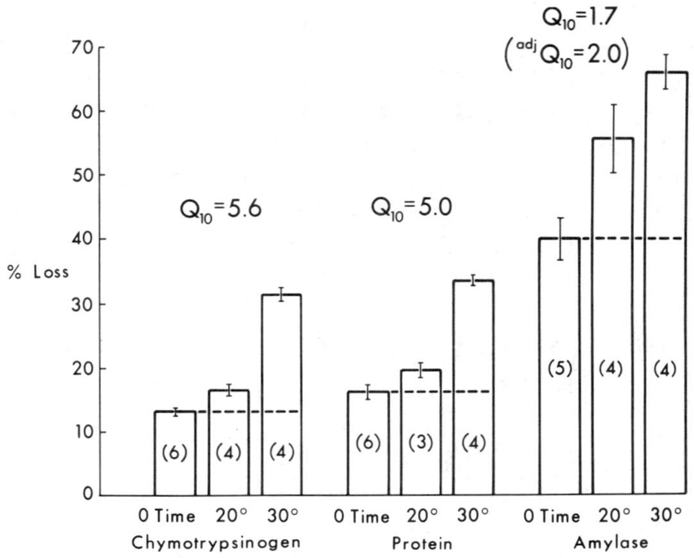

Figure 5 Q_{10} determination for chymotrypsinogen, protein, and amylase release. 0 Time refers to release prior to incubation; 20° and 30° refer to release after 20 min incubation in 20° or 30°C baths. The ratio of release during 30°C incubations (30°C release minus 0 time release, 30°C column above dashed line) to that released during 20°C incubations was used to determine the apparent Q_{10}. ( ) = number of trials; error bar is the SE. Adjusted Q_{10} for amylase derived from supernatant/pellet ratio, rather than percent loss, to compensate for the close approach to 100% loss. (Reprinted, with permission, from Liebow and Rothman, 1972.)

proteins between granule and medium is essentially temperature independent (Rothman, 1971; Burwen, 1972). This behavior is that expected for transport by facilitated diffusion. If a lytic event were involved, both kinetic parameters would show approximately the same temperature sensitivity.

11.4 THE UPTAKE OF ENZYME INTO GRANULES

Could release be reversed? That is, could the uptake of enzymes into partially depleted granules be demonstrated? Both the uptake of radiolabeled enzyme (Liebow and Rothman, 1972) and the bulk reuptake of previously released enzyme (Liebow and Rothman, 1976) was shown (Figures 6 and 7). Uptake had a substantial time dependence (Figure 6) and was wholly reversible, at least when endogenous material was used (Figure

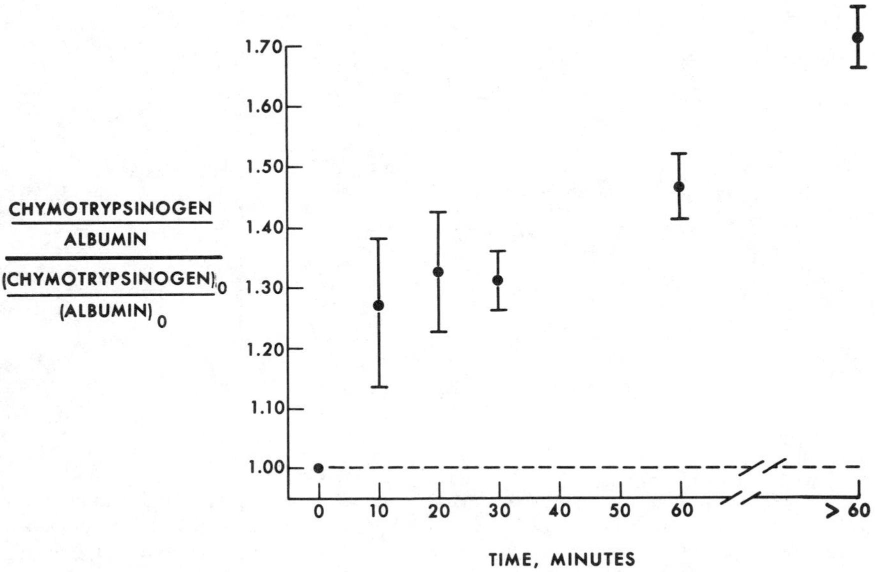

Figure 6 The change in the ratio of exogenously added [^{3}H] α-chymotrypsinogen (bovine) to ^{131}I-labeled albumin (human) with time, in granules previously about 50% depleted of chymotrypsinogen. Increase in the ratio indicates chymotrypsinogen uptake by granules. 0 time after separation is $^3H/^{131}I = 1.0$ by definition. Dashed line indicates no uptake. Moving out from 0 time, $n = 9, 4, 5, 5$ and 6; > 60 represents pooled data (1 at 75 min, 1 at 90 min, and 2 at 120 min). (Reprinted, with permission, from Liebow and Rothman, 1972.)

7). That is, an initial state could be reestablished after the dilution of a suspension of granules merely by reconcentrating the enzyme in the medium (by dialysis). Therefore, enzyme could be taken up into partially depleted granules, and this uptake had the characteristics of a time-dependent equilibrium, as did enzyme release from granules.

11.5 THE APPEARANCE OF ISOLATED GRANULES AFTER THE RELEASE OF ENZYME

What do granules look like after they have released a portion, or even all of their contents? Is their visual appearance consistent with the hypothesis that we have been discussing? Ermak extracted granules from tissue homogenates and compared undiluted sediments of granules to granules from the same starting material subjected to varying degrees of additional

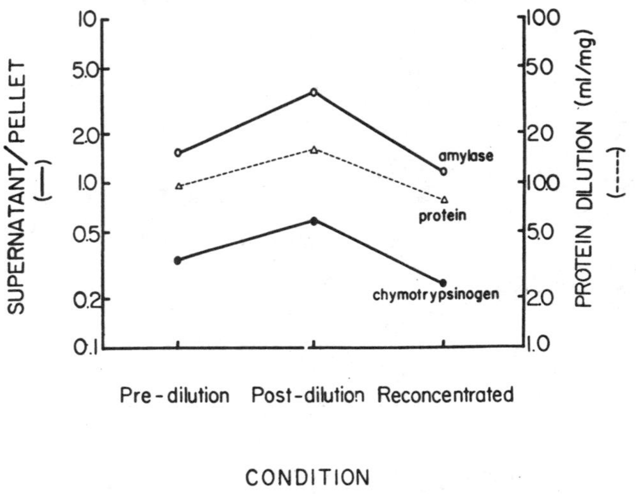

Figure 7 Redistribution of enzyme upon dilution and reconcentration. The distribution of enzyme between supernatant and pellet is shown by the solid line and circles for chymotrypsinogen (●) and amylase (○) in three conditions: before diluting the granule suspension (predilution), after dilution of the suspension, but before reconcentration (postdilution), and after reconcentration by dialysis (reconcentrated). The absolute degree of dilution and reconcentration is shown by the dashed line and triangles. Its units are on the vertical axis on the right and are the inverse of protein concentration, that is, protein dilution given in ml/mg of protein. The logarithmic scale for both ordinates is equal, although the absolute values are different. (Reprinted, with permission, from Liebow and Rothman, 1976.)

dilution (Ermak and Rothman, 1978). He found that, as a function of dilution, the usual electron-dense or opaque zymogen granule (Figure 8) gave way to apparently intact but internally honeycombed or "reticulated" granules of the same approximate size (Figure 9). That is, granule appearance changed as a result of enzyme release. Granules did not merely decrease in number in an all-or-none fashion, with all those that remained having the "normal" opaque appearance, as a lytic view of enzyme release would predict. The visual, as well as the kinetic evidence, indicated that the graded release of enzyme occurred from individual, seemingly intact, granules. Ermak also demonstrated directly that at least large segments of the membrane surface of individual granules were without apparent tears or schisms in their fabric by looking at three-dimensional images of thick sections through granules under the transmission microscope (Figure 10). Furthermore, the honeycombed appearance of the granule indicated that material was lost from throughout the whole of its interior and not from particular localized sites, or even generally from the surface towards the

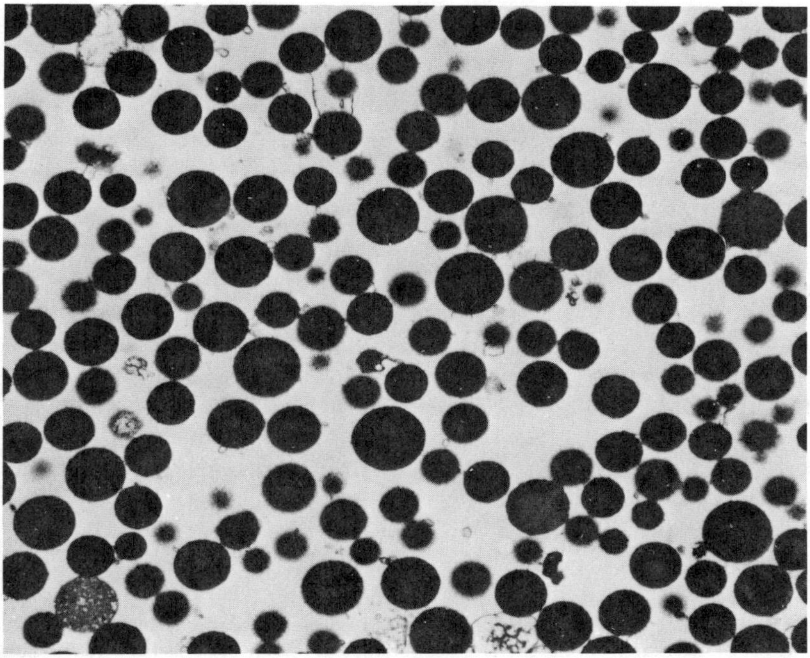

Figure 8 Electron micrograph of a final zymogen granule pellet (600 × *g*) illustrating the morphology of granules before experimental treatment (that is, dilution). The granules have circular profiles and a relatively homogeneous, dense content. Magnification, 9000×. (Reprinted, with permission, from Ermak and Rothman, 1978.)

center. This reticulated form is consistent with the requirement of the kinetic data that all or almost all of the enzyme within the granule be involved in the equilibrium with the external medium.

Moreover, the changes in appearance were a function of dilution, both in regard to the types of granules seen and their relative frequency. As granule suspensions were made more dilute, the number of dense granules decreased, at first being primarily replaced by reticulated granules, which upon further dilution were replaced by increasing numbers of virtually empty granules of the same approximate size and general appearance (Figures 11, 12, and 13). Empty granules accounted for about 50% of the total population in the most dilute suspension studied. It should be noted that the concentrations of granule suspensions used in these morphologic studies were chosen to match those employed in the kinetic studies discussed above.

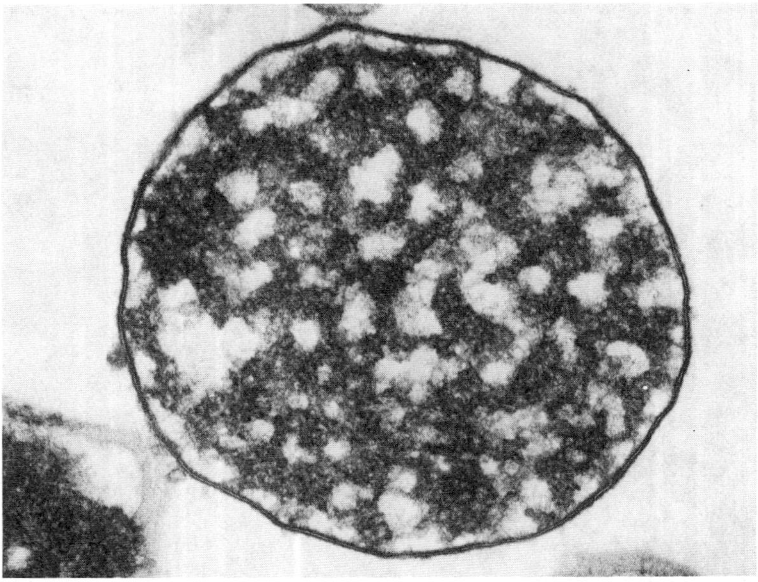

Figure 9 Reticulated granule containing a meshwork of electron-dense material. Reticular strands make attachments to the granule membrane and outline spherical spaces and larger polymorphic spaces. Content is also lost directly beneath the granule membrane. Magnification, 75,000×. (Reprinted, with permission, from Ermak and Rothman, 1978.)

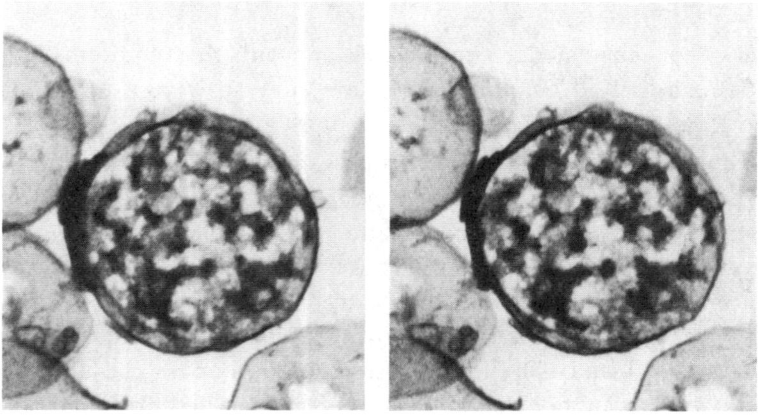

Figure 10 Stereo pair of electron micrographs of 0.3-μm-thick sections of reticulated and empty granule profiles. The meshwork forms a three-dimensional reticulum. Membranes are seen in section and en face. Magnification, 22,000×. The three-dimensional images can be seen with a pair of stereo glasses. (Reprinted, with permission, from Ermak and Rothman, 1978.)

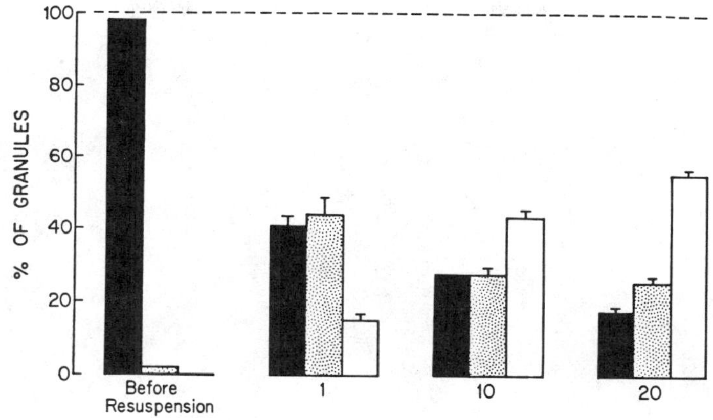

Figure 11 Change in the percentage of solid (black), reticulated (stippled), and empty (white) granule profiles before resuspension and as a function of suspension volume. Before resuspension, $n = 741$; 1 ml, $n = 1186$; 10 ml, $n = 1328$; and 20 ml, $n = 1214$; no error bar signifies a standard error of less than 1%. (Reprinted, with permission, from Ermak and Rothman, 1978.)

Thus, the microscopic exploration of the granule gave "form" to kinetic and thermodynamic observations of enzyme release. Granules appear internally structured by interlaced strands of variable thickness containing bound enzyme that border honeycombed lacunae, into which enzyme is released and from whence enzyme leaves the granule across its membrane. In this way the protein contents of the granule appear to have access to the medium.

Although these observations are reassuring and offer important support for the kinetic studies with isolated granules, they provide a very different visual picture from that discussed in Chapter 9 for changes in granule structure in situ as a result of the release of enzyme during active secretion. In that case, the size of granules decreased, but their appearance otherwise remained essentially unchanged; that is, opaque granule interiors were still predominantly observed. There are several things to be said about this difference. In the first place, although we have not accurately evaluated the size of isolated enzyme-depleted granules in a quantitative fashion, they do not appear to be substantially reduced in size, and if this is indeed the case, then it must be that such reductions in size do not occur in vitro but require a cellular environment.

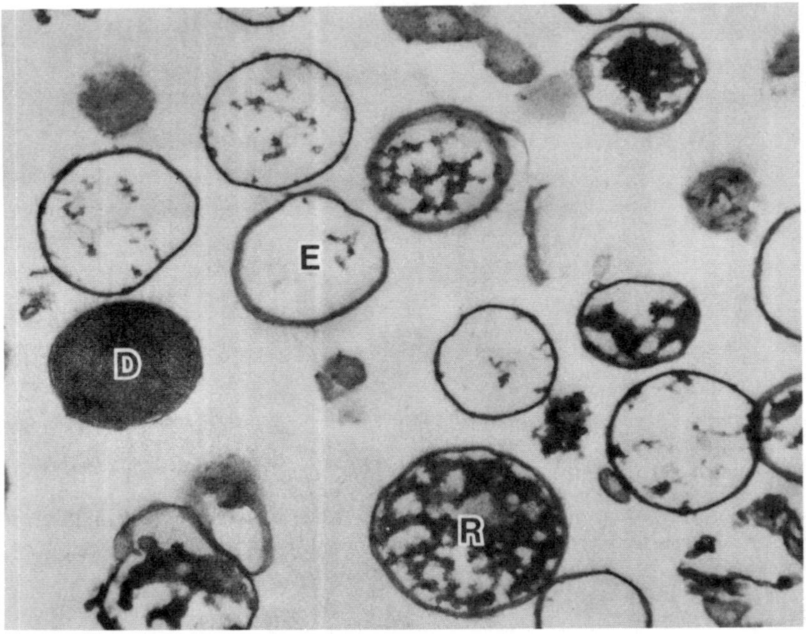

Figure 12 Thin section through a granule preparation diluted in a large suspension volume (20 ml) showing dense (D), reticulated (R), and empty (E) profiles. Empty granules contain filamentous material. Magnification, 25,000×. (Reprinted, with permission, from Ermak and Rothman, 1978.)

Conversely, as I have just said, granules in the cell do not appear to become, at least in major part, reticulated or empty during active secretion. This may merely reflect the fact that granule size can be decreased instead, but it should also be borne in mind that the relative density of granules in pancreas after active secretion has not been quantitatively assessed. Density may change, even greatly, but the granules may still be sufficiently opaque so that these changes cannot be distinguished visually in the usual electron microscopic thin section. Nor does opacity necessarily equate with protein content. Opacity may be retained in situ but reflect some other aspect of granule morphology.

Furthermore, it is not really correct to say that reticulated forms are not seen in situ. They most certainly are, but are thought of as "forming" zymogen granules (that is, condensing vacuoles) and not "emptying" granules, although the two obviously are not distinguishable on a visual basis. The appearance of some, though not all condensing vacuoles, is

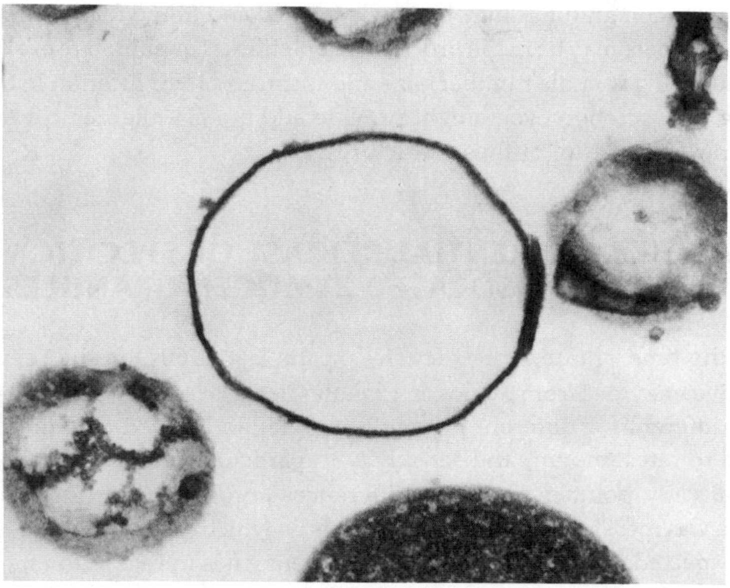

Figure 13 Totally empty granule profile. Magnification, 50,000×. (Reprinted, with permission, from Ermak and Rothman, 1978.)

essentially identical to the reticulated granules that Ermak produced in vitro by causing enzyme release (Figure 9). The evidence that is usually offered in support of the view that condensing vacuoles are forming zymogen granules is twofold. In pulse-label experiments, labeled protein appears to accumulate in these vesicles before any substantial accumulation occurs in the mature granule population (as estimated autoradiographically); thus the hypothesis that one gives rise to the other. However, if enzyme has access to these granules from the cytoplasm (that is, if we do not assume a priori that it can only arrive there by means of a vesicular mechanism as the paradigm proposes), then partially empty granules (condensing vacuoles) might well equilibrate more readily with labeled enzyme than full or mature granules and give the same kinetic pattern whether they are emptying or filling.

The other argument in favor of the condensing vacuole as a forming zymogen granule is its geographic location within the cell. Condensing vacuoles tend to be found in greatest abundance near the Golgi stacks that are thought to give rise to them. They are, however, also found, although in lesser abundance, in other areas of the cell, even near the apical cell membrane. Of course, both filling and emptying may occur; some partially filled zymogen granules may be newly formed and filling for the first time, whereas others may be in the process of depletion. A quantitative evaluation of condensing vacuole number, size and form, and their location in the cell during the secretion cycle might provide additional enlightenment about the types and nature of these structures.

11.6 THE DIFFERENTIAL RELEASE OF SPECIFIC ENZYMES FROM ISOLATED ZYMOGEN GRANULES

From the time of our earliest studies, in the late 1960s, on the release of digestive enzymes from zymogen granules, it was clear that release tended to be different for different molecular species and varied among them in regard to rate, amount, and sensitivity to particular release effectors. As I have already pointed out, such differences argued against the idea that release was merely caused by granule lysis, in which case one would have at least expected it to be similar for different enzymes in regard to rate, if not necessarily in regard to sensitivity to particular release effectors or to amount if intragranular or surface binding produced differential distributions. For some release phenomena, essentially the whole population of granules and their contents was shown to be involved, and for this reason, as I have already noted, differential surface desorption or the differential release of enzyme due to differences in intragranular binding affinity from a subpopulation of broken granules did not seem to provide a sufficient explanation.

Some observations of the differential release of different enzyme types are more convincing as evidence for membrane transport than others, but perhaps the most convincing case, and one that is difficult to explain as being other than the result of the membrane transport of the released enzymes, comes from relatively recent work by Grendell (Grendell and Rothman, 1981). We were interested in determining how end-products of digestive reactions, such as glucose and amino acids, produced the selective secretion of different digestive enzymes, a phenomenon for which we had evidence in situ (Rothman, 1974, 1976; Grendell et al., 1984). We thought that the effect might be expressed most simply and directly at the zymogen granule level, with end-products mobilizing specific enzymes from granule stores.

We initially studied two end-products, glucose and lysine, for their effect on the release of two enzymes, amylase and trypsinogen, that are involved in their respective liberation from food polymers in the gut. The results were striking. The major end-product of carbohydrate digestion, glucose, led to the release of amylase from isolated zymogen granules at glucose concentrations in the range normally found in plasma, but did not cause the release of trypsinogen (Figure 14). Conversely, the amino acid, lysine, led to trypsinogen, but not amylase, release and was also effective at levels of this amino acid found in plasma (Figure 15). Thus, a clear, indeed, an absolutely, selective distinction was observed that had potential physiological meaning. Moreover, release could be shown to involve a large percentage of the contained enzyme stores; that is, the effect was also quantitatively substantial.*

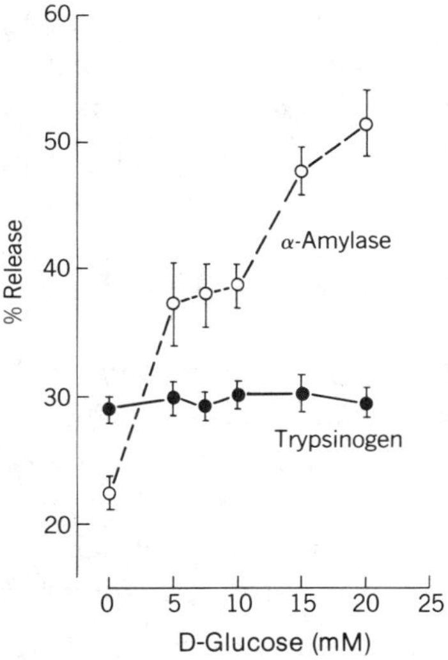

Figure 14 Percent of total α-amylase (--- ○ ---) and trypsinogen (— ● —) released into soluble phase from homogenates of pancreatic tissue as a function of the concentration of D-glucose. Both enzymes were measured in the same experiments. Points represent means, and error bars SE; $n = 3$. (Reprinted, with permission, from Grendell and Rothman, 1981.)

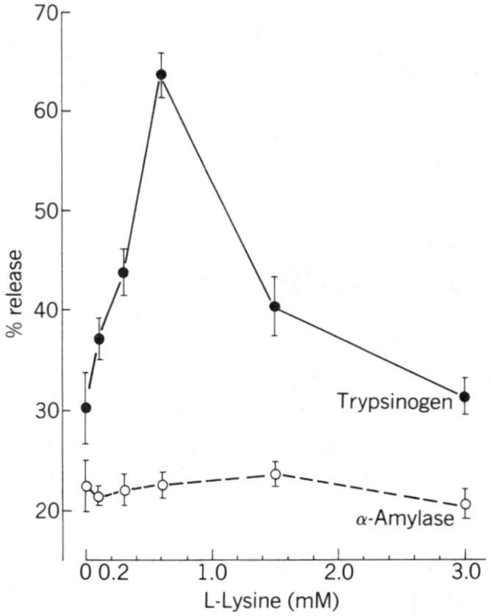

Figure 15 Percent of total α-amylase (--- ○ ---) and trypsinogen (— ● —) released into the soluble phase from homogenates of pancreatic tissue as a function of the concentration of L-lysine. Both enzymes were measured in the same experiments. Points represent means, and error bars SE; $n = 3$. (Reprinted, with permission, from Grendell and Rothman, 1981.)

*Recently, we have been able to demonstrate that glucose-1-phosphate and glucose-1,6-diphosphate produce essentially the same effect as glucose. However, we have been unable thus far to reproduce our original glucose observations, and are at the moment at a loss to explain the problem. In addition, we have studied the effect of other amino acids and have

Can such an effect be explained without invoking membrane permeability; that is, can it be explained in terms of granule lysis, and the differential surface or interior binding of different enzymes? In the first place, we would have to propose that all release effectors at millimolar, and lower, concentrations (see footnote); concentrations normally found in plasma and cells, are lytic agents. Moreover, for lysine, since the release effect is lost at concentrations above 1.0 m*M* (and is gone at 3.0 m*M*) (Figure 15), one would have to propose additionally that although low concentrations of this amino acid produce lysis, higher concentrations do not. Of course, there is no reason to think that these substances produce granule lysis, no less that they all do, no less differentially; nor, moreover, is there reason to expect them to have such effects in the range of concentrations that are normally found in plasma. But most significantly, there is no reason to expect, for example that glucose (or glucose-1-phosphate) would cause lysis that leads only to the release of amylase, whereas lysine causes lysis that leads only to the release of trypsinogen, and merely by chance at that.

11.7 STUDYING MEMBRANE PERMEABILITY IN SITU

Thus, the evidence indicates that the zymogen granule membrane is permeable to its contained secretory proteins. Can such a permeability be corroborated in situ, and can a similar permeability be shown for the plasma membrane of the secretory cell? With these questions in mind, Liebow and I performed a series of experiments in which we added radioactively labeled chymotrypsinogen to slices of pancreatic tissue in vitro, and followed enzyme uptake by the tissue (Liebow and Rothman, 1974). Time-dependent uptake of chymotrypsinogen was observed (Figure 16), and was far greater than that seen for a control protein, radiolabeled albumin. Washing the tissue to remove labeled protein from the interstitial spaces accentuated the difference between chymotrypsinogen and albumin uptake. The ratio of chymotrypsinogen to albumin uptake was 3:1 in unwashed tissues, but rose to 20:1 after a simple washing procedure that removed most of the albumin (Table 1).

---

found additional examples of differential release. For example, tryptophane produces the release of chymotrypsinogen and trypsinogen from particulate stores, whereas virtually no amylase release occurs in the presence of this amino acid. L-Arginine is effective in releasing trypsinogen, but neither chymotrypsinogen nor amylase. The effective concentrations of these amino acids are once again in the range found in plasma.

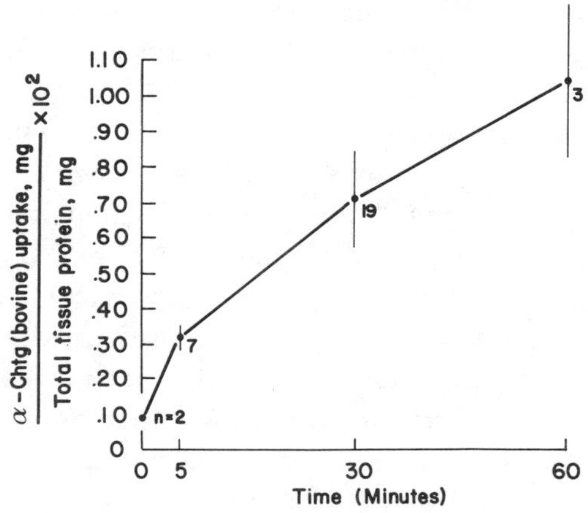

Figure 16 Uptake of exogenous bovine [^{3}H] chymotrypsinogen (Chtg) into pancreatic tissue versus time. Chymotrypsinogen uptake is corrected for albumin space. Uptake is normalized to quantity of tissue dividing by total tissue protein. Incubation was at 37°C. Numbers refer to number of experiments and error bars represent standard error of the mean. (Reprinted, with permission, from Liebow and Rothman, 1974.)

TABLE 1. Effect of Number of Rinses on Relative Amounts of [^{3}H] Chymotrypsinogen and [^{131}I] Albumin in Tissue Slices

Number of Rinses	(^{3}H/^{131}I) Tissue[b] / (^{3}H/^{131}I) Bath
0	3.0 ± 0.3 (9)
2	7.8 ± 0.6 (26)
4	17.0 ± 0.9 (4)
2 Krebs-Ringer's and then two additional Krebs-Ringer's rinses containing 1 mg/ml cold bovine chymotrypsinogen	20.0 ± 1.6 (4)

Values are means ± SEM. Numbers in parentheses are *n*.
[a]10 ml 4°C Krebs-Ringer's solution per 100 mg tissue sample.
[b]^{3}H in bovine chymotrypsinogen and ^{131}I in albumin.
Source: From Liebow and Rothman, (1974).

Although the relatively large difference between chymotrypsinogen and albumin uptake suggested that chymotrypsinogen had entered the cells of the tissue, and was not merely restricted to the interstitial spaces, we sought additional evidence for uptake by extracting zymogen granules from tissue homogenates after incubating the tissue in the labeled proteins, and then determining whether or not radiolabeled protein had entered the zymogen granules. The answer was yes and the results provided strong additional evidence for the permeability of granule, as well as the plasma, membrane to digestive enzyme. After 60 min of incubation, not only was the exogenous enzyme found in the zymogen granule fraction of tissue homogenates, but the specific radioactivity, expressed as labeled enzyme per milligram of protein, was by far the highest in the granule fraction and represented an enrichment over the homogenate of more than 20-fold (Figure 17). (N.B.: The specific activity in the zymogen granule fraction also rose from approximately 0.2 to 2.0 from 5 to 60 min and was enriched relative to the homogenate overall as a function of time.) The ratio of labeled chymotrypsinogen to albumin in the granule fraction was 150:1; seven times the ratio for the homogenate overall. Thus, the evidence indicated that chymotrypsinogen was selectively taken up by the tissue and was accumulated in zymogen granules.

Furthermore, when the data were expressed in terms of chymotrypsinogen specific radioactivity; that is, the amount of labeled chymotrypsinogen per unit of total chymotrypsinogen, specific activity in the zymogen granules was about the same as in the bathing medium by 60 min (Figure 18). That is to say, isotopic equilibrium had been attained. Not only was exogenous chymotrypsinogen taken up by the cells, but it had fully equilibrated with the endogenous granule enzyme pool. This meant that complete exchange of this protein between granule and medium pools had occurred over the 60-min period of incubation.

A vesicle or endocytic uptake system of some sort cannot readily account for this observation. In such a case, isotopic equilibration would not be possible. Material taken up by granules in this way would always be diluted by the endogenous pool, and even to approach isotopic equilibrium such an uptake system would have to have moved a volume of material in an hour that is greater than the volume of the whole granule pool and even greater than that of the cell itself. This observation also demonstrates that it was indeed chymotrypsinogen that had equilibrated and not merely the label, either dissociated from the protein or reincorporated into the mixture of pancreatic enzymes. If either had been the case, then isotopic equilibration of the chymotrypsinogen pool would not have been possible, since much of the label would be in other substances (N.B., over 95% of the label was precipitated in trichloroacetic acid). Only a membrane transport process in which the enzyme presumably moved from bathing medium, via the cytoplasm, into the granule could seemingly account for what was observed.

This view was reinforced by the observation that the postmicrosomal supernatant fraction of the cell also essentially attained isotopic equilibrium

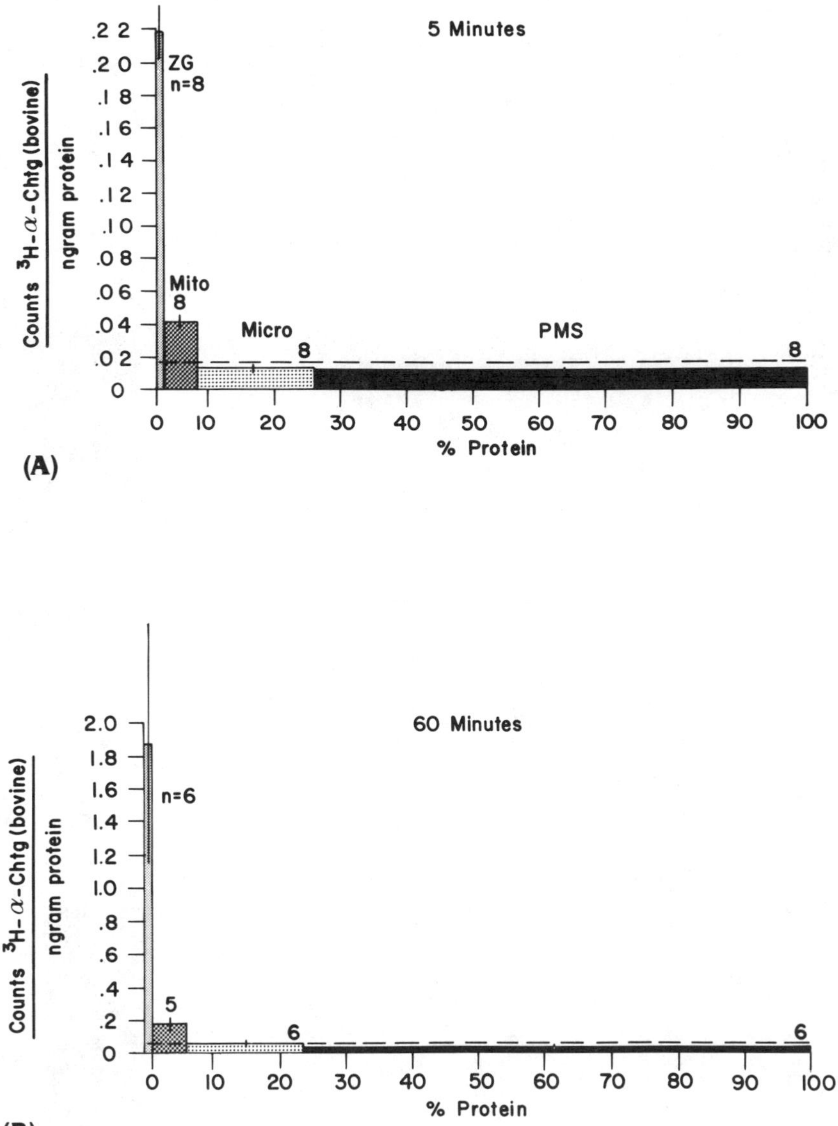

(A)

(B)

Figure 17 Specific radioactivity of [^{3}H]chymotrypsinogen (Chtg) (cpm/ng protein) in different cell fractions. Fractions are: ZG (zymogen granule), Mito (mitochondria), Micro (microsomes), and PMS (postmicrosomal supernatant). Specific radioactivity is plotted against percent of total protein recovered (after washes) in these fractions. Dashed line is sum of counts collected in fractions divided by sum of protein collected. Numbers are number of individual experiments from which means were determined. Error bars are standard error of the mean. (A) Fractionation after 5 min of incubation; (B) after 60 min of incubation. (Reprinted, with permission, from Liebow and Rothman, 1974.)

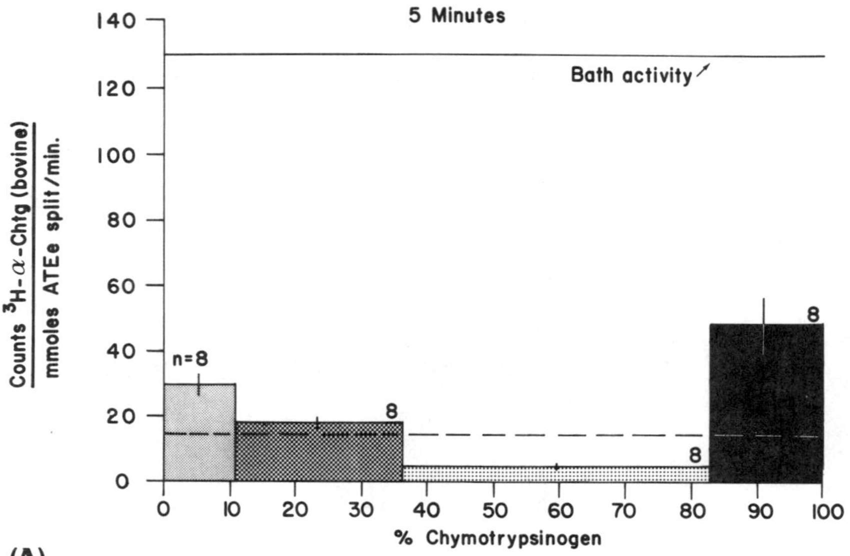
5 Minutes
Bath activity
Counts ³H-α-Chtg (bovine) / mmoles ATEe split/min.
140
120
100
80
60
40
20
0
n=8
8
8
8
0
10
20
30
40
50
60
70
80
90
100
% Chymotrypsinogen

(A)

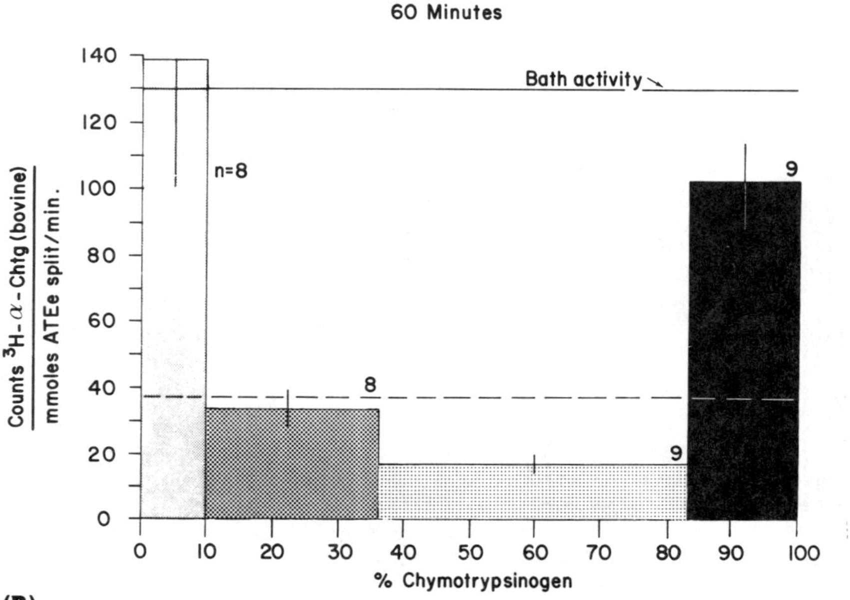
60 Minutes
Bath activity
Counts ³H-α-Chtg (bovine) / mmoles ATEe split/min.
140
120
100
80
60
40
20
0
n=8
8
9
9
0
10
20
30
40
50
60
70
80
90
100
% Chymotrypsinogen

(B)

with exogenous chymotrypsinogen (unlike mitochondrial and microsomal subfractions) (Figure 18), and indeed for short periods of incubation (for example, 5 min) had a higher chymotrypsinogen specific activity than the zymogen granule fraction itself (Figure 18).

11.8 MEMBRANE PERMEABILITY IN OTHER CIRCUMSTANCES AND ACROSS OTHER MEMBRANES

There are a variety of other observations, of both a direct and indirect nature, that support the notion of the membrane transport of these proteins, not only across granule and apical acinar cell membrane, but across other biological membranes as well. Although I will not consider these observations in any depth here, and must leave it to the reader to ferret them out in the primary literature, I would like to touch on two, both as examples and because I think that they are noteworthy.

As we discussed in the last chapter, the basolateral membrane of the acinar cell is permeable to at least some digestive enzymes, by an equilibrating process suggestive of membrane transport, and not vesicular processes. In addition, digestive enzyme added to the interstitial fluid bathing the tissue can cross the epithelium and be collected in pancreatic juice. This process of transpancreatic transport has been characterized to some degree in work by Liebow and Isenman using the organ culture preparation of rabbit pancreas that was described in the last chapter (Liebow and Rothman, 1975; Isenman and Rothman, 1979).

Transport across the epithelium has been shown to involve the movement of the proteins across the secretory cell in a concentration-dependent fashion. For example, it was found that the addition of exogenous amylase to the bathing medium led to a concentration-dependent increase in amylase secretion up to rates of the order of maximal endogenous rates (about an

Figure 18 Specific radioactivity of [^{3}H]chymotrypsinogen (Chtg) (cpm in terms of chymotrypsin activity in fractions). Fractions are from left to right: zymogen granule, mitochondrial, microsomal, and postmicrosomal supernatant. Counts per minute bovine [^{3}H]α-chymotrypsinogen per ATEe esterase activity (total chymotrypsinogen) are plotted against percent esterase activity found in each fraction. Dashed line represents specific radjoactivity of sum of represented fractions. Numbers are number of experiments and error bars are standard error of the mean. Solid line represents bath, or tracer, specific radioactivity. (A) 5 min incubations; (B) for 60 min. (Reprinted, with permission, from Liebow and Rothman, 1974.)

order of magnitude above control or unstimulated rates), while at the same time, and also in a concentration-dependent manner, inhibiting the secretion of previously labeled endogenous amylase (by up to 70–90% for exogenous amylase concentrations that produced magnitudinal increases in overall amylase secretion) (Isenman and Rothman, 1979) (Figure 19). These, as well as other similar, observations provided evidence that exogenous enzymes can be taken up by secretory cells, mix with endogenous enzyme pools, and compete with them for exit from the cell. In another experiment, Isenman incubated the gland in a medium containing a substantial concentration of chymotrypsinogen for 2 hr prior to adding a cholinergic agonist to augment secretion. Under these circumstances, the peak chymotrypsinogen response to the cholinergic agonist was enhanced more than three fold, compared with tissues that had not been previously incubated with the enzyme (Figure 20).

In the second example, Götze, studying amylase transport across the intestine, mounted pieces of ileal epithelium as a membrane to separate two halves of an incubation chamber. He found that: (1) amylase transport occurred at about twice the rate in the mucosal to serosal (gut lumen to blood) (mucosal/serosal, MS) direction as in the serosal to mucosal

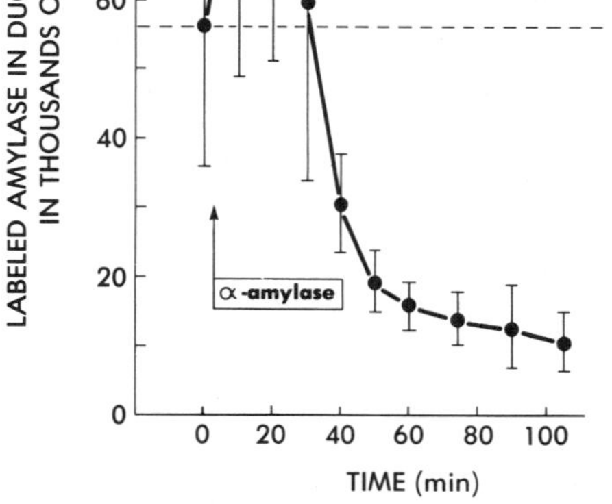

Figure 19 Effect of the addition of 1 mg/ml α-amylase to the bath on the amount of radioactive amylase in ductal secretion collected from rabbit pancreas in organ culture. Data are expressed as mean cpm in amylase/10 min (± SE). Value at time 0 is cpm in amylase in secretion for 10 min control prior to the addition of α-amylase to the bath and 230–240 min after the addition of [^{3}H]leucine to the medium. $n = 8$ at 0–40 min; 7 at 50 and 60; and 3 at 75, 90, and 105 min. (Reprinted, with permission, from Isenman and Rothman, 1979.)

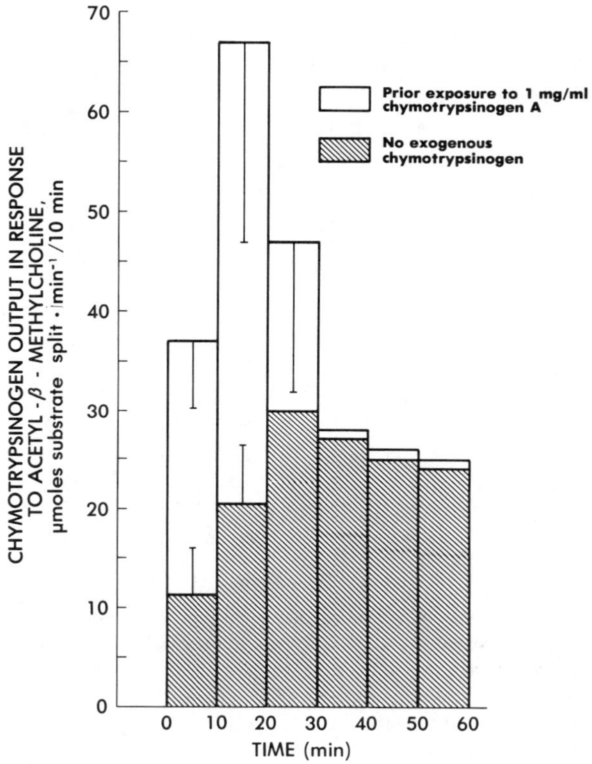

Figure 20 Effect of acetyl-α-methylcholine chloride (1 mg/100 ml bath volume) on chymotrypsinogen output from rabbit pancreas in organ culture with and without prior exposure (2 hr) to 1 mg/ml bovine chymotrypsinogen A in bathing medium. Acetyl-β-methylcholine chloride was added at time zero. $n = 5$ at 0–40 min; 3 at 40 and 50 min; and 4 at 50–60 min for prior exposure condition. $n = 7$ for control data. The difference between chymotrypsinogen pretreated and control groups was $p < 0.01$ at 0–10 min (two tailed t test), and $p < 0.05$ at 10–20 min (Mann-Whitney rank test). (Reprinted, with permission, from Isenman and Rothman, 1979.)

direction; (2) transport (MS) was inhibited by more than 75% in the absence of oxygen (Figure 21); and (3) amylase transport across the epithelium was two to three orders of magnitude greater than for the 5000–6000 dalton carbohydrate, inulin (Figure 22). Thus, in a well-defined system, specific energy-dependent amylase transport was demonstrated across other than acinar cell membranes.

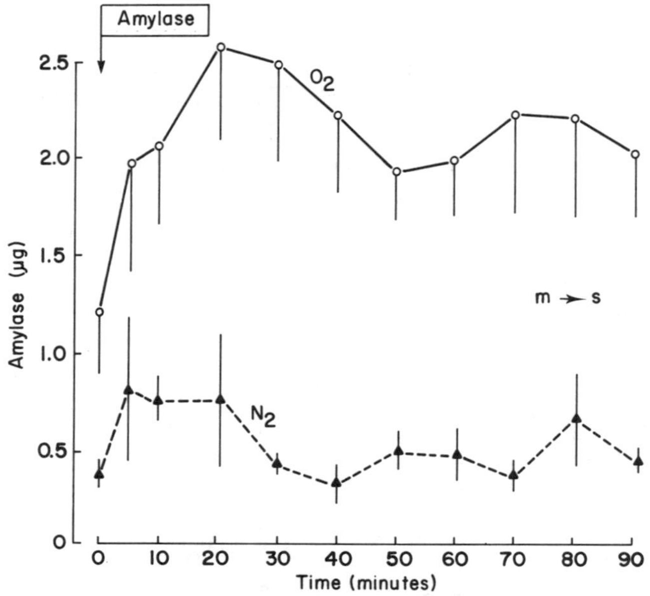

Figure 21 The effect of metabolic inhibition on mucosal to serosal ileal amylase transport in vitro (Ussing chamber). The release of amylase into the serosal medium after the addition of amylase (at time 0) was determined in the presence of a 95% N_2–5% CO_2 gas mixture (▲, n = 6) and compared with the results obtained with a 95% O_2 and 5% CO_2 mixture (○, n = 6). The serosal medium was replaced after each sampling period. (Reprinted, with permission, from Götze and Rothman, 1978.)

11.9 SUMMARY

Of course, all of the experiments that I have just discussed can in one fashion or another be interpreted in ways to accommodate them to paradigmatic precepts. However, I believe that this evidence, particularly when viewed in light of the other studies that I have discussed (and shall discuss further on), presents a convincing case for the existence of the membrane transport of these substances.*

Perhaps we can use the following parallel as a litmus test of our preconceptions about protein transport across membranes. How would you view the evidence for uptake and release, for the equilibrium behavior of the

*For further consideration of the question of granule membrane permeability, and attempts to explain such observations in terms of the vesicle paradigm, see Liebow (1985) and Rothman and Liebow (1985).

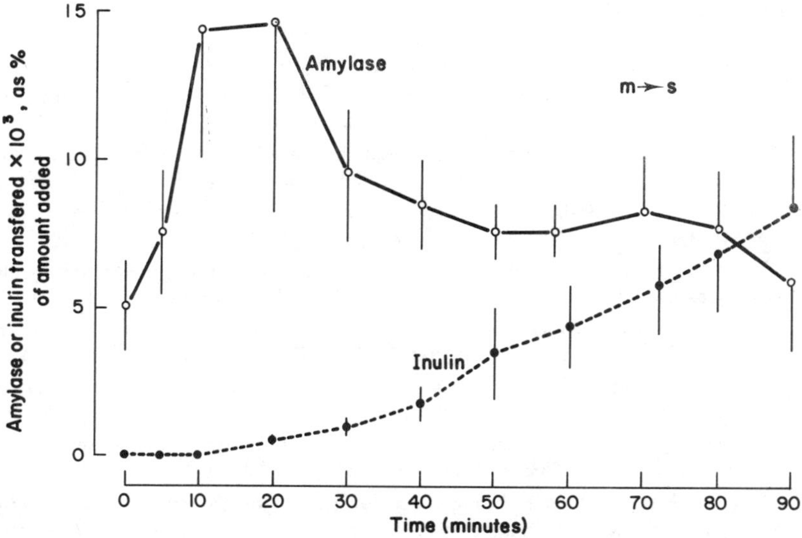

Figure 22 Relative amounts of amylase (○) and [^{14}C]inulin (●) appearing in the serosal medium following their simultaneous addition to the mucosal surface. Amylase data were corrected for endogenous release (n = 6). The serosal medium was replaced after each sampling. At 10 min amylase activity, as percent transferred, was 0.014 ± 0.0035 SE and inulin 0.00007 ± 0.000025. At 5 min amylase activity, as percent transferred, was 0.0075 ± 0.002 SE and inulin was 0.000013 ± 0.000013. (Reprinted, with permission, from Götze and Rothman, 1978.)

system, and so forth, if it were for a small sugar rather than a large protein molecule? Would evidence for membrane transport of the sort presented here be more convincing in this case, or perhaps even self-evident? If so, then the reason for this view must come more from a priori assumptions about the likelihood of the membrane transport of proteins, as opposed to sugars, than the evidentiary questions.

In any event, the existence of the membrane transport of digestive enzyme from interstitium to cytoplasm to granule, and vice versa, requires the existence of a cytoplasmic pool of digestive enzyme and therefore the evidence for the permeability of the granule membrane, as well as much of the other evidence presented in this chapter, is also evidence for the existence of a natural pool of digestive enzyme in the cytoplasm of the intact cell. Indeed, if the granule membrane is permeable, such a pool must exist. But in addition there is a range of other evidence that bears on this issue, which we will now consider.

SELECTED REFERENCES

1. Burwen, S. J. Thesis, Harvard University, 1972.
2. Burwen, S. J. and S. S. Rothman. Zymogen granules: Osmotic properties, interactions with ions, and some structural implications. *Am. J. Physiol.* **222**:1177–1181, 1972.
3. Ermak, T. H. and S. S. Rothman. Internal organization of the zymogen granule: formation of reticular structures in vitro. *J. Ultrastruct. Res.* **64**:98–113, 1978.
4. Götze, H. and S. S. Rothman. Amylase transport across ileal epithelium in vitro. *Biochim. Biophys. Acta* **512**:214–220, 1978.
5. Grendell, J. H. and S. S. Rothman. Digestive end-products mobilize secretory proteins from subcellular stores in the pancreas. *Am. J. Physiol.* **241**:G67–G73, 1981.
6. Grendell, J. H., H. C. Tseng, and S. S. Rothman. Regulation of digestion. I. effects of glucose and lysine on pancreatic secretion. *Am. J. Physiol.* **246**:G445–G450, 1984.
7. Hokin, L. E. Isolation of zymogen granules of dog pancreas and a study of their properties. *Biochim. Biophys. Acta* **18**:379–388, 1955.
8. Isenman, L. D. and S. S. Rothman. Transpancreatic transport of digestive enzyme. *Biochim. Biophys. Acta* **585**:321–332, 1979.
9. Liebow, C. Protein transport through zymogen granule membrane: reconsideration after a decade, in, *Nonvesicular Transport*, S. S. Rothman and J. J. L. Ho, Eds. John Wiley & Sons, New York, 1985.
10. Liebow, C. and S. S. Rothman. Membrane transport of proteins. *Nature* **240**:176–178, 1972.
11. Liebow, C. and S. S. Rothman. Transport of bovine chymotrypsinogen into rabbit pancreatic cells. *Am. J. Physiol.* **226**:1077–1081, 1974.
12. Liebow, C. and S. S. Rothman. Enteropancreatic circulation of digestive enzymes. *Science* **189**:472–474, 1975.
13. Liebow, C. and S. S. Rothman. Equilibration of pancreatic digestive enzymes across zymogen granule membranes. *Biochim. Biophys. Acta* **455**:241–253, 1976.
14. Rothman, S. S. The behavior of isolated zymogen granules: pH-dependent release and reassociation of protein. *Biochim. Biophys. Acta* **241**:567–577, 1971.
15. Rothman, S. S. Association of bovine α-chymotrypsinogen and trypsinogen with rat zymogen granules. *Am. J. Physiol.* **222**:1299–1302, 1972.
16. Rothman, S. S. The molecular regulation of digestion: bond-specific and short-term. *Am. J. Physiol.* **226**:77–83, 1974.
17. Rothman, S. S. The independent secretion of different digestive enzymes by the pancreas. *Am. J. Physiol.* **231**:1847–1851, 1976.
18. Rothman, S. S., S. Burwen, and C. Liebow. A zymogen granule: Intragranular organization and its functional significance, in, *Advances in Cytopharmacology*, Vol. 2, B. Ceccarelli, F. Clementi, and J. Meldolesi, Eds. Raven Press, New York, 1974, pp. 341–348.
19. Rothman, S. S. and C. Liebow. Permeability of zymogen granule membrane to protein. *Am. J. Physiol.* **248**:G385–G392, 1985.

CHAPTER 12

The Cytoplasmic Pool

A cytoplasmic pool of digestive enzyme that is the direct, or immediate, source of protein transported out of the acinar cell is a requirement of the equilibrium theory of secretion. Conversely, the cisternal packaging–exocytosis theory, in all of its forms to date, excludes the existence of such a pool, namely, new protein is separated from the cytoplasm of the cell as it is being synthesized and this separation is continuously maintained by its enclosure in and transfer by means of special membrane-bound structures from the cisternal spaces of the rough-surfaced endoplasmic reticulum (RER) to the zymogen granule, and then out of the cell. Thus, the existence, or lack thereof, of a cytoplasmic pool of digestive enzyme is another clear predictive difference between the two models; one requires its presence, the other its absence. For this reason, seeking evidence for such a pool has been, and continues to be, an important focus of investigation.

12.1 WHAT IS THE CYTOPLASM?

Before we consider whether or not a cytoplasmic pool of digestive enzyme exists, we should ask what we mean by cytoplasm? The cytoplasm, as the term is commonly used today, is the liquid matrix of the cell, the solvent phase, including its contained soluble materials (dissolved solutes) and dispersed (insoluble) particles, in which intracellular structures, or organelles, such as the nucleus and mitochondria, are suspended, and which is the intracellular compartment in direct contact with the extracellular environment across the cell membrane.

This is a descriptive definition of cytoplasm—a description that has changed with changing times and evidence—not part of a logical system defining and subdividing the parts of a cell. For example, some insoluble material ("dispersed particles") is considered cytoplasmic, whereas other insoluble particles are viewed as being separate from the cytoplasm ("intracellular structures"). Yet both are part of a continuum of particles that is not naturally separable as a function of their particular quality. This is not to say that there are not features that distinguish one group of suspended particles from the other. Indeed, their classification is based on differences in their nature, usually differences in size, complexity, and function. However, this classification is informal and is not always consistent, and is sometimes merely arbitrary or a matter of convenience.

In general, large objects, at least in cellular terms, often, but not always, membrane-bounded, that are usually, but not necessarily, chemically heterogeneous, particularly those that are known to perform special cellular functions, such as highly organized organellar structures (for example, the mitochondrion), are not viewed as being part of the cytoplasm, but as being separate from it, and suspended in it. Particles that are considered part of the cytoplasm, on the other hand, tend to be smaller and less complex in terms of both structure and function; and include all soluble material, insoluble homogeneous protein polymers, heterogeneous aggregates of protein, and lipoprotein micelles. A useful rule of thumb in considering whether or not a particle is cytoplasmic is whether it is larger and more complex than the ribosome or smaller and less complex. That is, the ribosome forms a convenient boundary for these empirical distinctions in today's terms. Although the ribosome's chemical content and physical organization have yet to be established fully, it is a particle of about 150 Å in diameter, in the main comprised of a variety of different proteins (perhaps as many as 25 or so) that appear to be designed to carry out a single type of reaction, the formation of the peptide bond.

Although one can classify intracellular elements with greater logical clarity, and this may be useful to do, as I have said, the concept of cytoplasm is not part of such a classification. Rather, it is the result of a continuing experimental clarification over the years of what the cell contains. For example, before the electron microscopic era, one could successfully argue that in many cells whatever was not found in the nucleus was part of the cytoplasm. Little else was observable. Partly for this reason there are two different cytoplasms that are not necessarily congruent in practice; the anatomical cytoplasm, a geometrical space within the cell, and the biochemical cytoplasm, its molecular substance.

The anatomical cytoplasm is of course based on the appearance of the cell under the microscope. When we look at a microscopic image of a section through a cell, the cytoplasm is the material contained within the cell membrane, excluding large structures that are able to be seen at a given level of magnification. That is, the cytoplasm is the volume of the cell that is not occupied by large definable structures at a particular level of magnification and resolution. As our capability to resolve small particles has improved, the volume of the cell that is considered cytoplasmic has been reduced by their exclusion, and hence today's cytoplasm occupies a smaller percentage of the cell volume than the cytoplasm of old.

Under the electron microscope, the cytoplasm appears as an inhomogeneous ground substance. It excludes the interior matrices of various intracellular structures, such as the nucleus, and in practice includes all particles that are free standing, not contained within larger structures, that are as small as or smaller than the size of the ribosome, 150 Å diameter. Although the volume of the cell occupied by the anatomical cytoplasm may vary widely from one cell to another, we can estimate that it occupies 50–80% of the cell volume in an "average" cell.

A full description of the biochemical cytoplasm would be a list of the nature, concentrations, physical forms, and relationships among substances contained in the anatomical cytoplasm. Ideally, the biochemical and anatomical cytoplasm should be identical in the sense that the substance of one is contained in the volume of the other; but as I have already suggested, this ideal is far from being realized on the basis of current knowledge.

Our understanding of the chemical composition of the cytoplasm has in great part been derived from attempts to extract directly and collect its contents, and then to analyze them chemically. This has, for the most part, been accomplished by physically separating the cytoplasm from the rest of the cell. Usually, cells are broken open in one fashion or another (cell and tissue homogenization), their contents exposed to varying degrees of centrifugal force, and the material that cannot be, or is not, sedimented is considered the "cytoplasmic fraction"; that is, the fraction of the homogenate that contains the cytoplasm.

Since this is dependent upon, among other things, the degree of applied centrifugal force, it is not surprising that as the power of our centrifuges has increased, and we have been able to sediment smaller and smaller particles, that what is included in the biochemical cytoplasm has become more and more restricted, much as with the anatomical cytoplasm. Today, materials that do not sediment readily at about 100,000 times the force of gravity are usually thought of as being cytoplasmic, although this is certainly not a firm and fast line of demarcation. Such 100,000 $\times$ g supernatants contain a large percentage of our "particle of division," the ribosome, but are most often free of substantial contamination with membrane-bound particles.

12.2 WHAT IS A CYTOPLASMIC PROTEIN?

It has been said, although not as often recently, that breaking open a cell to extract its parts, represents an act of some conceit by scientists that they can learn about the workings of an intact cell after they have destroyed its natural structure and relationships. Conceit or not, one can learn about cells in this fashion, and the effectiveness of such techniques, as well as their historical importance in the development of our knowledge of the cell, as I have pointed out, cannot properly be denied. From the early separation of the nucleus and with it the genetic apparatus, to the separation of the ribosome that carries out protein synthesis, this approach has been eminently rewarding. However, even if one does not think, like the vitalists of old, that breaking open the cell cannot provide any substantive information about its function, it is equally clear that studying parts of the cell isolated from their natural environment (and they can only be isolated by destroying their natural environment), must limit what we can infer about in situ events merely from knowledge of isolated cellular parts. Of course, to understand wholly cellular processes we must know the dynamical and thermodynamic relationships between parts, physical and chemical, that are extant within the whole cell.

Thus, the central problem with cell fractionation technology as we know it today, is that isolation itself limits what we can learn (at least in the absence of what one biology text calls a "deblenderizer"). Compounding this presently irresolvable difficulty is the fact that the most effective and common methods of cell fractionation replace the intracellular cytoplasmic environment that normally bathes the organelles with one that is very unlike it. Three major changes have commonly been made: (1) as the tissue is homogenized, the parts of the cell are diluted in a volume that is usually more than an order of magnitude greater than that within the cell; (2) the suspending medium is usually a simple, nonionic medium, unlike the cytoplasm which is chemically complex, contains relatively high concentrations of charged solutes (~150 m*M*), and includes a substantial concentration of protein; and (3) separation is carried out in the cold (usually 4°C), whereas cells, at least of homeotherms, are normally at 36–39°C.

There is every reason to think that these circumstances could alter, even dramatically alter, the distribution of substances between particulate and soluble phases of the homogenate, or between different particulate phases. For example, if a substance were maintained at a relatively high concentration in a particular organelle by a membrane transport process, then the dilution of the medium outside the organelle, the reduction in temperature, and altering the concentration and gradients of many other substances across the membrane of the organelle might well alter the distribution of the substance of interest. It might even lead to such substantial release from the organelle that we could mistakenly derive the impression that this substance was mainly in the cytoplasm. Conversely, in the same way, the conditions for separation might increase the access of a substance to an organelle from which it is normally excluded and we might derive the opposite erroneous impression; namely, that it is normally found in this particular structure.

To the degree that redistribution can occur, we have no ready means of determining either its occurrence, direction, or extent without an independent method of measuring the presence of the substance in different parts of the cell in situ. And indeed, if we had such knowledge, cell fractionation for this purpose would probably be unnecessary, since the knowledge would already be ours. Therefore, cell fractionation as a tool for determining the true intracellular distribution of substances is of limited usefulness, *as long as* the material of interest may be redistributed as a result of our experimental manipulations.

For example, one would not attempt to determine the intracellular distribution of the sodium or potassium ion by cell fractionation procedures. This would be fruitless with current technology because we would expect their distribution to be greatly altered when the cell was broken open, and in ways that we could not readily, if at all, assess. This of course does not mean that we could not study the sodium and potassium transport properties of a particular extracted subcellular structure.

However, for substances thought to be held firmly in a compartment, such as proteins within a secretion granule, held by an impermeable membrane, or contained within the membrane itself by essentially irreversible forces, separation would not alter the natural distribution of the substance as long as the organelle itself was not damaged by the procedure. Thus, if all of a particular protein were recovered in an ultraspeed supernatant, or conversely none were found there, then the conclusion that the protein is naturally contained in the cytoplasm, or restricted from it, would be drawn based on the assumption that no shift in its distribution had occurred as a result of breaking open the cell.

Most cytoplasmic and noncytoplasmic proteins have been catalogued in this fashion based on the assumption that no substantial redistribution of material occurs as a result of the separation procedure. This assumption is probably correct in many, perhaps most, cases, but is most likely incorrect in some.

In any event, it certainly is reasonable to *hypothesize*, even if we cannot rigorously demonstrate, that if all or virtually all of a material is found in the cytoplasmic fraction or is excluded from it, that this distribution reflects the state of affairs in situ. However, what are we to make of a situation in which the protein is found both in particulate and supernatant fractions of a tissue homogenate? If our assumption that no redistribution occurs is justified, then would we have to conclude that the particular protein is found naturally in both cellular compartments, and moreover, if we subdivided the particulate material and found it in a variety of subcellular sediments, then would we have to conclude additionally that it was present in all of them?

Of course this is not necessarily the case, nor is it necessarily or even usually, the conclusion that is drawn. It is commonly proposed that the material occurs naturally in one intracellular compartment, and that its presence elsewhere is an artifactual result of cell fractionation procedures. This redistribution, however, is not usually considered to be the result of a "natural" redistribution as with the sodium ion. As I have said, it is usually held that such a redistribution does not occur for proteins. Rather, the presence of a protein in an "inappropriate" compartment is most often viewed as being the result of an "unnatural" redistribution; for example, due to damage to an organelle during homogenization leading to the subsequent solubilization of a protein component. Such a proposal of an "unnatural" redistribution may be bolstered by knowledge that damage to subcellular organelles may well occur during homogenization, and in some cases clearly does.

Similarly, it may be believed that material found in a particulate fraction of a tissue homogenate is artificially derived from a natural cytoplasmic pool of this protein. In this case, it may be thought that an "unnatural" adsorption to the surface of the particulate material has occurred, as opposed to the protein being naturally associated with the fraction or its association being the result of a "natural" redistribution.

Nevertheless, a redistribution may either be "natural," or "unnatural," or even both. Nor does the occurrence of a redistribution of whatever kind necessarily mean that an organelle does not contain the material as a normal constituent as well. Of course, the point is that merely by observing a distribution of material in cell fractions, we cannot infer its causes in a rigorous fashion, that is, whether due to a "natural" redistribution (as a result of changes in gradients, temperature, and so forth), an "unnatural" redistribution (due to breakage or adsorption, and the like), or whether it reflects no redistribution at all. Nor, if more than one of these causes is involved, can we determine to what quantitative extent each accounts for what is observed.

These explanations of course remain hypotheses to be tested and verified in other ways. Discounting the presence of the substance in other than its primary location as being artifactual is particularly problematic. This can perhaps be appreciated by an example. Let us say that we recover most of a certain enzyme involved in oxidative metabolism in a eukaryotic cell in the particulate phase of the homogenate, and find only a minor component in the cytoplasmic or supernatant fraction. We might even know that the enzyme is concentrated in the mitochondrion, and have a good idea of how it functions chemically in concert with other enzymes in that organelle. We might also know that mitochondria were damaged to some extent during their separation from the cell. On this basis we might feel quite justified in drawing the conclusion that the small amount of this enzyme recovered in the supernatant fraction was merely an "unnatural" artifact of separation and that its sole natural home was the mitochondrion. However, such a conclusion could be incorrect. Many mitochondrial proteins are known to be synthesized on free ribosomes and released from them into the cytoplasm prior to their uptake into the mitochondrion (Schatz and Mason, 1974; Ho, 1985). That is, a natural cytoplasmic pool of such a protein exists. Its size might be quantitatively trivial or it might be substantial, depending upon the rate of its synthesis relative to the rate of its uptake into the mitochondrion and the size of the cellular pool overall. But whatever its size, a natural pool would exist. This of course would not mean that all of this particular protein found in the high-speed supernatant would have been part of this natural intracellular pool. Some, even most, might be derived from the redistribution of the protein, either "natural" or "unnatural," from the particulate phase. Without additional information, we cannot distinguish one possibility from the others, no less quantitatively.

Thus, in considering the meaning of an observed distribution of a particular protein in fractions of cells derived from cell and tissue homogenates we need to be cognizant of the following:

1. We should not assume that an observed distribution is the natural one, the one within the cell, on the basis of the distribution alone.
2. But neither should we assume that it is not.
3. We should not assume that the presence of a protein in a particular fraction is artificial merely because this is possible.
4. We should not assume that protein is only redistributed upon homogenization in "unnatural" ways, and that a "natural" protein redistribution cannot occur, without specific evidence for this view.
5. We should not assume that there is only a single natural location for a protein in the cell, nor that there is any particular number of natural compartments on an a priori basis.
6. We should not assume that a protein is necessarily either cytoplasmic or noncytoplasmic, because it may be both.
7. Even when the great preponderance of a protein is found in a particular fraction, we should not assume that this is the natural and sole location of the protein in the cell if we have no other evidence upon which to base our case.
8. And I would add, the distribution of a molecule between fractions may change with physiological state.

12.3 THE CONCEPT OF SPECIFIC ACTIVITY

One way in which scientists have attempted to overcome some of the limitations of cell fractionation as a technique for determining the location of a protein in the cell has been to borrow the concept of specific activity from protein chemistry. I have already considered this concept in a particular context in Chapter 7. Changes in the specific activity of an enzyme, that is, changes in the ratio of the activity of a particular enzyme in a sample to the total amount of protein in that sample, is used to determine the relative degree of purity of the sample and to indicate whether or not an attempt at purification has been successful. If it has, the specific activity increases, because the quantity of other proteins contaminating the sample will have been reduced.

In applying this idea to cell fractionation, one compares the specific activity of a protein in a particular cell fraction to that found in the homogenate (that is, in the cell overall). If the particular fraction has a higher specific activity than the cell, then the inference is made that the material is concentrated in the major cell component comprising the fraction. On the other hand, a lower specific activity leads to the opposite inference, that is, the substance is a contaminant and not a natural constituent of the fraction. This approach has been useful in identifying compartments in which substances appear to be concentrated, particularly if the compartment occupies a small percentage of cytoplasmic volume (and therefore large differences between the specific activity of the fraction and the cell can be established).

On the other hand, this approach has its limitations. A low concentration of a substance in a compartment does not necessarily mean that it is an unnatural inclusion. It may indeed be present naturally at a low concentration. Moreover, specific activity expresses the activity of an enzyme in terms of the protein content of the fraction, and although this is a convenient measurement, it is not the proper one. If we wish to know the concentration of a substance in a particular cellular compartment, then of course we need to know the volume of that compartment. Frequently, however, the volume is unknown (or unknowable) and protein content is a convenient and useful substitute. In using it however, we make the assumption that the protein content of each and every cellular compartment is related to its volume at the same, fixed ratio, and this is undoubtedly incorrect.

12.4 THE ACINAR CELL AND THE INTRACELLULAR LOCATION OF THE DIGESTIVE ENZYMES

Now let us look at the situation for the digestive enzymes of the pancreas. How are they distributed in the various subcellular fractions of pancreatic tissue homogenates? Is all, or almost all, of the enzyme found in the sedimentable fraction, in one vesicle or another, as the cisternal packaging–exocytosis model proposes, or is there a "soluble" pool of enzyme? Using the literature as our guide, we could, on the one hand, say that almost all of the material is recovered in the sedimentable fraction. However, on the other, citing other experimental work, enzyme and animal species we could just as well say that nearly all is soluble and therefore of cytoplasmic origin. Published estimates have varied from about 85% sedimentable to about 85% in the supernatant, and most values range between these two extremes, including those in which an approximately 50–50% distribution is seen.

What sense can we possibly make of these differences? Most are not differences in the measurement of the same enzyme in the same animal species, but rather differences between different enzymes or different animals, or both. Moreover, usually the same, or similar, techniques are used to homogenize and separate the material. Therefore, the differences may not be explicable in terms of differences in methods. If we were instead considering a diverse group of intracellular enzymes, one could not conclude that all of the different proteins were either particulate or cytoplasmic, or even solely one or the other, based on such data.

However, when the secretory proteins of the pancreas have been considered, it is commonly thought that all of the various proteins should have the same in situ distribution, and therefore observed differences must reflect artifacts of redistribution and not natural variations. This view is based on the belief that the mechanism for the intracellular transport and disposition of these proteins does not distinguish between them; namely, they are handled in concert or in parallel.* Thus, the same distribution should be seen for all 20 different proteins, and deviations from that distribution must perforce be artifactual, and moreover, according to the paradigm, must be artifacts of the "unnatural" kind, since the possibility of a "natural" redistribution of protein as a result of their crossing membranes is excluded a priori.

But how can we determine what is actually the real distribution and what is the artifact? Here the hypothesis itself often acts as the guide in a pattern of circular reasoning. The paradigm proposes that all secretory protein is excluded from the cytoplasm, and therefore all material recovered in the

*This view appears to be changing. See below and Chapter 13 for a discussion of nonparallel transport.

supernatant fraction must be artifactually derived, and studies that demonstrate in one species, or for one enzyme, or another, more material in the supernatant must be flawed, and less adequate than those that find less of a particular enzyme in this fraction. Of course, since most of these observations are derived from essentially the same cell fractionation methodology, there is no way that we can distinguish between the adequacy of one or another such study. That is, one has no clear and unbiased way of evaluating the differences and their causes. Nor can we know if they reflect an "unnatural" redistribution, a "natural" redistribution, no redistribution at all, or if material is redistributed, to what extent and as the result of what particular cause or causes. Merely on the basis of such observed subcellular distributions themselves, we have no way of determining their cause or causes, and we certainly have no right to consider arbitrarily one observation superior to another merely on the basis of our belief in a particular hypothesis.

Perhaps because of its widespread acceptance, the mere ability to imagine an artifactual explanation for an observation that runs counter to the vesicular view of secretion is often sufficient interpretive justification. That is, it is not necessary to prove that a particular artifact is responsible for what is observed. It may be sufficient merely to know that it is possible that it could be. Such an argument is bolstered, if, as is often the case, we have good reason to suppose that such an artifact occurs at least to some extent or another, although we do not know to what extent and hence do not know whether it can account for the observation. For example, one can argue that damage to enzyme-containing organelles during homogenization or sedimentation produces an observed distribution, although one has no real knowledge that that is the case, or if it is, that it accounts for what is observed quantitatively. Or, the animal can be blamed for interspecies differences, and the conclusion drawn that one species or another is not satisfactory for studying secretion, although we have no information other than the fact that the data do not fit our hypothesis upon which to form such a judgement.

Furthermore, just as the presence of enzyme in the supernatant fraction of cells can be dismissed as being the result of an unnatural redistribution of protein from particulate fractions, one can also explain the absence of an enzyme in a particulate fraction that is derived from a cellular organelle thought to contain it, in the same way. For example, the rough endoplasmic reticulum (RER) is apparently a closed system of membrane-bounded spaces. It is thought that the internal or cisternal spaces of the reticulum are a major site for the storage of digestive enzyme and of secretory proteins in general. However, as I discussed in Chapter 7, when one collects the microsomal fraction of homogenates, which is predominantly comprised of vesicles formed from the endoplasmic reticulum, only small amounts of digestive enzyme are found associated with it. Moreover, of the material recovered in that fraction, the great majority is either associated with the ribosomes or adsorbed to the surface of the membrane of the vesicle (or otherwise readily removed from it), and is apparently not within the vesicle itself. The specific activity of digestive enzymes in the microsomal fraction, even including ribosomal and adsorbed enzyme, is usually relatively low; about one-third to two-thirds of the specific activity of the whole homogenate in the rat. [N.B.: This is a low value for a large compartment (~20% of cell volume), some 90% of which enzyme is even so not attributable to an intravesicular location.] Thus, there is very little evidence based on the fractionation of the endoplasmic reticulum that its internal spaces contain a substantial pool of digestive enzyme. Nevertheless, despite this "negative evidence," one can merely argue that the reason that the specific activity is low is because most of the enzyme contents of the cisternal spaces are released when the reticulum is broken open during homogenization (that is, it is lost as a result of an "unnatural" redistribution of material) (see Chapter 7 for further discussion of this question). If, nonetheless, one finds little enzyme in the supernatant fraction, then we can additionally assume that the released material had been adsorbed onto the surface of particulate cell fractions (thereby also explaining the "adsorbed" enzyme).

In a study by Scheele et al. (Scheele et al., 1978) which attempted to decipher these relationships, a variety of complex experiments and interpretations are presented which, after all is said and done, rest on two a priori assumptions: (1) there is no cytoplasmic pool, and (2) there can be no natural redistribution of these proteins (namely, membranes are not permeable to them). With these assumptions in hand, they are then able to conclude from their experiments that exogenous material that associates with particulate fractions is simply due to adsorption, and that endogenous material found in the supernatant fraction is due to its "unnatural" release from particulate fractions.

None of this is to say that these hypotheses and interpretations are necessarily incorrect, nor that the presence of enzyme in the supernatant fraction of homogenates is proof of a cytoplasmic pool, but rather to note a pattern of reasoning in which it is thought sufficient to merely imagine an artifactual explanation for an observation that is not consistent with the paradigm in order to discount it. Of course, it is no more justified to say that the presence of digestive enzyme in the supernatant fraction is the result of an "unnatural" redistribution of material, than it is to say that the relative sparcity of enzyme in that fraction in another case is the result of an "unnatural" loss from a natural cytoplasmic pool due to adsorption to particulate fractions in the nonionic homogenization medium. Indeed, these are not mutually exclusive occurrences. Both release and adsorption may occur, from and into a natural cytoplasmic pool, and as the result of a "natural" redistribution of material between fractions as well as an "unnatural" one. We need to be able to distinguish between these possibilities, and, moreover, in a quantitative fashion.

To do this, we must first know to what extent the contents of a natural compartment are lost, during cell fractionation, for example, from the cisternal spaces of the RER. And to know this, we must know in the first instance (in situ) what substances it contains and at what specific concentrations. Such attempts, for the cisternal spaces of the RER, have been made, but with at best unclear, if not wholly ambiguous, results thus far (see Chapter 7 and below for a discussion of the anatomical approach to the question of a cytoplasmic pool).

In addition to the cisternal spaces of the RER, enzyme recovered in the supernatant fraction of tissue homogenates may also be derived from the contents of zymogen granules broken during homogenization. As with the endoplasmic reticulum, the question is not merely whether granule breakage occurs, but if it does, to what extent it accounts for the observed enzyme content of the supernatant fraction of the homogenate? Dandrifosse (1985), in an interesting attempt to deal with this question, compared the number of zymogen granules in two pieces of tissue. In one, he measured the number of granules per unit volume in the intact cell. He homogenized the other, and then sedimented the whole homogenate and measured the number of granules per unit volume of sedimented homogenate. The number of granules per unit volume was essentially the same in both cases, and he concluded that homogenization did not damage a substantial number of granules. Of course, this does not mean that the granules did not lose their contents as the result of a "natural" (permeability), as opposed to an "unnatural," redistribution of digestive enzyme.

The conclusion that should be drawn from my discussion to this point is not to affirm the view that cell fractionation studies are useless, but to reinforce the need to appreciate the difficulty in using this approach to determine the natural distribution of substances in the cell. Most importantly, it is hard, if not impossible, to show in this fashion that the presence of a substance in a particular fraction is wholly artifactual (or natural) unless one has prior quantitative knowledge of the natural distribution of the substance in situ in the first instance. The cell fractionation approach is often more useful when we turn the question around and ask positively and qualitatively if a substance is a natural constituent of a particular compartment to one extent or another? I will now consider two examples of this approach taken from my laboratory's early work on protein transport.

12.5 CHANGES IN THE DIGESTIVE ENZYME CONTENT OF THE SUPERNATANT FRACTION AS A FUNCTION OF PHYSIOLOGICAL STATE

Our first approach, in the mid-1960s, to the question of a cytoplasmic pool using cell fractionation techniques was to look for changes in the digestive enzyme content of the supernatant fraction of tissue homogenates as a function of changing physiological state. We had observed that the gastrointestinal hormone, cholecystokinin, caused a nonparallel secretion of two enzymes; that is, it favored the secretion of one relative to the other, in particular it favored trypsinogen (Tg) secretion over chymotrypsinogen (Chtg) secretion from the isolated rabbit pancreas (Table 1) (Rothman, 1967; Rothman and Wilking, 1978) (see the next chapter for further discussion of this observation and of nonparallel transport in general). At the time, we thought that this might indicate the segregation of different enzymes or mixtures of enzymes in different zymogen granules, and that we might see an inverse change in the ratio of these two enzymes in the granules remaining in the cell after stimulation. If the secretion of one enzyme was favored, then perhaps the granule pool would be preferentially depleted of this enzyme. When we performed these experiments (Rothman, 1970), we found that the ratio of the two enzymes in the granule fraction had not been altered (Δ Tg/Chtg $= 0.20 \pm 0.50$ SE), nor was any significant change seen in their proportions in the microsomal (RER-derived) fraction (Δ Tg/Chtg $= 0.34 \pm 0.45$) (Figure 1). The ratio was, however, changed in

TABLE 1. Trypsinogen to Chymotrypsinogen Ratio in Subcellular Fractions of Rabbit Pancreas and in Secretion from the Rabbit Pancreas in Vitro

Source	Trypsinogen/Chymotrypsinogen
Zymogen granules	3.3 ± 0.4
Microsomes	3.6 ± 0.5
Postmicrosomal supernatant	9.2 ± 1.0
Unstimulated secretion	6.1 ± 0.4
CCK-stimulated secretion	15.8 ± 3.5

Values are means $\pm$ SE, $n = 12$ for cell fractions and $n = 6$ for secretion.
Source: Taken from Rothman (1970).

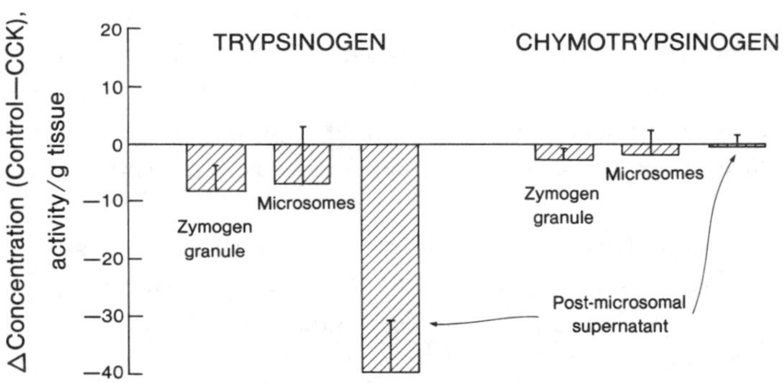

Figure 1 The effect of cholecystokinin-pancreozymin (CCK) on the relative distribution of trypsinogen and chymotrypsinogen (s) in certain cell fractions from the rabbit pancreas. Values are mean $\pm$ SE, $n = 8$ for all groups. Animals were killed 15 min after the injection of CCK (2.0 Ivy dog units/kg body weight). Values are the differences between paired control and treated rabbits. Only the value for trypsinogen in the postmicrosomal supernatant differed from its paired control, $p < 0.005$. (From Rothman, 1970.)

the postmicrosomal supernatant fraction; the fraction that contained the contents of the cytoplasm (Δ Tg/Chtg $= 2.35 \pm 0.77$). Moreover, it had changed in the predicted direction; that is, in the opposite direction to that observed in secretion, with its trypsinogen content being depleted preferentially (Figure 1).

Two other aspects of the observations reinforced the view that at least a portion of the supernatant fraction's content of digestive enzyme was due to a natural cytoplasmic pool. In the first place, the ratio of trypsinogen to chymotrypsinogen in the supernatant was dissimilar from that seen in either the zymogen granule or microsomal fraction, the two major putative sources of its enzyme contents in artifactual terms (Table 1). Furthermore, the enzyme ratios in both microsomal and zymogen granule fractions were similar to each other, thus excluding the possibility that the contents of the supernatant reflected a composite of loss from both of these pools. Therefore, their presence in the supernatant fraction could not have merely been due to their release from other parts of the cell as the result of an "unnatural" redistribution, unless there had been differential release, or differential readsorption of the two proteins, or both. Only in this case could the observed proportions be explained within the context of the paradigm, that is, in the absence of a natural cytoplasmic pool. This possibility seemed unlikely because the two enzymes, trypsinogen and chymotrypsinogen, were very similar; of almost the same molecular weight, size, shape, surface charge (isoelectric point), and ~50% homologous in terms of amino acid composition (Hartley et al., 1965; Iacino et al., 1980). Therefore, extensive, *but nevertheless nonspecific,* differential release or readsorption would have had to have occurred.

In addition, the ratio of the two enzymes in secretion was quite different from that found in either the microsomal or zymogen granule cell fraction. Indeed, if anything, it more closely resembled that in the supernatant (Table 1). Of course, according to the exocytosis model, the enzyme contents of secretion should closely match, indeed should be identical with, that in the zymogen granule pool.

Thus, the enzyme content of the supernatant fraction changed in a unique fashion in response to a physiological stimulus in a manner that suggested that a natural cytoplasmic pool of digestive enzyme was a source of the secreted material. In a second study, we tried a somewhat different approach. We produced pancreatic hypertrophy by feeding rats egg-white trypsin inhibitor in their food (Rothman, 1970). Feeding rats, and other animals, substantial quantities of trypsin inhibitors produces a considerable increase in the mass of the pancreas within a few days to a week. This increase is characterized by, among other things, a two- to four-fold increase in the concentration of digestive enzymes stored in the tissue. Where is the additional protein stored? Is there an increase in the concentration of digestive enzyme within the zymogen granule pool? When we performed this experiment, we found that there was virtually no change in the amount of three enzymes in the zymogen granule fraction (Table 2), and only a small

TABLE 2. Effect of Dietary Egg-White Trypsin Inhibitor on Enzyme Contents of Zymogen Granule Fractions[a]

	Trypsinogen (10) μmoles Substrate Split·min^{-1}·gland^{-1}	Chymotrypsinogen(s) (10), μmoles Substrate Split·min^{-1}·gland^{-1}	Amylase (9), mg Substrate Split·min^{-1}·gland^{-1}
Normal	7.6 ± 1.0	51 ± 9	758 ± 151
EWTI	7.4 ± 0.8	49 ± 6	806 ± 130

[a]Numbers in parentheses = *n*. EWTI = 4% egg-white trypsin inhibitor diet. Values are means ± SEM.

Source: Taken from Rothman (1970).

increase for two of the three in the microsomal fraction (Table 3). Between 80 and 100% of the increase in the enzyme content of the tissue was traced to the supernatant fraction (% nonnuclear activity increase, Table 3). Indeed, the specific activity of the enzymes in the supernatant was almost as high, and higher in one case, as in the zymogen granule fraction (Figure 2).

Thus, almost all of the increased concentration of digestive enzyme in the tissue was attributable to an increase in the enzyme content of the cytoplasm-containing fraction. The zymogen granules did not appear to be the source of the additional material in the supernatant, at least to any appreciable extent, because a proportional increase in its enzyme content was not observed. It was possible, however, that the cisternal spaces of the RER might have been the source of the enzyme recovered in the supernatant if virtually all of its contents had been lost during homogenization.

To consider this possibility, we plotted the enzyme content of the supernatant versus that of the microsomal fraction from normal rats and from rats fed trypsin inhibitor. In normal rats, the two variables were correlated; that is, the more enzyme that there was in the supernatant, the more there was in the microsomes (Figure 3). If the endoplasmic reticulum had been the source, then we would expect the higher values seen in the supernatant fraction after trypsin inhibitor feeding to fit the same function, with data points merely being placed further away from the origin. However, this was not the case and although the relationship was maintained in that the two variables were still correlated, feeding trypsin inhibitor altered the slope of the function so that there was considerably

TABLE 3. Effect of Dietary Egg-White Trypsin Inhibitor on Enzyme Activity Recovered in Microsomal Fraction and Postmicrosomal Supernatant[a]

	Trypsinogen (10), μmoles Substrate Split·min^{-1}·gland^{-1}		Chymotrypsinogen(s) (10), μmoles Substrate Split·min^{-1}·gland^{-1}		Amylase (9), mg Substrate Split·min^{-1}·gland^{-1}	
	Microsomes	p-m sup	Microsomes	p-m sup	Microsomes	p-m sup
Normal	29 ± 2	138 ± 24	204 ± 22	558 ± 57	6,328 ± 1,230	17,061 ± 3,607
EWTI	36 ± 3	252 ± 21	352 ± 39	1,386 ± 206	5,422 ± 1,098	32,147 ± 6,815
P	0.05	0.005	0.01	0.005	ns	0.01
% Nonnuclear activity increase	6	91	13	80	0	100

[a]*Numbers in parentheses* = *n*. EWTI = 4% egg-white trypsin inhibitor diet; p-m sup = postmicrosomal supernatant. Values are means ± SEM. Probability values compare normal vs. EWTI; ns = not statistically different.

Source: Taken from Rothman (1970).

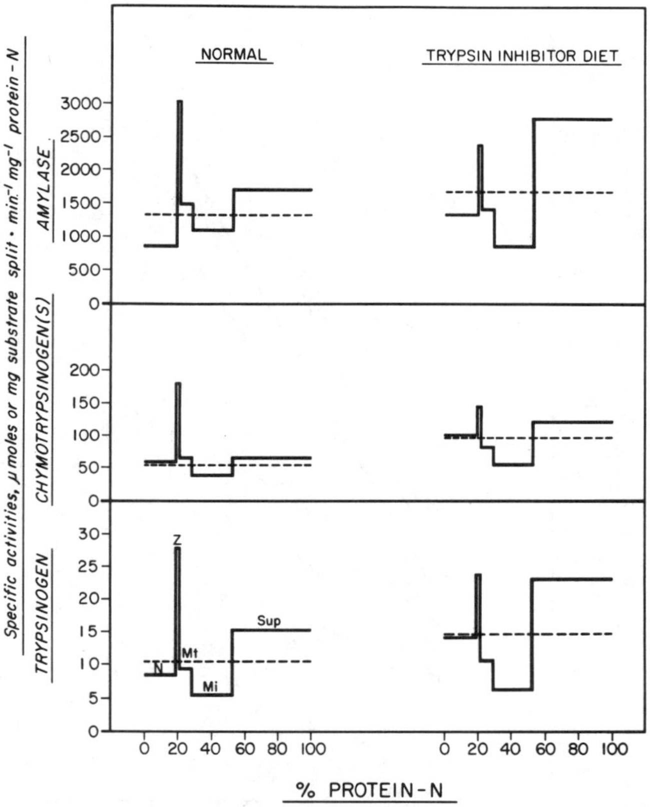

Figure 2 Comparison of the specific activity of amylase, chymotrypsinogen (s), and trypsinogen in various subcellular fractions of pancreas from normal and egg white trypsin inhibitor-fed rats (EWTI). (N) Nuclear fraction; (Z) zymogen granule fraction; (Mt) mitochondrial fraction; (Mi) microsomal fraction; (sup) postmicrosomal supernatant fraction. Dashed line refers to specific activity of the whole homogenate. Data are means of 10 experiments for both normal and EWTI-fed animals, except for amylase where $n = 9$. (Reprinted, with permission, from Rothman, 1970.)

more enzyme in the supernatant fraction for a given amount in the microsomes (Figure 3). That is, feeding trypsin inhibitor led to a preferential increase (relative to zymogen granule and microsomal fractions) in the enzyme content of the supernatant fraction. Thus, the additional enzyme recovered in the supernatant appeared to be due to the cytoplasm contained therein.

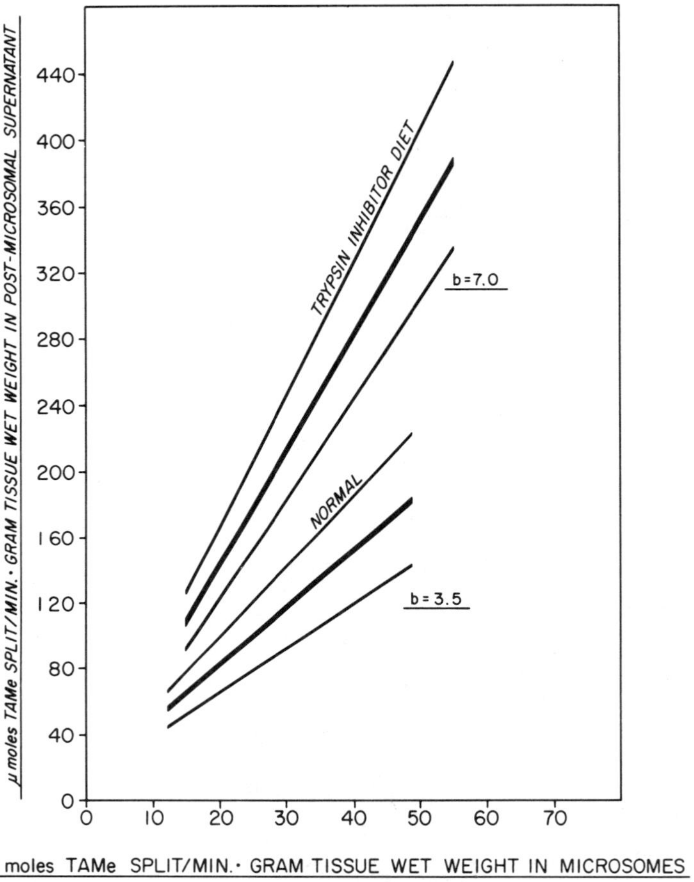

Figure 3 Trypsinogen in microsomal versus postmicrosomal fractions for normal and EWTI-fed rats (expressed as TAMe esterase activity per gram tissue wet weight). Goodness of fit for both lines, $p < 0.01$. Intervals around each line give 95% confidence limits. Difference between slopes, $p < 0.001$, $n = 15$ for each line. (Reprinted, with permission, from Rothman, 1970.)

12.6 THE ANATOMICAL APPROACH

As I discussed above, cell fractionation, the "biochemical" approach of determining the location of a protein in different subcellular compartments, has numerous problems. Indeed, in the final analysis, if conclusions drawn from such data are to be unambiguous, other in situ evidence is generally necessary. Investigators have been well aware of this problem, and as a result have sought other means, either independent or corroborative, of determining the location of specific proteins within the cell. A common approach has been to combine the anatomical and the biochemical by attempting to measure the chemical content of particular "intact" intracellular spaces in tissue sections. As I have already discussed in Chapter 7 in reference to the endoplasmic reticulum, two procedures have been used in pancreas. In one, the protein is labeled with an amino acid that contains the isotope tritium, and its presence over different parts of the cell determined from radioactive emissions emanating from the protein. This, of course, is autoradiography. The second approach is cytochemical and conceptually quite similar. In it a complex or precipitate is formed between an antibody to a specific protein and the protein itself (the antigen). One then determines the location of the protein by looking for the precipitate in tissue sections under the microscope. Nowadays, the antibody often is first coupled to an "opaquing agent" such as gold or ferritin to make the complex more easily seen in the electron microscope.

Both of these approaches have their particular strengths and weaknesses. For example, a photographic emulsion can be exposed to an isotopic emission at a substantial lateral distance from the site of its origin within tissue, and as a result the "resolution" of autoradiography is limited. For this reason, it has not been possible to determine whether or not a particular emitting protein is within the cisternal spaces of the endoplasmic reticulum or outside of it, or whether or not emitting proteins are contained within or outside of the small vesicles that have been proposed to account for intracellular protein transport. Although the antibody method has greater resolution, producing as it does small opaque particles at the site of precipitation, it still can present similar problems of resolution.

Furthermore, in autoradiography the period of time chosen for exposure of the emulsion to the labeled tissue, and a host of other technical details, such as the particular emulsion chosen, can have an important and often uncertain bearing on what is observed, particularly when quantitative issues are involved. Indeed, one can make grains appear or fail to appear essentially at will by these choices, especially if the emitting protein is present at a low concentration in a given pool relative to another. Moreover, the investigator, often, with perfectly good intentions, by these choices accentuates pools that contain the material at high concentrations, to make the evidence "clearer," by reducing the "backround"; namely, pools that contain the protein at a low concentration. There are similar problems with the antibody technique, which is further complicated by the fact that it can only indicate the location of the protein on the surface of a tissue section, where binding takes place, and at which site protein loss and redistribution are most likely to occur, rather than within the tissue section.

This raises an additional problem that both techniques share, that although not technically limiting, has made many studies (and certainly those that have been performed on the pancreas) uncertain and ambiguous as a means of determining the natural distribution of proteins in various parts of the cell. I believe this is particularly so when proteins that are known to be secreted; that is, that in one way or another cross membranes are studied. The problem concerns the handling of the tissue from the time that it is exposed to the marker (antibody or isotope) to the time that it is viewed in the microscope.

To prepare labeled tissue for viewing under the microscope, the sample is often washed in some fashion or another to remove "excess" label or antibody. It is then fixed, washed again, dehydrated in organic solvents, embedded in plastic, and sectioned. To know that the material seen in the microscope bears an accurate resemblance, in terms of concentration and location, to that found normally in situ, we must know what has been lost or redistributed, and from where within the cell, during this preparation for viewing so that we can correct for it; or alternatively, we must know that nothing has been lost, or that whatever is, is either quantitatively trivial, or lost equivalently from all parts of the cell. Moreover, we cannot assume that all proteins and all tissues behave the same way in this regard, or perhaps even more significant here, that even the same protein in different compartments of the same tissue behave identically.

Perhaps when intact tissue is washed after its exposure to a marker, it is thought that no loss or redistribution of protein can occur, because such loss presupposes the existence of membrane permeability to protein. However, to assume that it cannot occur, for example, in studies on pancreatic protein secretion, is to assume that the cisternal packaging–exocytosis or vesicle model is correct while performing experiments that are presumably designed to test the validity of the hypothesis. And even this assumption of impermeability cannot be made during the preparation of thin tissue *sections* for viewing where no membrane barrier exists.

As I have said, not only does the potential for the loss of material have to be evaluated, but more difficult to deal with, loss may occur in a differential fashion from one cellular compartment to another in the same tissue. For example, in the pancreas, the digestive enzymes are found at a relatively high concentration in the zymogen granule, and moreover, in a bound state. Whereas, the same proteins in the cytoplasm may be present at a lower concentration and in soluble, not bound, form. Thus, it would not be surprising if protein was more readily lost from the cytoplasm than zymogen granules. This is a particularly troubling problem with the antibody technique, in which case the liability of a few molecules at the surface of the tissue section might substantially alter the apparent distribution.

Whether for the sake of editorial brevity, because the proper controls are very difficult to perform, or because they are thought unnecessary, the treatment that a tissue receives after its exposure to a marker is often not noted in detail, nor is it often reported how and to what extent these treatments affect the concentration and distribution of particular proteins within the cell. Such knowledge is, however, crucial, particularly if we are interested in quantification, and not merely in noting some compartments in which a substance appears to be found. In the absence of this evidence, we can only assume that whatever we do to the tissue subsequent to its exposure to the marker has no effect either on the concentration or distribution of the substance. And of course, such an assumption, not based on evidence, is unwarranted.

This does not mean that these techniques are useless or hopelessly inadequate otherwise. For example, it has been shown visually that radioactively labeled secretory protein is in the zymogen granule, and that antibodies to a variety of these proteins complex with the protein contents of the granule (Kraehenbühl et al., 1977; Bendayan et al., 1980). The problem arises when one attempts to make quantitative estimates of the relative concentrations of substances found in various parts of the cell, particularly if one expects to be able to exclude the presence of a particular protein from a given location.

I will now briefly consider two relatively recent studies that have used the antibody technique to evaluate the compartmental distribution of secretory proteins in the acinar cell, as examples of the problems that one faces in such studies, as well as examples of the assumptions that may underlie such attempts at compartmental analysis (I have touched on these studies already in Chapter 7). In the first study (Kraehenbühl et al., 1977), the investigators demonstrated that ferritin-labeled antibodies to several digestive enzymes bound to sections of zymogen granules. However, they reported little in the way of opaque particles over either the cytoplasm or the cisternal spaces of the endoplasmic reticulum. The presence of particles over the zymogen granules was certainly expected, but what were they to make of their absence over the cytoplasm and cisternal spaces? Their conclusions are instructive. Should the investigators conclude that digestive enzymes are not normally found in either the cytoplasm or the cisternal spaces, or should they conclude that their technique was faulty and resulted in the loss of protein from both of these compartments? They answered this question in opposite ways for the two compartments, cytoplasm and cisternal space. Since they apparently believed that digestive enzymes are normal constituents of the cisternal spaces of the RER, in accord with the paradigmatic view of things, they concluded that the absence of opaque particles over the cisternal spaces must have been due to the protein having been washed out during tissue preparation. On the other hand, for the exact same tissue, treated in exactly the same fashion, they concluded that the relative sparcity of label over the cytoplasm indicated that digestive enzymes are not normally found in this cellular compartment. Indeed, the presence of some particles over the cytoplasm, led to the conclusion that they must have been an artifactual *addition.* Of course, one cannot appropriately apply one standard of evaluation to one compartment, and another to another in the same tissue, merely because it fits one's hypothesis. In this way what begins as an attempt to test a hypothesis does nothing of the sort, and the evidence is merely used to reinforce prior notions. The plain fact is that the investigators had no way of knowing (at least as one can judge matters based on published evidence) whether the materials had been lost or not, and if so, from where and to what extent.

In the other study (Bendayan et al., 1980), gold-conjugated antibodies were used. In this case, the investigators *found* label over both the cytoplasm and the cisternal spaces. Once again the conclusions are instructive. Did they conclude that enzyme is normally found in both places, cytoplasm and cisternal spaces, disagreeing with the apparently contradictory observations of the other group? No indeed. They drew exactly the same conclusions. Label found anywhere in the basolateral region of the cell (in the cytoplasm, as well as the cisternal spaces, and in great part apparently associated either with the ribosomes or the membranes of the RER; see Chapter 7), indicate the natural presence of digestive enzyme in the cisternae of the RER; whereas, particles over the cytoplasm are artifactual.

Thus, two studies with apparently contradictory observations led the investigators to the same conclusions; digestive enzyme is found in the cisternal spaces and excluded from the cytoplasm. Furthermore, the same observation in each study, the presence or absence of label over both cellular compartments, led to contrary conclusions depending upon which compartment was being considered. Such is the strength of paradigms.

In the latter study (Bendayan et al.), the conclusion that the cytoplasm did not contain enzyme was apparently based on the fact that the concentration of particles over this space (particles per unit area) was much lower than that found over the zymogen granules, although it was only about half that over the endoplasmic reticulum-containing area of the cell. A range of concentration differences was reported for different enzymes from about a one-third cytoplasm to zymogen granule ratio to a one-sixteenth ratio. But how does this lower concentration of enzyme in the cytoplasm bear on the conclusion that there is no cytoplasmic pool? Ad lib fed rats were studied at midmorning, at which time the cell should be relatively depleted of its contents of zymogen granules (rats being nocturnal), and although the volume density of granules was not measured, from the work of others, we can estimate that it was not likely to be more than 15% of cell volume at most. The cytoplasm of the acinar cell, on the other hand, occupies about 50% of the cell's volume. Therefore, if an enzyme is three times as concentrated in the zymogen granule, distributed in a volume that is about one-third that of the cytoplasm, then approximately equal amounts of protein would be contained in both compartments. At the other extreme, a ratio of one-sixteenth, about 20% of the enzyme would be in the cytoplasm. If for the sake of estimation we assume that only small amounts of enzyme are in other particulate fractions of the cell, then the cytoplasmic pool would contain between 20–50% of total tissue enzyme according to this histochemical evidence. These numbers fit the percentage of the enzyme activity of the tissue that is recovered in the supernatant fraction of homogenates quite well. If half of the digestive enzyme associated with particulate material is not in the zymogen granule, then these values would of course be halved, or 10–25% of the total in the cytoplasmic pool; values that are also consistent with some cell fractionation data.

Thus, the observed values are not far off from what we would predict from a straightforward interpretation of cell fractionation values. We can also estimate the expected ratio in another way. From the experiments on the flow dependence of protein secretion that I discussed earlier (Chapter 10), we can conclude, in the equilibrium model's terms, that the digestive enzyme concentration in the cytoplasm is either roughly equal to that found in secretion or is less. It does not appear to be significantly higher. Thus, the concentration of protein in secretion can be used as a maximum estimate for the concentration of digestive enzyme in a cytoplasmic pool. The concentration of protein in unstimulated secretion collected from anesthetized rats is about 10 mg/ml of fluid, or about 3–7% of the concentration of protein in the zymogen granule (Ho and Rothman, 1983), or equivalent to a cytoplasm to granule concentration ratio of one-tenth to one-thirtieth at most [that is, if the concentration of enzyme in the cytoplasm is the same as in secretion, and not less, and if the estimates of granule protein concentration are not too low (otherwise the ratios would, of course, be lower yet)]. Thus, once again the predicted values are in the range reported with antibody cytochemistry or even lower.

I would like to consider briefly two other points about this study before moving on. The first is that the distribution of particles raised to different enzymes in different compartments was not parallel, that is, the proportions varied from one compartment to another. Yet the evidence does not indicate the intracellular segregation of different enzymes in different vesicles, RER cisternae, and so forth. The authors note this, but feel that they cannot draw any conclusions from their data, because they have not actually measured the rate of transport of the proteins. However, such a measurement is not necessary, and the observation is one of state and not rate, in which case one must ask, "How did the differential distribution come about?" That is, if we take their evidence for the distribution of proteins as being quantitative, as they claim, and that no artifactual redistribution occurred, then the nonparallel transport of the various proteins must account for their observed asymmetric distribution, which in the absence of special enzyme-specific or selective vesicles can only be accounted for by their independent movement by means of membrane transport.

The second point is related to the question that I raised earlier about this type of study; that is, "to what extent is protein lost during the extensive procedures required to prepare the tissue for microscopy?" The data that are presented indicate that about 300 particles are found over each square micron of zymogen granule area; that is, if we sum the values for the individual enzymes. If we estimate conservatively that this reflects the amount of digestive enzyme in a volume of the granule equal to a granule cross-section times the diameter of a single protein molecule (let us say, about 50 Å), then, since the great majority of secretory proteins were estimated in this study, we can estimate as well the protein concentration of the granule based on these numbers, and knowledge of average granule volume (0.32 μm^3 in the fasted rat). When one does this, an average estimate of approximately 20,000 molecules per granule is arrived at, or about 1–2% of the chemically estimated value. We can make a similar calculation for the digestive enzyme content of the whole cell summing the particle count for all of the enzymes over all parts of the cell, and then comparing the digestive enzyme content of the cell arrived at in this fashion to that determined chemically (about 40% of the total protein content of the cell) (Ho and Rothman, 1983). When this is done (using a cell volume of 1000 μm^3), we find similarly that the value obtained from the particle count is less than 1% of the chemically estimated value.

We can explain this discrepancy in two ways; most of the protein is lost from the surface of the tissue during the preparation of the section, or only a small percentage of the molecules at the surface bind to the antibody. Regardless of which is the case, the measurements of particle density cannot be considered quantitative unless we know which is responsible and to what exact quantitative extent accounts for the observed discrepancies.

In the end, this anatomical approach has as many, if not more, problems than the biochemical approach of cell fractionation, particularly when our goal is to deal with quantitative issues, such as the distribution of a particular protein or group of proteins between one cell compartment and another. Perhaps these problems can be overcome, and the use of quick-freeze techniques holds some promise, but in the final analysis, we must be able to account, mole for mole, for the protein that was in the tissue initially, both in terms of loss and redistribution. In the absence of such specific knowledge, these techniques cannot be used in a truly objective fashion.

12.7 THE PHYSIOLOGICAL CYTOPLASM

Thus far, we have considered anatomical and biochemical approaches to identify the digestive enzyme contents of the cytoplasm, each of which presents its own particular set of problems. The biochemical approach, cell fractionation, requires that we break open the cell, and thereby introduces the potential, unknown in quantity and character for the redistribution of protein. The anatomical approach, cytochemistry, also has the problem of redistribution, but introduces, in addition, the potential for the loss of protein during the preparation of tissue sections for viewing. There is a third approach to identify the digestive enzyme content of the cytoplasm that avoids the problems of redistribution and loss. We can call it the physiological approach, because it evaluates the question of a cytoplasmic pool of digestive enzyme in whole functioning cells and tissues, and uses physical analysis to help understand the nature of the system. We have already discussed examples of this approach in several different experimental contexts; for example, in the last chapter, I described studies in which we measured the uptake of labeled digestive enzyme into tissue in vitro, or earlier on, I discussed the kinetic relationship between the disappearance of granules from the cell during augmented secretion and the output of enzyme into the duct system, as means of evaluating the existence of a cytoplasmic pool.

These were merely different experimental applications of a kinetic compartmental analysis applied to the question of a cytoplasmic pool. This physiological approach is hardly new, but it has suffered somewhat from a lack of interest recently, perhaps because of the notion that more direct approaches, such as the structural (anatomical and biochemical) ones that I have discussed, have the advantage of being more "direct," breaking open, as it were, the "black box," and are sufficient unto themselves. Of course, the physiological approach is nevertheless still necessary. Most notably, it does not require that we deal in extrapolations from the parts to the whole, since it, in the first instance, provides a description of an intact process, in much the same way that anatomy, in the first instance, provides a description of objects. As such, it catalogues the behavior of the system, which in the end our models and hypotheses must both be derived from and consonant with. Indeed, as I discussed at the outset of this volume, somehow such knowledge must be ours if our models and hypotheses are to have any claim on reality whatsoever.

The cytoplasm is, as I have noted, in part, defined as the intracellular pool in direct contact with extracellular space across the cell membrane. It is through this intervening compartment that free-standing intracellular structures, such as the mitochondrion and nucleus, have access to the extracellular world. Thus, in this view, the cytoplasm is the middle or intervening compartment in a three-compartment system through which substances pass as they move between intracellular structures and the environment. This model has been the paradigm for the movement of molecules in and out of various intracellular spaces since the time that it became known that such spaces and structures existed and that biological membranes were semipermeable. However, as we have been discussing, significant exceptions to this three-compartment model for transport have been proposed over the years; in terms of the present discussion, for the processes of secretion (neurotransmitters and proteins) and protein transport more generally. In these cases, instead of a three-compartment system (intracellular structure–cytoplasm–extracellular space), movement is usually thought to involve only two compartments (intracellular structure–extracellular space). Transport is accomplished by means of direct, although transient, connections between intracellular structures, such as secretion granules, and the extracellular space.

If a cytoplasmic pool of digestive enzyme exists, then we should be able to demonstrate that secretion occurs via a three-compartment system. Or conversely, if it does not, then a two-compartment model should suffice to describe the secretion process. As we have already discussed in Chapter 10 for one line of investigation, a two-compartment model fails, and the evidence in hand requires the interposition of a third, presumably cytoplasmic, compartment in the secretory pathway.

Isenman, Ho, and I have performed another series of experiments from the descriptive perspective of compartmental analysis that bears on this question, and that came about as the result of an unexpected observation made over a decade ago concerning the kinetic character of protein secretion, and perhaps, as it turns out, of secretion processes in general. The cisternal packaging–exocytosis model, in addition to proposing that secretion results from the direct transfer of material between two compartments, is, from a compartmental perspective, a series process. That is, protein is supposed to move sequentially through a number of compartments separated by membranes; from the ribosome where it is made, eventually to the duct lumen. These compartments are arranged in series with each other in much the same way as resistances in an electrical series; as current flows through one to the next, so does protein move from compartment to compartment, except that in the case of proteins movement is thought to be irreversible.

If movement through a single series of compartments leads to the secretion of digestive enzyme, then if we label secreted proteins with radioactive amino acids at the site of their synthesis, we should be able subsequently to identify the labeled protein in the various putative compartments in the proposed linear sequence in a timed-ordered fashion, thereby tracing their path from ribosome to duct lumen. Evidence for such a system had been adduced, at least in part, from studies in which radioactive amino acids were applied to the tissue for a short period of time (a "pulse"), and the location of the label traced as a function of time in different cell fractions. If the interpretation given these "pulse label" studies was correct, that is, a linear sequence of membrane-bound compartments accounts for intracellular transport, then if instead of a short discontinuous "pulse" of radioactive amino acids, we added the labeled precursors to the medium bathing the gland continuously, the following should occur. In addition to the sequence of compartments being traced by the sequential appearance of labeled protein in one and then the next compartment, the approach to isotopic equilibrium (namely, reaching the maximal attainable or steady-state concentration of labeled protein in each compartment, at which time each pool would be wholly comprised of protein made subsequent to the addition of the radioactive amino acids to the medium) should also occur in the same unique sequence.

If we were to hasten the attainment of, or approach to, the steady-state or to isotopic equilibrium, by adding a stimulant of protein secretion to increase the rate of turnover of secretory protein, then this should be reflected in an increase in the rate of attainment of steady-state radioactivity in the various compartments; expressed finally in the fluid collected from the secretory duct.

However, when we performed such an experiment we found instead that there was a sharp *depression* in the specific radioactivity of protein (radioactivity per milligram of protein) in secreted fluid under these circumstances; not as we expected, a continuing, even sharper, stimulus-induced increase, as the paradigm predicted (Rothman and Isenman, 1974) (Figure 4). This meant that secretion was not derived from a single precursor pool at all, but from a minimum of two intracellular pools. The stimulant had called forth secretion from an older, less labeled, pool of digestive enzyme, thereby reducing specific radioactivity, whereas in its absence, secretion occurred preferentially from a more rapidly labeled (higher specific radioactivity) intracellular pool. It was not clear from this observation alone whether both pools had direct access to the pancreatic duct, or whether one was an indirect source, as in a three-compartment model for secretion. But it did occur to us that the cytoplasm might be one of the source pools, and we felt that the zymogen granules were likely to be the other. Since the time that we made these observations it has become clear that many, if not all secretory systems display similar behavior, being derived from at least two intracellular pools whose contents can be called forth differentially [see *Nonvesicular Transport* (eds. S. S. Rothman and J. J. L. Ho), John Wiley & Sons, New York, 1985].

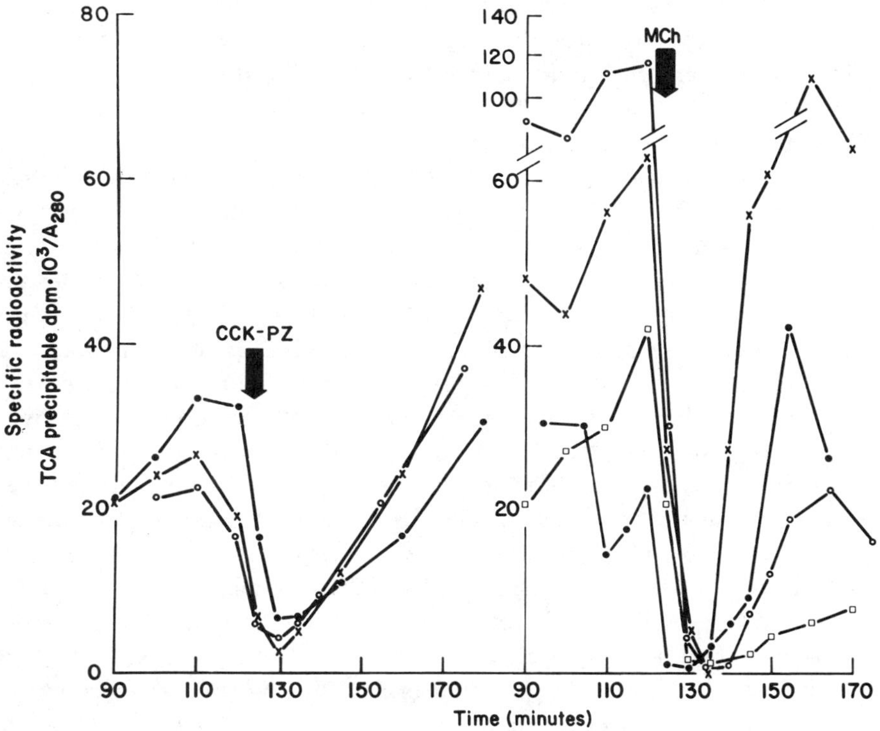

Figure 4 Effect of the cholinergic agonist methacholine chloride (MCh) and cholecystokinin-pancreozymin (CCK-PZ) on specific radioactivity of secreted proteins. Four MCh and three CCK-PZ experiments are shown. (Reprinted with permission, from Rothman and Isenman, 1974.)

It was of course possible to explain this observation within paradigmatic borders, that is, in a purely vesicular context, by adding a new layer of complexity to the cisternal packaging–exocytosis concept; new processes and events could be proposed. We could say that in the unstimulated pancreas new granules are preferentially secreted as a result of their preferential movement through the cell to its apex, or preferential binding to "exocytosis sites" on the membrane, or both, vis à vis older granules. Moreover, since some new granules would presumably not be secreted, because old granules must come from somewhere, this preferential behavior would either have to be lost over time, or would only apply to some newly formed granules. Beyond this, the preferential effect would be "canceled" by the stimulant. This of course would mean that the stimulant would have to be able, in some fashion or another, to eliminate the selective movement or binding of the new granules. However, there was no evidence of such events, although one certainly could not exclude the possibility that they might occur.

Alternatively, we could envision selectivity as being between cells, rather than granules, and design a parallel model of differentiated cells to account for this observation. Once again, beyond the kinetics themselves, there was no evidence that such a group of differentially responsive cells existed in the pancreas, no less that behaved in this fashion. Furthermore, in the paradigmatic view of the secretion process, the idea of a selective exocytosis of new granules seemed without clear functional meaning or purpose. Most of the material that is secreted is, even under these circumstances, derived from old granules, despite the selectivity, and very little in the way of new protein is normally secreted within an hour or so of it being manufactured either in the unstimulated or stimulated state. That is, the quantitative contribution of such a bypass would be minimal in the short run. Moreover, as I have said, since this original observation, similar effects have been reported by others, not only for the pancreas, but for many other secretory cells, including neurons (for example, "rapidly turning-over granules" are thought by some to account for much of acetycholine release), and thus this kinetic pattern may well be a general one for secretion, for which we have to seek a general functional explanation.

But this additional complexity only represents the tip of the iceberg. As we explored the nature of the secretory response, attempts to use paradigmatic models to explain our observations became more and more forced, and thus presented a more demanding challenge to the traditional view of the process. A cholinergic agonist not only reduced the specific radioactivity of protein in secretion, but actually depressed the secretion of labeled protein in absolute terms (Figure 5); that is, not only would one have to propose that the preferential release of new granules was eliminated by the stimulant, but that the stimulant had actually inhibited the exocytosis of such granules. Inhibition was about 75% at the nadir on the average, and thus the stimulant would have had to have essentially prevented the secretion of new granules, whose release had been preferred prior to its application.

Thus, our exocytosis system would have to release the contents of new granules selectively in the absence of a stimulant (or secretion would have

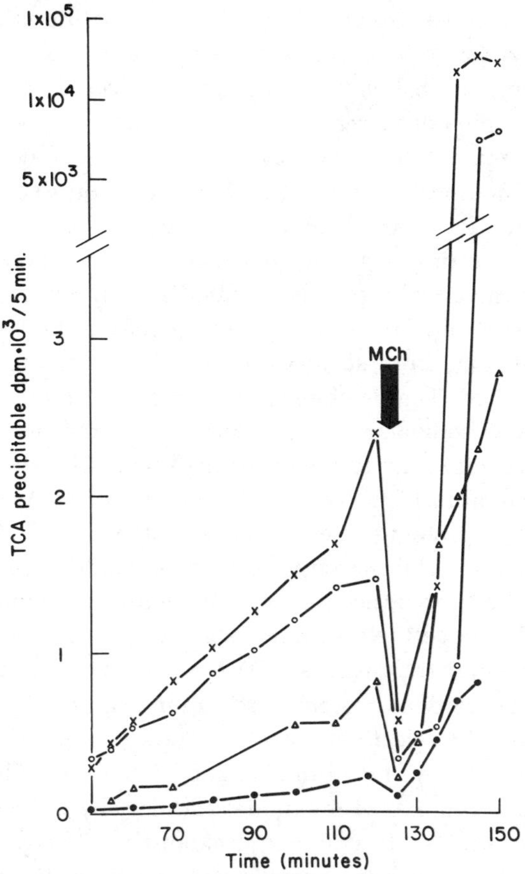

Figure 5 Effect of methacholine chloride (MCh) on the amount of new (labeled) protein secreted by rabbit pancreas in short-term organ culture. Four experiments are shown. (Reprinted, with permission, from Rothman and Isenman, 1974.)

to come from a particular group of rapidly turning-over cells), although the quantitative contribution of this group of granules (or cells) would nevertheless be minor. In addition, the preferential effect would not only have to be eliminated by the stimulant, but the exocytosis of new granules (or the secretory activity of rapidly turning-over cells) would have to have actually been inhibited.

But the complexity does not end here. An additional detail of the response to the cholinergic agonist was that the depression in the specific radioactivity of secreted protein, as well as the absolute inhibition in the secretion of new (labeled) protein, was seen *prior* to the development of augmented secretion. That is, the stimulant produced these effects before protein secretion overall increased (Table 4). How can one explain this observation in terms of the exocytosis model? At first glance this may not seem to require modification of what we have just proposed, but that is not so. Because inhibition was observed prior to the onset of augmented secretion, that is, during a period in which protein secretion remained unaltered, or at the control or unstimulated rate, the stimulant had: (1) inhibited the secretion of new protein; (2) increased the secretion of old protein; and (3) done both so that the overall rate of protein secretion remained unchanged. That is, the increase in the secretion of old protein was quantitatively equivalent to the decrease in the secretion of new protein—quite a complex undertaking for an exocytosis system, done for no apparent purpose, and which effect would only last for a very brief period of time before overall protein secretion was elevated.

The last observation in this group of experiments added yet another layer of complexity. The gastrointestinal hormone cholecystokinin (CCK) produced a similar but nevertheless quite distinct effect on secretion from that produced by the cholinergic agonist. Like the cholinergic drug, it decreased the specific radioactivity of protein in secretion (Figure 4), and hence appeared to "cancel" the selective secretion of new protein. However, it did not inhibit the secretion of new protein in absolute terms (Figure 6). Thus, we have to modify our exocytosis model additionally to include different actions of the two agents. The cholinergic agonist would favor the secretion of old granules, and inhibit the secretion of new granules, whereas CCK would favor the secretion of old granules, but not inhibit the secretion of new granules (or in terms of cells, the cholinergic agent would inhibit secretion from the rapidly turning-over group of cells, whereas CCK would not).

Although we could not at the time exclude such vesicular explanations, there was, and remains, no substantive evidence that such things occur. Moreover, these seemingly complex observations could be explained rather simply if one of the two precursor compartments was the cytoplasm and if the other, in all likelihood, the zymogen granules, only had access to ductal fluid indirectly, that is, via the cytoplasm. In this case, we would merely be dealing with questions of competition between individual labeled and unlabeled protein molecules for exit from the cell; the time, nature, and

TABLE 4. Overall Protein Secretion Relative to Inhibition of New (Labeled) Protein Secretion: Five Minutes after Methacholine Chloride (MCh) Administration[a]

	Percent Inhibition of Total Protein Secretion (120–125 min Compared with Mean Output for Prior 60 min)	(−MCh)–(+MCh), %	Percent Inhibition of Labeled Protein Secretion (120–125 min Compared to dpm/5′ for Prior 10 min)	(−MCh)–(+MCh), %
−MCh	18 ± 19 (4)[b]		+17 ± 16 (4)[b]	
+MCh	18 ± 10 (4)	0	71 ± 7 (4)	−88

[a]−MCh refers to preparations left untreated at 120 min. Numbers in parentheses are the number of experiments.
[b]Values are means ± SE.

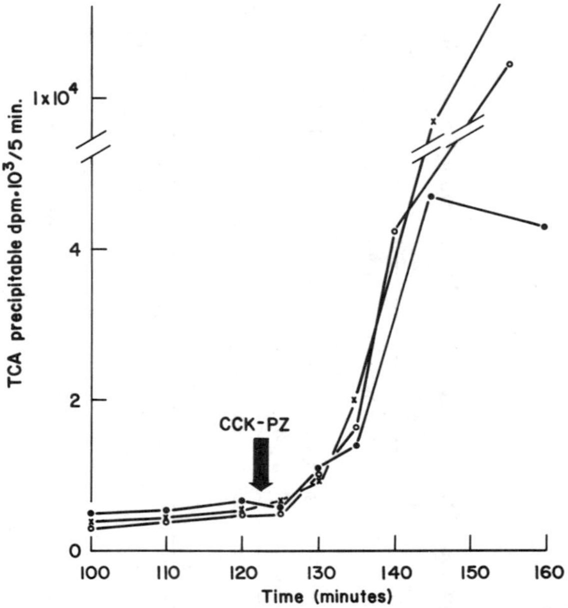

Figure 6 Effect of cholecystokinin-pancreozymin (CCK-PZ) on the amount of new (labeled) protein secreted by the rabbit pancreas in short-term organ culture. Three experiments are shown. (Reprinted, with permission, from Rothman and Isenman, 1974.)

extent of such competition being determined by the relative concentrations of new and older protein at the secretory site. The various effects that we observed would simply be kinetic characteristics of the secretory process, and would not have to have any particular functional meaning or purpose or require new or additional mechanisms. Competitive inhibition at the plasma membrane introduced before an overall increase in protein secretion (permeability) would, of course, inhibit the secretion of labeled protein and increase the secretion of old protein in exact proportion, leaving the total amount secreted unaltered. This simple explanation was sufficient and no ad hoc mechanism was needed to account for this effect. The difference between the relative mix of labeled and unlabeled protein found in secretion in the unstimulated versus the stimulated state would simply be due to differences in the mix present at the apical membrane of the cell; the stimulants acting to mobilize older stored enzyme from the zymogen granules. The differences between the two hormones could simply be explained in terms of differences in the timing of the increase in secretion (permeability) across the apical membrane, as well as by stimulus-induced differences in the relative concentrations of enzyme from the two pools (more highly labeled and less highly labeled) at the site of secretion. Thus, although one can certainly imagine explanations for the results in terms of various complex forms of selective exocytosis, they can be explained in a simple fashion by a three-compartment model for the secretion process, in which zymogen granule contents only have access to the duct lumen by means of the cytoplasm. Furthermore, the observations in their own right were fully consistent with a three-compartment model for the secretory process, and reinforced other evidence for its existence.

12.8 THE OSMOTIC EFFECT

Although one cannot, at least readily, imagine cells displaying competitive inhibition relative to each other; that is, being able to account for the results that I have just discussed in terms of competition between cells for the release of their enzyme content, perhaps the exocytosis model can be saved by viewing these results as being due to competitive interactions between old and new granules in the same cell. That is, couldn't the competition merely be between granules, rather than molecules? If this were the case, wouldn't the same explanations apply? For example, we could introduce the idea that the stimulants "mobilized" the older granules and made their exocytosis more "competitive," than prior to the addition of the stimulant, or conversely, made the exocytosis of new granules less "competitive." In this way, by altering their relative competitive position for exocytosis, couldn't we explain the decreased specific radioactivity prior to the occurrence of an overall increase in protein secretion as being the result of prior changes in the competitive position of new versus old granules? Or might not different stimulants augment or inhibit the potential for the exocytosis of one type of granule or another to a greater or lesser extent? That is, we can imagine an analogous model for competitive inhibition in which granules, rather than molecules, would be involved. And, indeed, such a possibility cannot be excluded on the basis of these observations alone, although admittedly we have to introduce the concept of specific granule mobilization or inhibition. In any event, we attempted to distinguish between these two explanations (granule versus molecule) for the observed competitive inhibition by making use of another aspect of the secretory response to cholinergic stimulants that we had observed using the rabbit pancreas in organ culture as the experimental system.

For older protein molecules to competitively inhibit the secretion of new ones in a molecule-by-molecule fashion (to a greater extent after the addition of a stimulant, than before its addition), the concentration of old protein at the site of secretion would have to be increased, at least in relative terms, by the stimulant. Thus, the stimulant should have mobilized older protein molecules from storage pools, presumably from within zymogen granules. Since the protein inside the granule is primarily in a bound state, its release into the cytoplasm would transiently increase cytoplasmic osmolarity, at least in the apical region of the cell. Such an increase in osmolarity might in turn reduce secretory flow in one of two ways; either water would move back into the cell due to the development of an osmotic gradient, or the flux of water out in the first instance would be reduced. Although the release of protein from the zymogen granules into the cytoplasm would only produce a small increase in intracellular osmolarity, even a small change is capable of affecting secretory flow in a relatively dramatic fashion. That is, flow rate is a very sensitive osmometer, recording minor (~1%) and short-lived osmotic fluctuations within the cell.

If such an osmotic effect could be observed coincident with the competitive inhibition of labeled protein secretion, this would provide support for the view that competitive inhibition was between molecules, and that protein had been mobilized into the cytoplasm from zymogen granules; and conversely that competition did not reflect alterations in the relative rates of exocytosis of different types of zymogen granules.

However, although it was possible that such an osmotic effect of protein mobilization might be seen, it was not a necessary sequelae of the release of digestive enzyme from zymogen granules into the cytoplasm. If the rate of mobilization of protein was slow enough, then the osmotic force and the resultant changes in flow might be too small to be observed. Or if the osmoles released into the cytoplasm were accompanied by parallel increases in the release of protein into the duct lumen, thereby moving the protein and not water, then the effect also would not be seen, since no increase in osmolarity would occur. Thus, a negative observation would not be instructive.

However, we were fortunate and the cholinergic agonist that we used, at the dose that we used, produced just such a transient depression in flow from the rabbit pancreas. If one was careful in preparing bathing solutions, a reproducible, sharp, although a transient, decline in the rate of fluid flow, of about 75% at the nadir, could be produced that occurred *coincident* with the reduction in specific activity seen after the application of a cholinergic stimulant to the gland (Ho and Rothman, 1980) (Figure 7; compare with Figure 5).

Thus, there was evidence that an infusion of osmoles into the cytoplasm occurred. The first question was whether or not the reduction in flow was of the order that we might expect; that is, was it quantitatively consistent with the hypothesis that it was due to the mobilization of osmoles from the

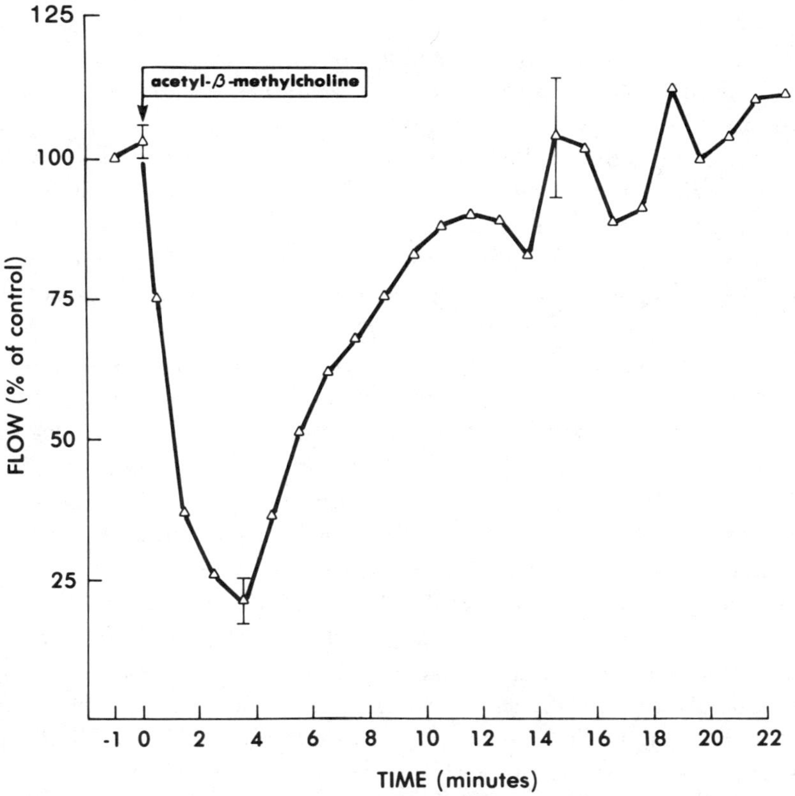

Figure 7 Effect of cholinergic agonist, acetyl-β-methacholine (1.0 mg/100 ml bath volume), on flow from cannulated pancreatic duct of rabbit pancreas in short-term organ culture. Stimulus was applied at zero time (arrow). Data are expressed as percentage of control flow rate (1 min before addition of stimulant) determined from 30-sec collection periods. Points, means; error bars, SE; $n = 7$. (Reprinted, with permission, from Ho and Rothman, 1980.)

granules in the form of secretory proteins. We would expect the amount of enzyme secreted in the early phases of the response to the secretatgogue to be of roughly the same order as the number of protein osmoles liberated into the cytoplasm just prior to its secretion (as estimated from the reduction in flow). When such a calculation was made, the reduction in flow in osmotic terms and the amount of protein subsequently secreted were quantitatively quite similar (Ho and Rothman, 1980).

Furthermore, if the reduction in flow was due to the mobilization of protein, then we would expect that variations in the reduction in flow seen from tissue to tissue, presumably a function of the amount of protein released into the cytoplasm, would be correlated to protein secretion or output at the onset of the secretory response. That is, the higher the concentration of secretory protein in the cytoplasm, reflected in the magnitude of the flow depression, the greater the subsequent rate of protein secretion at the onset of augmented secretion (due to the elevated concentration gradient). When these two variables, the reduction in flow and the initial output of an enzyme (amylase), were plotted against each other for different experiments, they were found to be highly correlated (Figure 8). A lesser flow reduction was accompanied by a lower amylase secretion, and a greater flow reduction indicated that amylase secretion was going to be greater. This was just as expected if the reduction in flow had been an osmotic sequelae of the mobilization of proteins into the cytoplasm.

Further evidence for the osmotic nature of the effect was sought by setting the osmolarity of the cells slightly higher or slightly lower than normal prior to the application of the stimulant. In this case, we should be able to alter the flow depression by biasing the system in advance; an increase in osmolarity should reduce the osmotic inhibition, and an increase should accentuate it. Both effects were seen (Figure 9).

We performed one additional set of experiments considering the behavior of the system and its consonance with our previous studies in which we measured specific radioactivity. In these earlier experiments, as I discussed above, the effects of the cholinergic agonist and the gastrointestinal hormone CCK were somewhat different in that CCK did not produce an absolute inhibition in the secretion of labeled protein, as the cholinergic agonist did. In addition, the decrease in specific radioactivity with CCK did not appear prior to the augmentation of protein secretion, but at approximately the same time. These observations suggested that CCK, unlike the cholinergic drug, increased the mobilization of proteins from

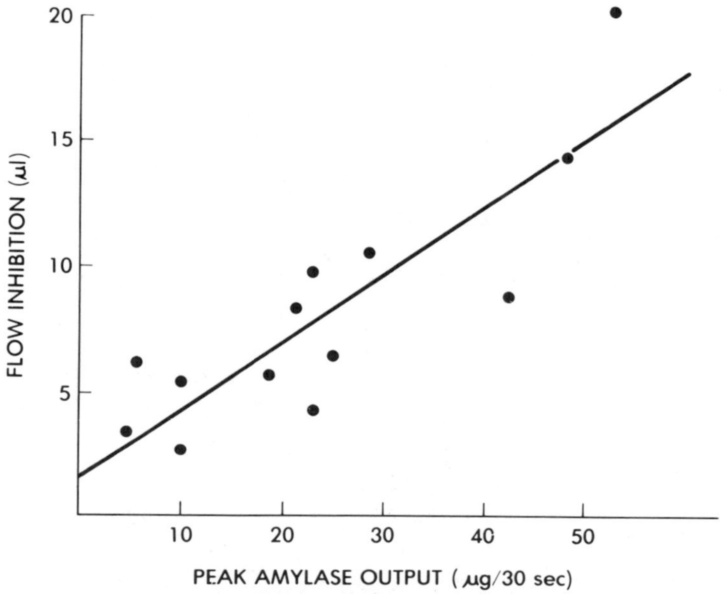

Figure 8 Relationship between the magnitude of flow depression and peak amylase secretion or output produced in response to addition of acetyl-β-methacholine (1.0 mg/100 ml bath volume). Total flow inhibition to nadir is plotted against peak amylase output for each individual experiment. Amylase is expressed as micrograms of crystalline α-amylase standard (730 IU/mg porcine pancreas, Worthington Chemical) per 30-sec collection period. Points were fit to the line by regression analysis ($y = 0.27x + 1.53, r = 0.853, n = 13$). (Reprinted, with permission, from Ho and Rothman, 1980.)

granule stores more or less simultaneously with increases in the flux of enzyme out of the cell; that is, the permeability of granule and cell membranes were more or less simultaneously increased in this case. In this way, although the decrease in specific radioactivity, indicating competition, would be seen, because there was also a simultaneous increase in the overall flux of protein out of the cell, there would be no absolute inhibition in labeled protein secretion. On the other hand, the cholinergic agonist would produce such an inhibition because it mobilized proteins from the granules prior to increasing protein secretion, and therefore, not only would competition occur, but there would be an absolute decrease in the amount of labeled protein secreted as well.

In light of these ideas, could we predict the osmotic behavior that we might expect as a result of CCK stimulation of protein secretion? If it was

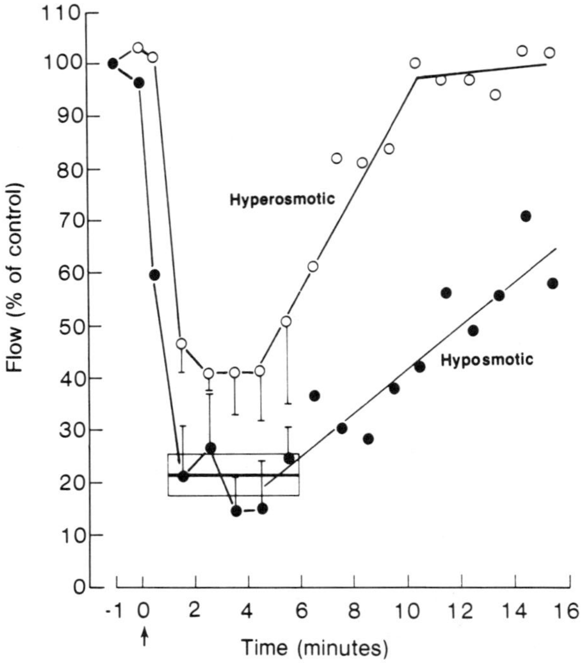

Figure 9 Effect of alteration of medium osmolarity on flow depression produced by acetyl-β-methacholine (1.0 mg/100 ml bath volume). The osmolarity of the bath was either increased (hyperosmotic) by 20 mosm/L (○) or decreased (hyposmotic) by 20 mosm/L (●) with NaCl and the gland equilibrated in the altered osmotic environment before addition of the cholinergic agent (arrow). Data are expressed as percent of control flow rate (1 min before addition of stimulant) determined from 30-sec collection periods. Points, means; error bars, SE ($n = 3$ for both groups). Rate at which flow returned to prestimulus levels was approximated by linear regression analysis (hyposmotic, $y = 4.2x + 0.04$; $r = 0.939$, $n = 12$; hyperosmotic, $y = 9.30x + 1.32$, $r = 0.969$, $n = 7$). Rectangle shows value of mean flow at nadir (dark horizontal line) $\pm$ SE (bracketing light horizontal lines) 3.5 min after addition of cholinergic stimulus to medium of normal osmolarity. Flow depression in hyperosmotic and hyposmotic states differed both in terms of nadir achieved and rate of return to control flow rates. (Reprinted, with permission, from Ho and Rothman, 1980.)

true that both an increase in protein secretion and the mobilization of protein from the granules occurred relatively simultaneously, then we would expect the osmotic consequence of CCK administration either to be less pronounced than seen with the cholinergic drug, or even absent if the increased concentration of granule-derived proteins was paralleled by an increase in the rate of protein secretion, so that no substantial osmotic force was generated by their release from the zymogen granule. What we found was that there was no discernible reduction in flow as a result of the addition of CCK (Figure 10).

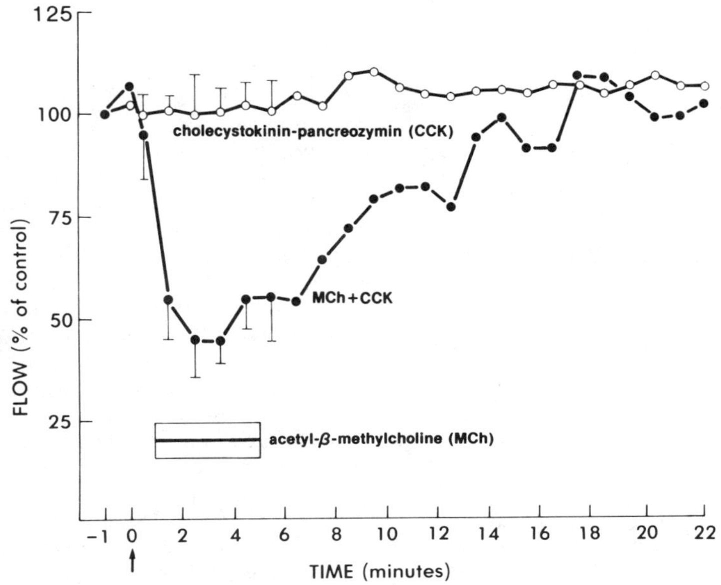

Figure 10 Effect of cholecystokinin-pancreozymin (CCK, 6.25 Ivy dog units/100 ml bath volume) on flow. The conditions are the same as those described in the previous figures. Either CCK alone or in combination with acetyl-β-methacholine (MCh, 1.0 mg/100 ml of bath) was added to the medium at time indicated by arrow (CCK alone, $n = 4$; MCh + CCK, $n = 6$). Rectangle shows value of mean flow at nadir (dark horizontal line) ± SE (bracketing light lines) 3.5 min after the addition of MCh alone to bathing medium. CCK had no effect on flow by itself, but attenuated the depression producted by MCh. (Reprinted, with permission, from Ho and Rothman, 1980.)

If our osmotic interpretation for the absence of an effect of CCK on flow was correct, then we would expect to be able to reduce the flow depression produced by the cholinergic drug merely by adding CCK along with it. CCK would increase the release of protein across the apical membrane earlier on, and therefore would truncate the protein buildup normally produced by the cholinergic drug. When this experiment was performed, the prediction was confirmed; CCK substantially reduced the flow depression produced by the cholinergic agonist (Figure 10). Finally, if these interpretations were correct, then having reduced the buildup of protein in the cytoplasm in this way, the rate at which protein secretion increases as augmented secretion first begins should be less in the presence of CCK than when the cholinergic agonist is present alone. Once again the prediction was borne out by the evidence and the rate of increase of amylase secretion was greatly diminished as augmented secretion began in the presence of both CCK and the cholinergic agonist (Figure 11).

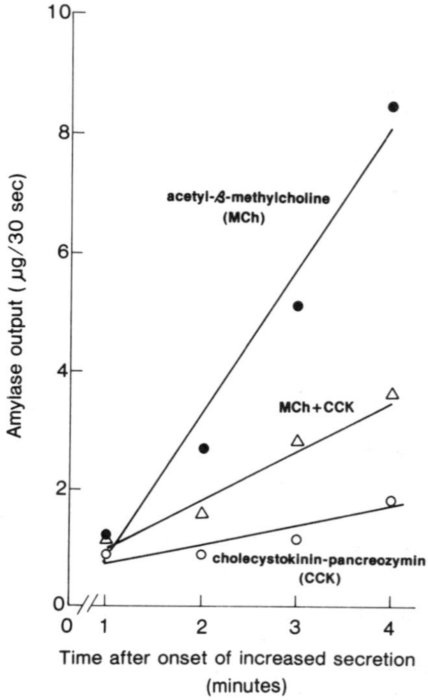

Figure 11 Effect of cholecystokinin-pancreozymin (CCK, 6.25 Ivy dog units/100 ml of bath) on rate at which amylase secretion initially increases after addition of acetyl-β-methacholine (MCh, 1.0 mg/100 ml of bath volume). First four collection periods after onset of augmented secretion are shown; in presence of MCh alone (●); CCK alone (○); and combination of both CCK + MCh (△). Amylase is expressed as micrograms of crystalline α-amylase standard (730 IU/mg, porcine pancreas, Worthington Chemical) per 30-sec collection period. Points, means of seven experiments for MCh, six experiments for MCh + CCK, and four experiments for CCK alone. Data were fitted to lines by linear regression analysis ($b_{MCh} = 2.43, r = 0.984; b_{MCh+CCK} = 0.87, r = 0.987; b_{CCK} = 0.31, r = 0.893$). Addition of CCK reduced rate of increase produced by MCh alone by more than half, although the overall response (over time) was not significantly altered. (Reprinted, with permission, from Ho and Rothman, 1980.)

Thus, it seemed that the flow reduction that we had observed with cholinergic stimulation was indeed a sequelae of the liberation of individual protein molecules from storage pools into the cytoplasm. I cannot explain this series of observations, in conjunction with those in which specific radioactivity was followed, by means of a vesicle system in any way that I think does not stretch, at least my, credulity.

12.9 SUMMARY

Although there may be some who will view these physiologic observations as being "indirect," as though that made them suspect, they are to my way of thinking in many ways more direct than the seemingly "direct" anatomical and biochemical approaches that I discussed above. More importantly, if the exocytosis model cannot explain or account for these results in an experimentally verifiable manner, then to my way of thinking these studies falsify the hypothesis, and require that we seek another model to explain the events of secretion.

SELECTED REFERENCES

1. Bendayan, M., J. Roth, A. Perrelet, and L. Orci. Quantitative immunocytochemical localization of pancreatic secretory proteins in subcellular compartments of the rat acinar cell. *J. Histochem. and Cytochem.* **28:**149–160, 1980.
2. Dandrifosse, G. From digestive enzyme secretion to bioelectrogenesis, in, *Nonvesicular Transport* (eds. S. S. Rothman and J. J. L. Ho), John Wiley & Sons, New York, 1985.
3. Hartley, B. S., J. R. Brown, D. L. Kaufman, and L. B. Smillie. Evolutionary similarities between pancreatic proteolytic enzymes. *Nature* **207:**1157–1159, 1965.
4. Ho, J. J. L. The signal hypothesis—therein lies a tail. II. Mitochondria, in, *Nonvesicular Transport* (eds. S. S. Rothman and J. J. L. Ho), John Wiley & Sons, New York, 1985.
5. Ho, J. J. L. and S. S. Rothman. Intracellular osmotic events accompanying protein secretion by the exocrine pancreas. *Am. J. Physiol.* **238:**G289–G297, 1980.
6. Ho, J. J. L. and S. S. Rothman. Protein concentration in the pancreatic zymogen granule. *Biochim. Biophys. Acta* **755:**457–466, 1983.
7. Iacino, D., G. A. Scheele, and C. Liebow. Secretory response of the rabbit pancreas to cholecystokinin stimulation. *Am. J. Physiol.* **239:**G247–G254, 1980.
8. Kraehenbühl, J. P., L. Racine, and J. D. Jamieson. Immunocytochemical localization of secretory proteins in bovine pancreatic exocrine cells. *J. Cell Biol.* **72:**406–423, 1977.
9. Rothman, S. S. "Non-parallel transport" of enzyme protein by the pancreas. *Nature* **213:**460–462, 1967.
10. Rothman, S. S. Subcellular distribution of trypsinogen and chymotrypsinogen in rabbit pancreas. *Am. J. Physiol.* **218:**372–376, 1970.
11. Rothman, S. S. Intracellular storage of exportable protein in functionally hypertrophied pancreas. *Am. J. Physiol.* **219:**1652–1657, 1970.
12. Rothman, S. S. and L. D. Isenman. The secretion of digestive enzyme from two parallel intracellular pools. *Am. J. Physiol.* **226:**1082–1087, 1974.
13. Rothman, S. S. and H. Wilking. Differential rates of digestive enzyme transport in the presence of cholecystokinin-pancreozymin. *J. Biol. Chem.* **253:**3543–3549, 1978.
14. Schatz, G. and T. L. Mason. The Biosynthesis of Mitochondrial Proteins. *Ann. Rev. Biochem.* **43:**51–87, 1974.
15. Scheele, G., G. Palade, and A. Tartakoff. Cell fractionation studies on the guinea pig pancreas. Redistribution of exocrine proteins during tissue homogenization. *J. Cell Biol.* **78:**110–130, 1978.

CHAPTER 13

Nonparallel Transport

There was a controversy in the early part of this century between the famous Ivan Pavlov in St. Petersburg and one of his more renowned students, B. P. Babkin, then at McGill University in Canada (Babkin, 1950). Pavlov and another student, A. A. Walther, had argued that digestion in the gut was regulated at the level of specific enzymatic events (Pavlov, 1910). They believed that the mixture of digestive enzymes secreted by the glands of the gastrointestinal tract, especially the pancreas, varied depending upon the particular contents of an ingested meal, and that this was the means used by the body to regulate specific digestive reactions; different proportions of food substrates led to the secretion of different, substrate-appropriate mixtures of enzymes to carry out digestion. Their hypothesis rested on certain observations that the proportions of different digestive enzymes found in pancreatic juice and in the intestines varied in a substrate-related manner, and perhaps equally, or so it seems, on Pavlov's belief that in its wisdom nature would provide such a system.

Babkin, on the other hand, could find no evidence for such regulation in his own work. He and his colleagues, looked for, but did not find, changes in the secretion of individual enzymes in the absence of changes in the secretion of others by the pancreas, as they thought should occur. All stimuli that they tested increased the secretion of all of the enzymes that they measured. As a result, they argued that the relative amounts of the different digestive enzymes secreted into the gut did not vary, but were the same or constant regardless of the content of the meal undergoing digestion, or the nature of the applied stimulus. Digestion was regulated solely as a function of the amount of the applied food load, and not its specific chemical content. Although the matter was not pursued to any great depth, and neither position was experimentally very secure, Babkin's prevailed, perhaps because it proposed less, and became the generally accepted view.

The term "parallel secretion" was coined to describe Babkin's concept. Variations in the overall rate of protein secretion by the gland (viz., the pancreas) would be accompanied by "parallel" changes in the rates of secretion of all of the different enzymes, thereby leaving their relative proportions unchanged or invariant. The term "nonparallel" secretion, on the other hand, was applied to any observation in which the proportions of different enzymes in secreted fluid varied as a function of state, time, or circumstance otherwise. These ideas of parallel and nonparallel secretion will be the subject of my discussion in this chapter. I will describe these concepts as they are used today, along with the mechanistic trappings that have been associated with them, as we consider modern evidence for the existence of nonparallel secretion.

13.1 SYNTHESIS VERSUS TRANSPORT

About 25 years after the seeming resolution of the controversy in Babkin's favor, it was demonstrated by Grossman, Greengard, and A.C. Ivy (Grossman et al., 1943) that the enzyme content of pancreatic tissue, and presumably the fluid product of the gland as well, could be altered in predictable ways by dietary means. Thus, Pavlov's view now appeared to be supported by the data; diet did affect the enzyme content of digestive fluid. However, these observations, as well as numerous substantiations in the years that followed, did not lead to the rejection of Babkin's concept of parallel secretion. Indeed, the notion was retained. The effects that were observed were attributed to changes in the relative rates of synthesis of the various enzymes by the secretory cell, rather than to their differential rates of transport out of the cell. Thus, although changes in the enzyme content of secretion did occur, in just the way that Pavlov had predicted, these changes were due to alterations in the rates of synthesis of different enzymes and not to their differential transport out of the cell from a previously formed pool in response to the contents of a particular meal.

In accordance with the parallel secretion concept, the secretion of all of the various enzymes was thought to be uniformly parallel, accurately reflecting the contents of the preexisting intracellular precursor pool of digestive enzyme. Only as the contents of that pool changed, as the result of changes in the rates of synthesis of different enzymes, would the proportions change in secreted fluid. Indeed, at all times, the enzyme composition of secretion would reflect accurately the composition of the precursor pool, and it would only vary in a dependent fashion in parallel with changes in the composition of that pool. Only when the synthesis of a new mixture of proteins had altered the composition of the precursor pool sufficiently would changes in the composition of the secreted mixture be observed.

The reasons for drawing the conclusion that synthesis was the cause of the change in the proportions of enzyme secreted was fourfold. Dietary changes led to changes in the enzyme content of the tissue. Indeed, initially all that was measured were the effects of diet on tissue enzyme content. Of course, alterations in a tissue's content of an enzyme might well reflect changes in that enzyme's rate of synthesis. Moreover, some changes in tissue content tended to follow in a sensical way the changes that had been made in the diet; for example, a high-starch diet might increase tissue amylase content. Although changes in the relative rates of transport of different enzymes might also change the composition of the cellular pool, the fact that some of the changes in tissue content were in the direction that one would expect (for a positive feedback system) if synthesis were responsible (in some, although not all cases, and for some, but not all, enzymes), and not in the opposite direction (which would suggest the selective transport of the enzymes out of the cell), reinforced the idea that synthesis acted to change the contents of the intracellular precursor pool. A third reason for concluding that synthesis was the cause, was that in the early studies on the subject, long periods of time (weeks to months) were used to demonstrate the effect of diet on tissue content. Such long-term effects seemed, at least on a prima facia basis, likely to reflect synthetic, rather than transport events, although the rate of turnover of digestive enzymes in the pancreas was in fact unknown, and indeed still remains to be estimated accurately in any situation, no less characterized fully as a function of state. More recent studies have made it clear that a few days generally suffice to produce substantial changes in tissue content, and such changes probably begin much sooner. Lastly, there was direct evidence that the rate of synthesis of at least some digestive enzymes could be increased by dietary means.

However, although these facts suggested the occurrence of synthetic changes, they did not demonstrate that such changes were either the primary or the sole effect of the diet, vis à vis transport, nor that changes in the contents of secretion were merely a dependent by-product of changes in synthesis. Changes in transport, particularly a comparison between the rate and extent of change in the composition of secretion, and the rate and extent of change in the proportions of different enzymes within the tissue, were not examined carefully if at all. On the basis of the available evidence, it could not be concluded that independent nonparallel transport effects did not also occur as a result of dietary manipulations. Indeed, it could not even be excluded that a transport effect was the primary, rather than a secondary, event.

This raises an important point about these studies and their interpretation. If it were true that secretion was uniformly parallel in the short run, reflecting the composition of the intracellular precursor pool, then the regulation of the digestion of single meals at the level of the digestive reactions themselves probably did not occur, because from what we know of the amount of enzyme stored within the tissue, it is previously stored material that is in great measure responsible for what is secreted subsequent to the ingestion of a particular meal, at least in many vertebrate species. Thus, these regulatory (synthetic) events would not in the main be for the meal undergoing digestion, but for subsequent meals, contrary to Pavlov's hypothesis. That is, in this view regulation would be predictive. If an animal is eating a high-starch meal, then the body draws the conclusion that the next meal will also be high in starch, and that it had better begin preparing. If several meals in a row were high in starch, then the effect would be reinforced, and over time the animal would adapt to a high-starch intake. This concept makes sense for grazing animals for example, who move into a field with a particular type of grass, stay there for some time, and only after having exhausted that particular food supply, move on to fresh grazing pastures, which may provide food with a different mixture of nutrients. Although such a system may certainly exist, and provide a selective advantage to the animal, for many animals and circumstances it not only would not seem to have such an advantage, but might even be disadvantageous, decreasing the efficiency of the digestion of the next meal if the prior meal was not a good predictor of the one that followed. And of course, it is even possible, in contrary creatures such as the human being, that what we eat in one meal, may predict that what we eat in the next will be different.

13.2 THE EFFECT OF FEEDING TRYPSIN INHIBITOR ON ENZYME SECRETION

More than fifteen years ago Wells and I performed experiments in which we considered the effects of diet on protein transport (Rothman and Wells, 1969). We chose a somewhat different approach from the one in which changes are made in the bulk composition of the diet, as for example, from a diet high in protein to one high in starch. We used identical, mixed diets, but added purified trypsin inhibitor (from the egg white) to the food fed one group of rats. It had been known for many years that feeding animals trypsin inhibitors produced pancreatic growth and an increase in the digestive enzyme content of the tissue. We thought that since trypsin inhibitors prevent tryptic hydrolysis, and sometimes chymotryptic proteolysis, that we might be able to see specific effects analogous to those seen when the gross composition of the diet was altered. Much to our surprise, we found that of the three enzymes whose presence in tissue we measured, trypsinogen, chymotrypsinogen, and amylase, the one that increased least was trypsinogen (Table 1). This either meant that the effect of trypsin inhibitor was not selective, increasing the synthesis of all of the enzymes without regard to kind, or that tissue content alone was not an adequate measure of what had occurred.

We thought that the pattern of change that was observed might reflect the selective transport of enzymes out of the cell, as well as changes in protein synthesis, and attempted to evaluate this hypothesis by measuring the secretion of the same three enzymes in fluid collected from the pancreatic duct of rats fed either a normal diet or one containing the trypsin inhibitor. The results were striking. The trypsin inhibitor-containing diet did not alter the rate of secretion of amylase or chymotrypsinogen at all, but had more than doubled basal (unstimulated) trypsinogen secretion (Figure 1; bars on left). The effect was even more pronounced when we gave a cholinergic agonist to increase the overall rate of protein secretion. In this case, trypsinogen secretion in response to the cholinergic agent was some five times greater in rats fed the trypsin inhibitor, than in controls. A lesser increase in the chymotrypsinogen response was also seen, about a doubling of the normal response to a set dose of a cholinergic agonist. However, trypsin inhibitor feeding had no effect on the extent of amylase secretion in the presence of the cholinergic agent (Figure 1; bars on right).

These experiments were among the earliest examples of nonparallel transport that we came across. They not only demonstrated that nonparallel

TABLE 1. Effect of Dietary Egg-White Trypsin Inhibitor on Pancreas Weight and Enzyme Content[a]

Diet	Body Wt (g)		Pancreas	Enzyme Content, μmoles or mg Substrate Split × min^{-1} $gland^{-1}$		
	Initial	Final	Wt (mg)	Trypsinogen	Chymotrypsinogen (s)	Amylase
Normal (6)	215	208	832	421	3,597	6,149
4% EWTI (6)	217	209	1,104[b]	570[c]	6,525[d]	12,976[b]
SEM			45	48	391	1,293
Δ%			+33	+35	+81	+111

[a]Values are means ± SEM. No. of rats in parentheses. P values refer to comparison between normal and EWTI fed.

[b]$p < 0.01$.

[c]$p > 0.05$.

[d]$p < 0.001$.

Source: Taken from Rothman and Wells (1969).

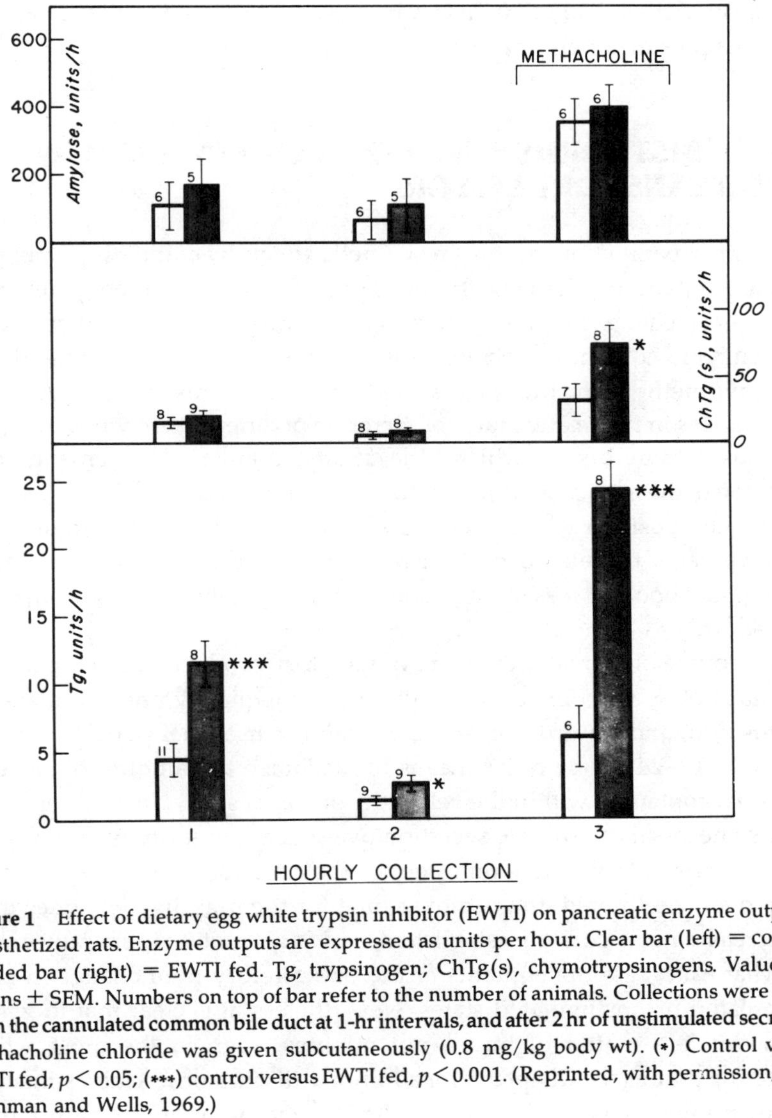

Figure 1 Effect of dietary egg white trypsin inhibitor (EWTI) on pancreatic enzyme output in anesthetized rats. Enzyme outputs are expressed as units per hour. Clear bar (left) = control; shaded bar (right) = EWTI fed. Tg, trypsinogen; ChTg(s), chymotrypsinogens. Values are means ± SEM. Numbers on top of bar refer to the number of animals. Collections were made from the cannulated common bile duct at 1-hr intervals, and after 2 hr of unstimulated secretion methacholine chloride was given subcutaneously (0.8 mg/kg body wt). (*) Control versus EWTI fed, $p < 0.05$; (***) control versus EWTI fed, $p < 0.001$. (Reprinted, with permission, from Rothman and Wells, 1969.)

transport occurred, but that dietary means could be used to alter selectively the transport of different enzymes.

13.3 DISTINGUISHING BETWEEN SYNTHETIC AND TRANSPORT EFFECTS

One can separate transport from synthetic effects by comparing changes in tissue content to changes in the secreted mixture of enzyme, or by comparing changes in the relative rates of protein synthesis to changes in the enzyme content of secretory fluid. However, the simplest and most effective method, and the one that has been most extensively used, is to look for changes in the relative rates of secretion of different enzymes over short periods of time during which little, if any, change in the enzyme composition of the intracellular precursor pool could occur. In this case, changes in the composition of the secreted mixture would reflect changes in the relative rates of release or transport of the different enzymes from a precursor pool of essentially constant composition, ergo nonparallel transport.

Of course, the greater the store of secretory product within the tissue at the start of such an experiment, the more adequately one can make this distinction, and toward this end it is useful to use brief periods of fasting (usually 12–24 hr) or other means to maximize the amount of secretory material contained within the cell at the outset of study. In addition, it is of course necessary to sample secretion over relatively short intervals of time in comparison to the time required for the contents of the storage pool to turn over. As I noted, the turnover time for the digestive enzymes in the pancreas is not accurately established, and as a result it is generally unclear at what exact rate the contents of the presecretory pool change in either stimulated or unstimulated states. Nevertheless, it is clear that in general turnover is a matter of hours, not minutes, even in the most actively secreting tissue (see Chapter 9 for additional discussion). A fair estimate is, I think, minimally between 5 and 20 hr, depending upon the nature of the variables. If this is so, then over a 30-min collection period, between 90 and 97.5% of what is secreted will have been manufactured prior to an imposed change of state, and changes in composition seen over this period of time cannot be attributed to new synthesis.

One can consider the measurement of turnover time from many points of view, for example, as I have already discussed, how much protein is secreted over such and such a period of time relative to the amount present in the tissue at the outset, and so forth. However, for the purpose of evaluating the cause of nonparallel secretion, the most useful approach is to determine how long it takes for new protein to find its way into secretion. If nonparallel secretion is observed before any protein manufactured subsequent to the application of the stimulant is secreted, then the effect must be due to the differential transport of previously manufactured protein. In the absence of heroic attempts to deplete storage pools, in virtually every report, in many species and under a variety of experimental and physiological circumstances, the appearance of substantial amounts of new protein in secretion is minimally about 20 min, and varies from about 20 to 75 min in most cases. In the rat, even after 3 or 4 hr of maximal secretion driven by a cholinergic agonist, and in the continuous presence of the stimulant, new protein is not found in substantial quantities in secreted fluid within an hour after the addition of a labeled amino acid to the bloodstream (see Figure 8 in Chapter 9).

Thus, the simplest and usually the clearest method at our disposal for distinguishing synthetic from transport causes of nonparallel secretion is merely to sample secretion over relatively short intervals of time, during which we would not expect new protein to be found in secretion in substantial quantities, and from which fact we can infer that the transport of previously synthesized protein is responsible for what is observed. Most observations of nonparallel transport that have been reported involve periods of sampling, 30–60 min or less, that are generally too short to be accounted for by changes in the overall composition of the precursor pool, and occur before there is substantial secretion of new protein. Thus, the possibility that such observations of nonparallel transport can be attributed to changes in the rates of synthesis of different enzymes can reasonably be put aside. This of course, does not mean that particular treatments do not lead to the preferential synthesis of a given protein or even to the preferential release of new protein. It is just that in such circumstances the observed nonparallel transport reflects in great part (that is, quantitatively) the selective transport of previously synthesized protein, and occurs in the absence of turnover of the whole precursor pool. Of course it should be borne in mind that even if selective release of newly synthesized protein occurs, this is still nonparallel or selective transport. In any event, most if not all rapidly produced changes in the proportions of enzyme in secreted fluid appear to be primarily the result of their differential transport, and not alterations in their rates of synthesis.

13.4 EXOCYTOSIS AND PARALLEL SECRETION

Parallel transport, in which the relative proportions of different enzymes in secreted fluid reflect their proportions in the intracellular precursor pool, is a prediction of the exocytosis model for secretion. To the degree that the notion of parallel secretion is correct, it lends support to the idea of secretion by the exocytosis of zymogen granules. If exocytosis accounts for secretion, then we would expect the proportions of enzyme in the granule pool and in secretion to be the same or to parallel each other, the latter being directly derived from the former as a result of the bulk transfer of mass.

This is the original notion of exocytosis that includes the following subsidiary hypotheses: (1) the enzyme contents of each granule are fixed once formed; (2) in a given state the contents of all granules are either exactly the same or vary randomly from granule to granule to form a normal distribution; (3) exocytosis is a probabilistic business in which a chance association of a secretion granule and a membrane site leads to exocytosis; (4) once exocytosis has occurred the contents of the granule are fully expelled into the extracellular environment; and (5) neither ordered differences in the enzyme content of different granules or cells, nor the means to express any differences that might occur by chance, exist.

These nations formed the original concept of exocytosis as a completely random, mass transport process in which the proportions of different enzymes in secretion would always reflect the contents of the precursor zymogen granule pool, and stimulants or inhibitors of secretion would act by either increasing or decreasing the frequency or rate of fusion of granule to cell membrane. In each and every circumstance the proportions of enzymes in secreted fluid would remain constant or unchanged over the short run, and would accurately reflect the composition of enzyme contained in the precursor granule pool.

Thus, parallel secretion and exocytosis are conceptually congruent; evidence that transport is parallel in all circumstances is support for secretion by exocytosis and evidence for exocytosis predicts parallel transport. Therefore, testing for the parallel transport of digestive enzyme also tests the exocytosis model of secretion, at least as the idea was originally conceived. Similarly, the existence of nonparallel transport argues against the exocytosis hypothesis. In this case, we must either propose a wholly different model for secretion, modify the view of exocytosis outlined above, or consider a combination of processes. In our work, my colleagues and I have proposed, and provided evidence for, a different construct, the equilibrium theory of secretion, whereas supporters of the exocytosis model have had to modify their hypothesis to attempt to account for nonparallel transport. I will first discuss the nature, and present some of the variety, of nonparallel transport phenomena, and then consider how such observations can be interpreted either in terms of the equilibrium model or modifications of the exocytosis construct.

13.5 THE CHARACTERISTICS OF PARALLEL TRANSPORT

Before we consider the data, what, in more specific terms, do we mean by parallel transport. Babkin and his colleagues appear to have had a relatively inexact definition by which they meant that there was merely a general coincidence in *responsiveness* in the secretion of different enzymes to all secretagogues. That is, all stimulants increase the secretion of all enzymes, although not necessarily in exact proportionate correspondence. When the idea of parallel transport is placed within the context of a mechanism for secretion, such as exocytosis, we can be more exact about what we mean and about what would be required. Thus, for the view of exocytosis that I outlined above, the secretion of a pair of enzymes should always covary, and increases in the secretion of one should be accompanied by coincident proportionate increases in the other under all circumstances of time, state, and rate. If we compare the rate of secretion of two enzymes, then we would expect, no matter what the circumstances or how the various samples were obtained, that a plot of the values for a variety of individual samples would form a straight line intersecting the axes at the origin. Thus, whenever the overall rate of secretion is increased or decreased, so would the rate of secretion of the two enzymes and in direct proportion to each other. As the overall rate of secretion approaches zero, so equally should the secretion of the two test molecules, and hence the line should regress through the origin; when one is not secreted, neither is the other. Furthermore, if we account for all measurement error and animal-to-animal variability in the proportions of the enzymes, then all of the points should fit the calculated regression line perfectly, that is, with a correlation coefficient of 1.0. To the extent that measurement error and animal-to-animal variability exist, the value should be less than 1.0, but whatever the correlation coefficient may be, it should be an invariant characteristic for the pair of enzymes that reflects the constancy of the secretory process and the sources of variation. That is to say, whatever the goodness of fit of the points to the line, it should be constant from circumstance to circumstance as long as the assay and the prior state of the tissue is not changed. Thus, transport is parallel whenever the relationship between the secretion of two or more proteins is constant despite changes in state or circumstance otherwise, and regardless of the overall rate of protein secretion. As a result, when the amount of one enzyme in a sample of secreted fluid is plotted against that of another, for the set of any and all such samples of secretion, a linear function should be formed that would both regress through the origin and have a constant variance from situation to situation that would be a particular and fixed characteristic of the specific enzyme pair (Figure 2).

13.6 THE FIRST OBSERVATION

My first attempt in the mid 1960s to evaluate the question of parallel secretion, produced an example of nonparallel transport (Rothman, 1967). It was measured in pure pancreatic fluid collected directly from the pancreatic duct in a defined in vitro experimental setting (the organ preparation of rabbit pancreas). Such reports had in the past often been

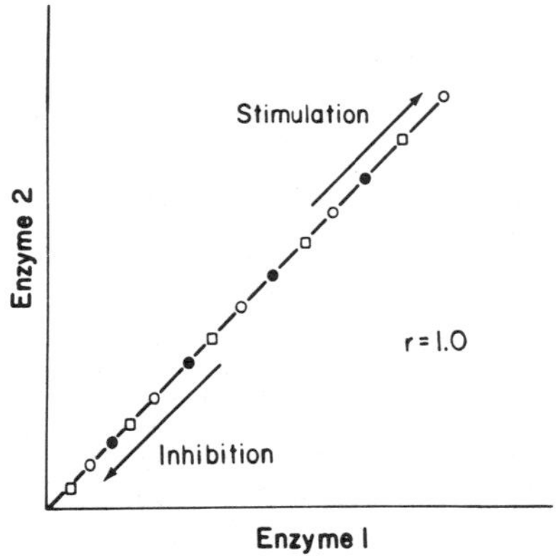

Figure 2 The relationship between the secretion of two enzymes predicted by a model that proposes the obligatorily parallel or linked secretion of different secretory proteins, as in the exocytosis of zymogen granules of mixed enzyme content. (○, ●, □) Secretion collected in three different unspecified states. In this model the secretion of any two enzymes would always fit the same line regardless of the rate of secretion or other situational factors. The line should intercept at the origin and if measurement error and precursor pool inconstancy had been accounted for, the data points should fit the calculated function with $r = 1$. (Reprinted, with permission, from Rothman, 1977.)

made on collected duodenal chyme, and, as a result, the question of whether or not the enzymes had actually been secreted in a nonparallel fashion, or whether their activity had merely been differentially altered by other substances in duodenal fluid after their secretion, had often been a point of uncertainty. Moreover, the measurements had always been made in situ, usually in awake subjects; thus, there was a great deal of variability from trial to trial and experiment to experiment. Our report was notable because trial-to-trial variability had been greatly reduced, although not eliminated, by the in vitro system; and of course, it was clear that whatever changes were observed were directly attributable to the fluid secreted by the cells of the gland.

However, the particular effect itself was somewhat surprising. In the first instance, nonparallel transport occurred as part of the response to the administration of a secretagogue, the gastrointestinal hormone cholecystokinin (CCK), that produced a general augmentation of protein secretion by the gland, and not in response to the presence of a particular foodstuff or product of digestion. It was not clear why this hormone should favor the secretion of one enzyme or another on regulatory grounds, unless it represented one of a series of such hormones, and there was no evidence for this at the time. Thus, it seemed that nonparallel transport did not necessarily have to be related to the regulation of digestion. It could be a consequence, even an apparently purposeless consequence, of the administration of a general stimulant of protein secretion. Perhaps, it was in some fashion, simply a by-product of the mechanism of secretion in which differences in the "permeability" of the cell to different molecules or differences in their available intracellular concentration produced by the hormone, led to their differential release.

Furthermore, the differences in the rates of secretion of different proteins were not absolute. That is, CCK did not increase the secretion of one protein while leaving that of another totally unaltered. Rather, it increased the secretion of both, but not in a proportional manner. Thus, an agent that produces nonparallel transport need not do so in an absolute fashion, augmenting the secretion of one enzyme, but not another. It might alter their rates of secretion *relative* to each other. The meaning of this, both in regard to the mechanism of secretion, and how secretion appears to be regulated, will be developed further below.

The particular observation of nonparallel transport produced by CCK in the rabbit was a disproportionate increase in the secretion of two proteolytic enzymes, trypsinogen and chymotrypsinogen, of very similar molecular weight, surface charge, shape, and amino acid sequence. It increased trypsinogen secretion to a proportionately greater degree, even though the concentrations of both enzymes in secreted fluid were elevated (Figure 3). (I have already touched on this particular observation in Chapter 12 in reference to the cytoplasmic pool of secretory protein.) On the other hand, when a cholinergic agonist, which also increased overall protein secretion, was tested instead of CCK, the increase in the secretion of both proteins was

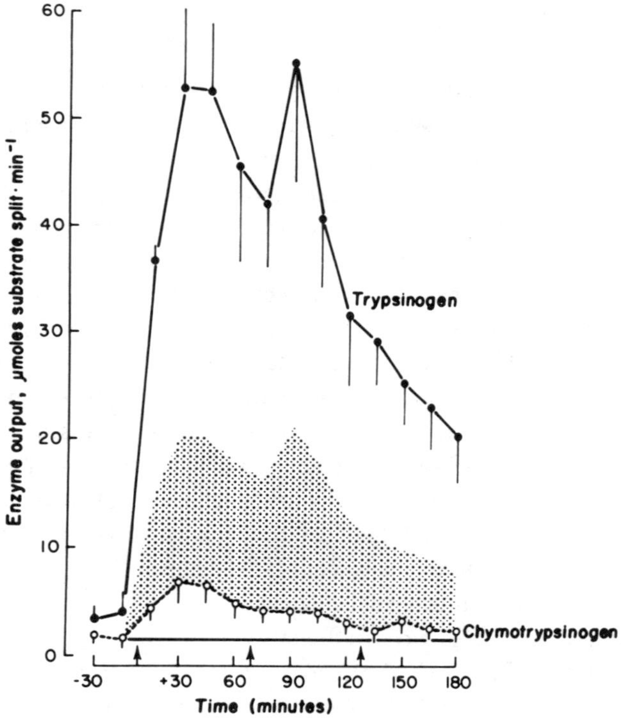

Figure 3 Trypsinogen and chymotrypsinogen secretion as collected from the cannulated duct of rabbit pancreas in organ culture in response to cholecystokinin-pancreozymin (CCK–PZ). Values are means ± SE for 16 preparations. The shaded area represents the amount of chymotrypsinogen secretion required for a response parallel or proportionately equivalent to the trypsinogen response. There is a large difference between the observed values (○) and the calculated parallel values (shaded area). The arrows on the *x* axis indicate that the medium and stimulant were replaced at the specified times. The two periods prior to the first arrow are control periods. The heavy solid line parallel to the *x* axis shows the control chymotrypsinogen output. Chymotrypsinogen secretion approached control values by 2 hr poststimulus, while trypsinogen output was still many times the control rate. (●) Trypsinogen; (○) chymotrypsinogen. (Reprinted, with permission, from Rothman and Wilking, 1978.)

more or less parallel (Figure 4). Thus, using the same biological system, experimental protocol and procedures, measurement techniques, and assays, one stimulant produced the secretion of a quantitatively different mixture of proteins than the other. The possibility that CCK had produced the secretion of another substance that had altered the activity of chymo-

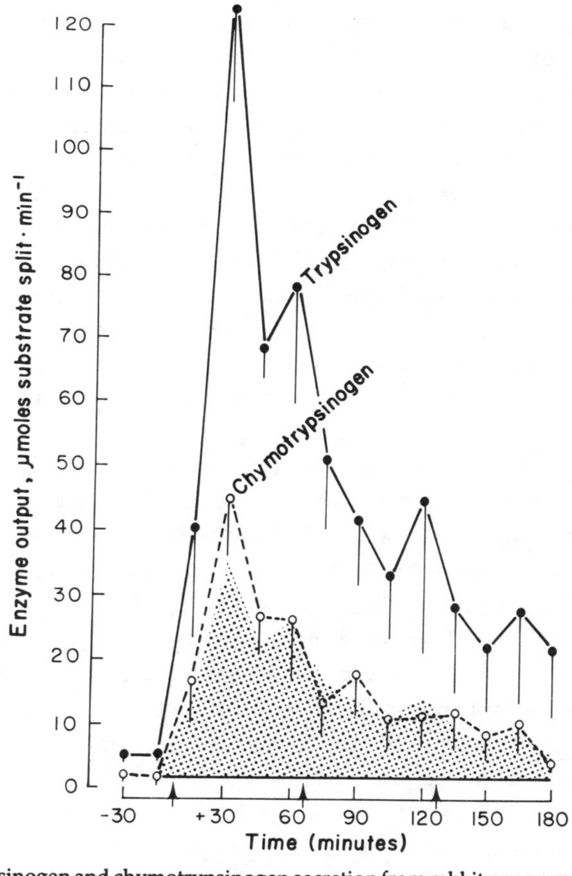

Figure 4 Trypsinogen and chymotrypsinogen secretion from rabbit pancreas in organ culture in response to acetyl-β-methylcholine. Values are means ± SE for seven preparations. The shaded area represents the amount of chymotrypsinogen secretion required for a response parallel or proportionately equivalent to the trypsinogen response. The arrows on the x axis indicate that the medium and the stimulant were replaced at the specified times. The heavy solid line parallel to the x axis shows the control chymotrypsinogen output. The observed chymotrypsinogen values approximated the shaded area or the calculated rate for a response parallel to trypsinogen. (●) Trypsinogen; (○) chymotrypsinogen. (Reprinted, with permission, from Rothman and Wilking, 1978.)

trypsin, thereby only giving the appearance of a lesser response, was tested, but no indication could be found that CCK-stimulated secretion contained a substance that either uniquely inhibited chymotryptic activity or activation, or conversely enhanced tryptic activity preferentially. The effect appeared to be due to a true difference in the rates of secretion of the two proteins.

At the time, I suggested three potential explanations for this phenomenon—the same three explanations that remain with us today, as we shall see. The first was that different zymogen granules contained different single enzymes or mixtures of enzymes and that CCK led to the selective exocytosis of certain granules. The second explanation was that the secretion of these proteins did not occur en masse, but directly and separately as a result of their movement across cell and granule membrane by membrane transport. Nonparallel secretion in this case would reflect differences in the permeability of the membranes to different proteins, differences in the concentration gradients for their movement, or both. This was the original proposal of equilibrium-based protein transport. The third hypothesis was a composite—the mixing of both types of secretion, vesicular and nonvesicular, producing the nonparallel effect (Figure 5).

There the matter rested for many years. We and others sought other examples of nonparallel transport, asking ourselves whether what we had discovered was a rare anomaly or a common occurrence that might have broader implications in terms of the mechanism of secretion and its regulation. We also began to try to distinguish between different models of secretion, as well as to define such models and the deductive predictions that one might make about them, in a more specific fashion. This led to a variety of experimental strategies and approaches, many of which I have already discussed. Those who did not believe in the existence of nonparallel transport either simply ignored the observation or discounted it, assuming it to be some sort of assay artifact, although no coherent explanation of how this could have been was put forth.

Even though almost a decade later no satisfactory alternative explanation had been proposed, at that time three groups felt the need to attempt to "reproduce" the original observation. In one study (Steer and Glazer, 1976), the investigators reported that they were unable to reproduce the effect, and concluded that the original observation must have been due to general problems with the activation of chymotrypsinogen (Glazer and Steer, 1977), although they did not explain how such a problem could have produced what was observed (particularly, the "better activation" with a

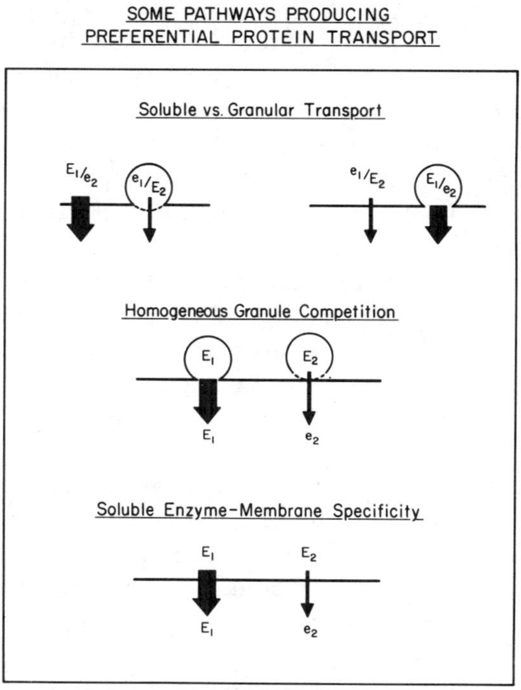

Figure 5 Some transport pathways that could produce "enzyme-specific secretion" by pancreatic acinar cells. (Reprinted, with permission, from Rothman, 1967.)

cholinergic agonist as the secretory stimulant).* Nor, more generally, did they explain why the extent of activation or level of activity was even relevant, as long as the measurements were made in a reproducible fashion and compared with each other for different conditions by means of standard curves or functions, as was of course the case. Beyond this, these investigators had chosen a dose of CCK that was ineffective, that is, it did not significantly augment protein secretion at all. Why they thought this a more satisfactory circumstance in which to seek the nonparallel effect of CCK than an effective

*Scheele and Palade (1975) also proposed that we underestimated the presence of chymotrypsinogen because of its adsorption to glass during sample preparation. They did not explain how a cholinergic stimulant and the CCK octapeptide, which produced much larger chymotrypsinogen responses for a given trypsinogen response, inhibit such adsorption, as they must if this hypothesis is correct. Moreover, this notion ignores the structural similarity between trypsinogen and chymotrypsinogen, most significantly the fact that the two proteins

dose has never been made clear, and it certainly seems as if they were not looking for the response under the most propitious circumstances. In later work (Steer and Manabe, 1979) however, they did increase the dose of CCK, and were able to produce a modest, although by our experience, poor secretory response to the hormone. They produced an average peak increase in enzyme secretion of about threefold, and were only able to maintain this augmented secretion for about 1 hr, whereas in our hands the response lasted at least 3 hr, and the peak was severalfold greater (Figure 3, Table 2). In any event, they were still unable to find the nonparallel effect. A comparison of their response and ours indicated that the chymotrypsinogen response was essentially the same (about a threefold increase), and therefore the discrepancy could not be attributed to the measurement of chymotryptic activity (for example, its autolysis or "incorrect" activation) (Rothman, 1981). The difference between the results was in the trypsinogen response; our response was some four times greater in relative terms than the one that they observed (Table 2). Thus, if one wished to attribute the differences between the two studies to an artifact of measurement, then it would have to have been due to their underestimation of trypsinogen (for example, inadequate or incomplete activation of high concentrations of trypsinogen).

The second group of investigators was no more successful than the first (Iacino et al., 1980). They attempted to bypass the potential for activation and activity artifacts by measuring the presence of the enzyme in secreted fluid using chemical separation methodology. The particular method that they chose involved labeling the proteins in situ with radioactive amino acids, separating the secreted proteins by two-dimensional disc gel electrophoresis, and then measuring the amount of radioactivity associated with each identified spot. Although this method sidestepped the issues associated with activity measurements, it had its own problems. Was it truly a satisfactory quantitative analysis? As a method, was it on a par, no less

---

have approximately the same isoelectric point in the rabbit. Why should one expect two proteins of equal size, similar shape, and surface charge to adsorb to glass differentially? Whatever plausibility one might attach to these ideas, the use of appropriate controls would both identify and correct for such an effect. If one develops standard curves to assess the presence of an enzyme, as we did, in which the measured activity is plotted against sample size, the occurrence of substantial adsorption (relative to sample size) would be reflected, depending upon its tenacity, in either a nonlinear function close to the origin, or a positive intercept on the sample size axis, or both. Neither was observed. But even if it had been, the appropriate use of a standard curve would correct for such an effect.

TABLE 2. Secretion of Trypsinogen and Chymotrypsinogen by Rabbit Pancreas in Vitro in Response to Presence of CCK-PZ[a]

	Steer and Manabe[b]		Rothman and Wilking[c]	
	1st Hr	2nd Hr	1st Hr	2nd Hr
Trypsinogen	314	211	1290	1165
Chymotrypsinogen	317	183	343	246

[a]Values indicate the response as a percentage of enzyme secretion during prior control hour. The dose of CCK–PZ (cholecystokinin-pancreozymin) in both sets of experiments was 2.5 μg/100 ml bath fluid.

[b]Data calculated from Table 1, Ref. 26.

[c]Data calculated from those used to prepare Figure 4, Ref. 23.

Source: Taken from Rothman (1981).

superior, to the very sensitive and reproducible measurement of enzyme activity? This question has yet to be satisfactorily answered, and hence it is not clear that it did not reduce the ability to make quantitative distinctions rather than enhance them.

Another problem, of which the authors were aware, was the need to label the whole storage pool prior to eliciting augmented secretion, so that the labeled protein properly traced the movement of protein overall. That is, it was necessary to be sure that tracer kinetics applied. In its absence one could only *assume* (incorrectly at that) that there was no difference in the movement of stored and newly manufactured protein, and this would of course in a sense be to assume the parallel transport hypothesis correct even as one was testing it. To overcome this problem, these investigators incubated the gland in vitro for many hours (about 7) prior to the application of the stimulant, although they apparently did not determine that labeling was at the steady state and that the old pool had actually turned over completely prior to the application of the stimulant. In any event, 7 hr in vitro is a long time for the biological preparation of pancreas that was used, which is generally on the edge of the useful, as seen by both its diminished secretory rate and responsiveness, by this time. Whatever problems there may have been with their approach, they were nevertheless once again unable to reproduce our result, although they did observe and report other examples of nonparallel transport.

Of course, it was possible that all of these reports were correct and that the differences between our observations and the others were real, for whatever reason, and attributable to natural causes, and not measurement problems. As the renowned physiologist Walter Cannon pointed out, it is not wise to assume that observations that you cannot reproduce are necessarily incorrect. We cannot always reproduce the exact conditions that others have applied. Although I cannot know with any certainty the reason for the failure of these investigators to reproduce our original observations, I think that the explanation may lie in the relationship between the secretion of the two enzymes at different levels of overall secretory activity. The expectation of these workers was no doubt that the change in ratio should be observed whatever the overall rate of secretion by the tissue might be. However, this need not be the case. Indeed, it is not the case. The function relating the secretion of trypsinogen to chymotrypsinogen in unstimulated conditions was nonlinear with a positive intercept (that is, trypsinogen secretion occurs, at least theoretically, in the absence of chymotrypsinogen secretion). I have redrawn the function (from Rothman, 1974) for secretion from the rabbit pancreas preparation in the unstimulated state (Figure 6).

If, in the simplest scenario, CCK merely moved points out along this curve, not changing its parameters, then we would find that close to the y-axis intercept the ratio would be high, decreasing as the rate of secretion was elevated, and subsequently increasing again as output or secretion values approached the apparent asymptote. Thus, CCK could, by increasing the rate of secretion, reduce or leave the ratio unaltered, as well as increase it, depending upon where our particular data points fit on this function. The average value for CCK-stimulated secretion found in our original experiments is shown in Figure 6 by the closed circle. It is beyond the limit of outputs observed in the unstimulated state (the end of the function as drawn) and displaced somewhat towards the trypsinogen axis, suggesting that the function's parameters had been altered, at least the apparent asymptote. We can look at the results of Steer and Manabe superimposed on this function. They produced roughly the same chymotrypsinogen response that we had, that is, a threefold increase over the unstimulated rate. If we assume that the control rate of secretion was approximately the same in both series of experiments (theirs and ours), then the open square estimates the enhancement that they observed, an approximately threefold increase over the control rate (open circle) if the points merely moved out the control function. In this case, the ratio would have decreased slightly in the presence of CCK from about 6.5 to 5.5. If the function was displaced somewhat toward the trypsinogen axis (as shown by the dashed line), then the ratio would have been unchanged.

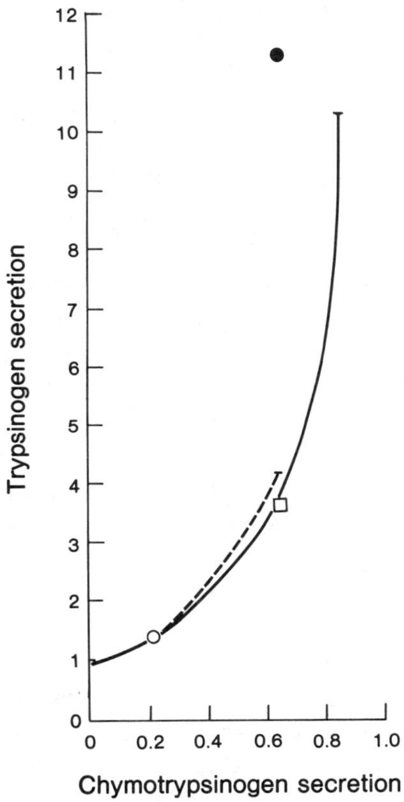

Figure 6 The relationship between unstimulated trypsinogen and chymotrypsinogen secretion from the rabbit pancreas in short-term organ culture (solid line) (taken from Rothman, 1974). The end of the function indicates the limit of observed values. (○) Estimate of the average control or unstimulated rate of secretion; (●) mean CCK stimulated value (taken from Rothman, 1967); (□) average CCK-stimulated output, based on the estimated control (○), taken from the experiments of Steer and Manabe, (1979); (----) function for which an approximately three- to fourfold increase in secretion would produce perfectly parallel responses for the two enzymes. See text for further discussion.

Of course, I cannot say that this is the reason for the differences in the observations, but it does at least offer an explanation that does not fall back upon the easy "complaint" that someone else's observations are flawed or erroneous, merely because you are unable to reproduce them. Moreover and more importantly, this analysis initiates us to the fact that we cannot assume that the relationship between the rate of secretion of different enzymes is merely due to changes in the slope of linear functions (as in Figure 2) that regress through the origin and that the question of nonparallel transport has, not surprisingly, its own descriptive intricacies, which we shall consider further below.

There is another possibility to be considered in evaluating these observations in comparison with the original ones. The preparation of CCK that we used in the mid 1960s was very impure relative to material available in the mid 1970s. Our CCK preparation, the purest generally available at the time, only contained CCK as some 0.2% of its total peptides, whereas the new (1970s) material contained 20% or more as CCK—a big difference in purity! Therefore, it was possible that our effect was due to a contaminant of CCK preparations that had been purged during the intervening years. It was with this thought in mind, as well as simply the desire to see if we could repeat our own observation (breaking the scientist's adage that one *never* attempts to repeat one's own experiment), that we, the third group, undertook to reproduce our observation with the new, purer CCK preparations that had become available (Rothman and Wilking, 1978). We performed a substantial series of experiments, collecting secretion every 15 min for 3 hr in the presence of the hormone, thereby providing us with a fair sampling of the situation.

We found, much to our delight, that even after 10 years the effect was still there, at least in our hands, and indeed was seen with even greater clarity than before. Trypsinogen secretion was augmented to a much greater extent than chymotrypsinogen secretion and the effect lasted for essentially the whole 3-hr sampling period (Figure 3). Indeed, augmented trypsinogen secretion continued over a 3-hr period, even as chymotrypsinogen secretion fell and became indistinguishable from control rates (Figure 3). We compared our activation and assay procedures for chymotrypsinogen to those used by others and found that the differences had no bearing on the results. In addition, we obtained an even purer CCK from Professor Mutt at Karolinska Institute and found that the purest available material (called "pure") also produced the nonparallel effect (Rothman and Wilking, 1978). Cholinergic stimulation was also tested again and found again not to produce nonparallel transport from this organ preparation for these two enzymes (Figure 4). Thus, whatever others had found, we were able to reproduce our observation and this was gratifying.

But we did one additional experiment; one could say, one too many. It had been observed that many of the actions of CCK could be obtained with its carboxy-terminal octapeptide. The chemically synthesized "octapeptide" was available, and we tested it, full well expecting to see the nonparallel effect once again. Instead, we observed a parallel response in which the secretion of both enzymes was increased proportionately (Figure 7). That is, for a given trypsinogen secretion, the octapeptide produced a greater chymotrypsinogen response than CCK itself. Therefore, it appeared that CCK or a lingering contaminant in CCK preparations might have *inhibited* the chymotrypsinogen response. In recent studies by Niederau, Grendell and the author (Niederau et al., 1985), we demonstrated that the presence of a CCK-receptor antagonist, proglumide, produces a dose-dependent selec-

Figure 7 Trypsinogen and chymotrypsinogen secretion from rabbit pancreas in organ culture in response to the COOH-terminal octapeptide fragment of CCK–PZ. Values are means ± SE for seven preparations. The shaded area represents the amount of chymotrypsinogen secretion required for a response parallel or proportionately equivalent to the trypsinogen response. Only one arrow is shown in this figure because the stimulant was added only at 0 time and the medium was not changed during the course of the experiment. The heavy solid line parallel to the *x* axis indicates the control chymotrypsinogen output. Unlike the whole CCK–PZ peptide, the octapeptide fragment produced a roughly parallel secretory response when trypsinogen and chymotrypsinogen secretion are compared. (●) Trypsinogen; (○) chymotrypsinogen. (Reprinted, with permission, from Rothman and Wilking, 1978.)

tive inhibition of chymotrypsinogen secretion in the same biological system, suggesting that it is possible specifically to inhibit chymotrypsinogen secretion at the level of the CCK receptor. I shall discuss these experiments later on.

13.7 OTHER CASES OF NONPARALLEL TRANSPORT

The parallel secretion model does not merely require that parallel secretion occur from time to time or circumstance to circumstance. It must always occur, in every circumstance that we measure, and for every possible condition, as long as transport, and not synthesis, determines the relative rate of secretion of the various proteins. Therefore, any single, clear case of nonparallel transport falsifies the hypothesis of parallel transport and requires any mechanistic model to be able to accommodate this fact. On the other hand, the view that transport is nonparallel does not require that it always be so. That would not make any sense, since the relative rates of transport of different proteins should always be parallel to each other in a given established state. It is only when state and circumstance change that nonparallel effects can come into play. That is, it is a "permissive" hypothesis in that whereas nonparallel transport may be seen in one situation or another, it is not a necessary consequence of each and every circumstance, or even each and every change of state. Thus, one can provide all of the necessary evidence for the hypothesis that nonparallel transport exists and rejection of the hypothesis of obligatory parallel transport with a single confirmed example of nonparallel transport, such as the one that I have just discussed.

Nevertheless, it is useful to seek, even accumulate, additional cases of nonparallel transport, because however reliable our single observation may be, the reliability of our conclusion is greatly improved with each new independent discovery of nonparallel transport. Many such examples have been published since the mid 1960s. And unless we wish to consider them all measurement artifacts of one sort or another, the existence of nonparallel transport has beyond any reasonable doubt been affirmed. I will not catalog all these examples here. Instead, I would like to discuss the particular series of nonparallel transport observations that have been made in my laboratory. They represent about two dozen independent, statistically validated examples, each demonstrating nonparallel transport for different enzymes in different biological settings or demonstrating different aspects of the phenomenon. They include examples of nonparallel transport that are either directly or indirectly related to the concentration of digestive end-products in the gut or blood, and therefore concern the regulation of specific digestive reactions in the intestine and are not solely "situational" examples of nonparallel transport, such as the CCK effect that I have just discussed. My major reason for discussing this group of experiments is to explore in some detail different descriptive and analytic aspects of the phenomenon of nonparallel transport; that is, its intricacies. What exactly do we mean when we say that we have observed transport that is nonparallel in one circumstance or another? One needs to understand the nature of these phenomena in the most complete manner possible in order properly to evaluate and consider the ability of different mechanistic models for the secretory process to explain them. In addition, as I have already mentioned, knowledge of these details is also crucial in helping us choose the appropriate means of identifying nonparallel transport in the first instance.

A look at the published literature might lead us to conclude that it has become easier to find nonparallel transport than it used to be. Why should this be so? Even though the measurement techniques, enzyme assays and the like, have improved, I do not believe that the increased number of reports of such phenomena can be attributed solely, if at all, to the nature of the assays. If this is not the cause, then to what can we attribute the increased frequency of such observations?

It has been the common view that gastrointestinal functions are rather variable as biological activities go. The frequency of light to which a given photoreceptor responds is well defined and reproducible within relatively narrow limits from receptor to receptor and trial to trial. A particular type of muscle cell will shorten to approximately the same degree time after time, and even from muscle fiber to muscle fiber under the same load. However, if one examines a gastrointestinal activity and makes a measurement several times, particularly in situ, we would not be surprised to find rather different values from trial to trial. Although this may not be true for all alimentary canal functions, it is certainly the case for secretion, in particular for protein secretion. If we collected a dozen cases of spontaneous secretion in awake or even in anesthetized animals, the rate of protein secretion might vary among the animals or even within an animal from trial to trial and period of collection to period of collection by as much as an order of magnitude. This even applies to animals whose past history and present state are the same, as best as we can know it. This is of course another reason that it has become common practice to fast animals for defined periods before studying secretion, and to study the process at roughly the same time of day, etc., as well as to study secretion in vitro. Despite attempts at standardization and however careful we may be, there nevertheless always remains a substantial variability in such processes in situ that we cannot avoid, or get rid of, because we do not know to what the variability is attributable.

The traditional method for determining whether or not nonparallel secretion has occurred is essentially the method that I presented above for the effect of CCK on protein secretion from the rabbit pancreas in vitro (Figure 3). The investigator averages output values for two or more enzymes at a particular period of time, that is, for a particular collection period, after a change in state and compares these values to control rates of secretion for the same enzymes. The question that is then posed is whether or not changes in the average rate of secretion of each enzyme subsequent to the imposed change of state have been equivalent or proportionate. If the changes in secretory rate were not proportionate, on the average, for different enzymes, then nonparallel transport occurred. If the average values changed in proportion for different enzymes, then transport was parallel. This method, although sometimes effective, is very insensitive because it does not take into account the large variability from trial to trial and animal to animal that I have just discussed. Instead it incorporates this variability into the calculated variance around the mean values; values that are then compared with each other statistically. Hence a large and overlapping variance between populations whose means may differ substantially is often seen.

Many, and perhaps most, cases of nonparallel transport are likely to be missed if this method of analysis alone is used. This has commonly been the case. The reason for this "insensitivity" can be easily understood by considering Figure 2. The line drawn in this figure describes the relationship between the secretion of two different enzymes. The points closest to the origin would be for animals, trials, or periods of collection when secretion is low and those furthest away for animals, trials, or periods of collection when secretion rates are high. This variability from animal to animal and time to time is, as I have said, incorporated into the variance or standard deviation around the mean or average value for each period and over time, and in a given situation can be said to reflect variability in secretory activity of unknown origin from trial to trial. One can look at the figure and consider the array of points and estimate an average rate for each enzyme, relative to the range of values on the graph. We can now consider in our mind another line of twice or half the slope, over approximately the same range of values in which case one of the two enzymes will be present at twice the concentration relative to the other. Since we have made the slope twice ourselves, we can be confident that this is indeed true. However, if we seek evidence for this known fact by averaging the values for the enzymes under two different conditions and comparing such averages statistically, we can see that although the average value would change, the mean values would have large standard deviations, and we might well not be able to "see" the effect in statistical terms despite the known change. The large and overlapping variance would mask the true difference between the two conditions. Thus, real differences can be lost in our attempt at statistical validation by parametric averaging techniques. For this reason such analysis is weak in this situation, and many nonparallel events, particularly if they are subtle, that is, involve changes of two- or threefold or even more (if one can call that subtle) may well be missed.

We can eliminate this source of variance completely, without knowing its origin, merely by comparing least squares or regressed functions calculated from plots of data points for one condition relative to another (as in Figure 2) rather than by comparing averages. This takes the variability from trial to trial into account and eliminates this source of variance from our statistical analysis, which is now a comparison between the parameters of two functions. In this way, our question can now be, "What is the ratio of one enzyme relative to another in secretion *for a given level of secretory activity?*" This simple device opens up, as we shall see, an abundant world of nonparallel transport events, that would without question never otherwise be noticed.

One can also improve upon averaging techniques by comparing average ratios of enzyme activity for any given trial, and in this way attempt to eliminate the variance in overall secretory activity by comparing only relative values. This is effective if, as in Figure 2, all of the points fit the line perfectly; that is, if there is no variance in the calculated function and hence no variation in the ratio from trial to trial. However, if substantial variation exists, then true differences may be missed if average ratios are merely compared parametrically (see discussion of chymodenin below). This would of course be a particular problem close to the origin, for example, for unstimulated secretion, where the two functions would converge. Once again, the use of correlation or regression analysis improves our ability to make such distinctions.

13.8 LYSINE, AN END-PRODUCT OF DIGESTION

The first time that this type of analysis was used was in studies on the effect of the amino acid lysine, added either to the intestinal lumen, blood, or extracellular fluid, on the relative rates of secretion of two proteolytic enzymes by the pancreas; one, trypsinogen whose active form splits proteins at lysine-containing peptide bonds, and the other, chymotrypsinogen, that does not. I will not consider the details of these experiments here and refer the interested reader to the primary article which is cited at the end of this chapter (Rothman, 1974).

However, briefly, when lysine was perfused through the duodenum of anesthetized rabbits at an effective concentration (Figure 8), the slope of the function relating the secretion of trypsinogen to chymotrypsinogen almost

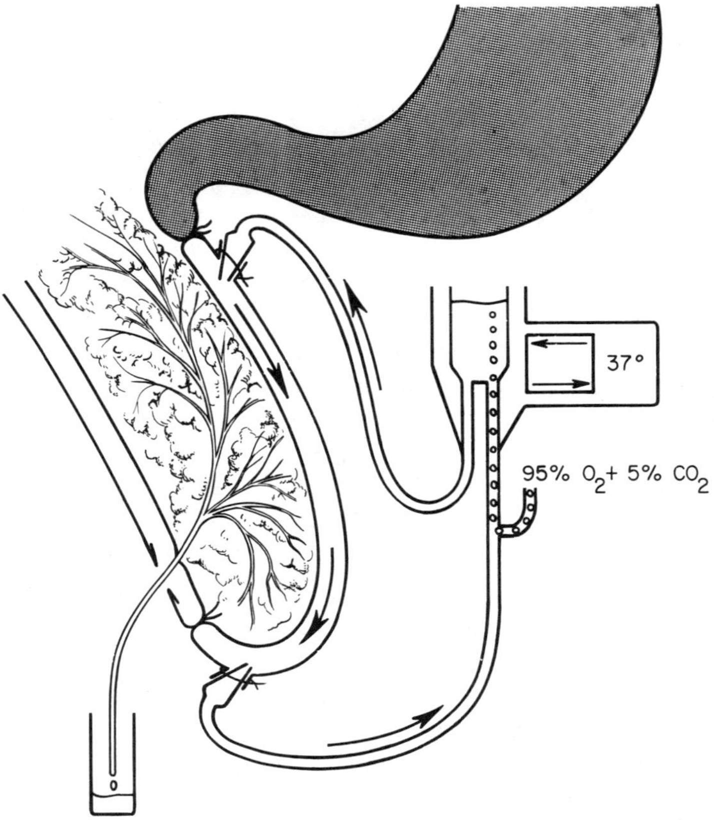

Figure 8 Rabbit mesomental loop and pancreas indicating perfused portion of duodenum and cannulated pancreatic duct. Total volume of circulating fluid was between 75 and 100 ml. (Reprinted, with permission, from Rothman, 1974.)

doubled; lysine had increased the secretion of trypsinogen relative to chymotrypsinogen, as compared to parallel controls (Figure 9). Similarly, in another series of experiments in which lysine was injected intravenously as a bolus, the slope of the function relating the secretion of the two enzymes was again increased, favoring trypsinogen secretion, this time by more than threefold (Figure 10). Lysine was also added to the medium bathing the rabbit pancreas in short-term organ culture, although at substantially higher concentrations. Once again, the relative rates of secretion of the two enzymes were altered, but in this case in the opposite direction, with

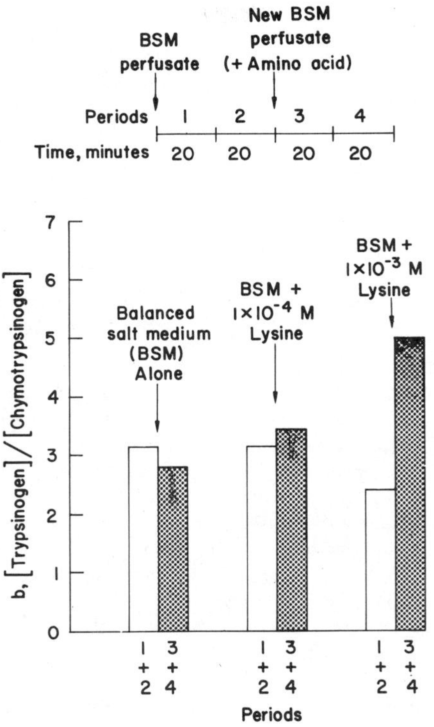

Figure 9 Effect of duodenal perfusion of lysine on relative concentrations of trypsinogen and chymotrypsinogen in secretion collected from the pancreatic duct of anesthetized rabbits. Design is shown at top of figure. Moving from left to right on x axis and comparing open bars to stippled bars: 1st pair, open ($b = 3.1$, $n = 18$, $r = 0.69$) versus stippled ($b = 2.8$, $n = 18$, $r = 0.76$), $t = 0.51$; 2nd pair, open ($b = 3.1$, $n = 12$, $r = 0.43$) versus stippled ($b = 3.4$, $n = 12$, $r = 0.46$), $t = 0.26$; 3rd pair, open ($b = 2.4$, $n = 12$, $r = 0.81$) versus stippled ($b = 5.0$, $n = 12$, $r = 0.72$), $t = 3.64$. n = number of observations. (Reprinted, with permission, from Rothman, 1974.)

chymotrypsinogen secretion being favored (Figure 11). It was also observed in these organ culture studies that in the control (unstimulated) state, secretion became more trypsinogen dominant over time (Figure 11).

Thus, the in situ effect of the end-product lysine was a positive feedback response favoring trypsinogen secretion, whereas in vitro it produced a negative feedback response, the end-product increasing chymotrypsinogen secretion relative to trypsinogen secretion. This difference, however, may not indicate a different effector modality for lysine in situ and in vitro but merely differences in the concentrations of the amino acid used. Recently, Grendell and I (Grendell and Rothman, 1981) demonstrated that lysine can cause the preferential release of trypsinogen (relative to amylase) from zymogen granules. The effect is one of positive feedback for lower concentrations of lysine (Figure 12), as in the in situ studies. However, when the concentration of lysine is increased to the levels used in the in vitro

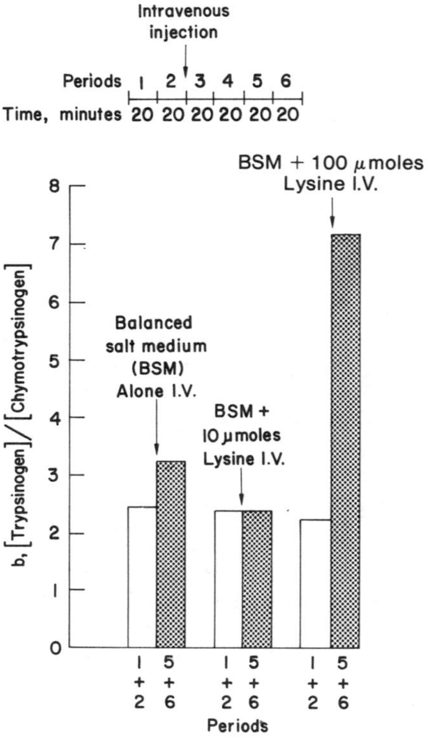

Figure 10 Effect of intravenous lysine on relative concentrations of trypsinogen and chymotrypsinogen in secretion collected from the pancreatic duct of the rabbit in situ. Design is shown at top of figure. Moving from left to right on x axis and comparing open bars to stippled bars: 1st pair, open ($b = 2.4, n = 12, r = 0.95$) versus stippled ($b = 3.2, n = 12, r = 0.90$); 2nd pair, open ($b = 2.4, n = 8, r = 0.99$) versus stippled ($b = 2.4, n = 8, r = 0.99$); 3rd pair, open ($b = 2.2, n = 8, r = 0.96$) versus stippled ($b = 7.1, n = 8, r = 0.98$), $p < 0.001$, n = number of observations. (Reprinted, with permission, from Rothman, 1974.)

organ culture study, a negative feedback is observed, in which case the effect of the amino acid on trypsinogen release is lost. Thus, a concentration-dependent switch from positive to negative feedback occurs.

Beyond their regulatory implications, these observations on the effects of lysine demonstrated the usefulness of regression or correlation analysis in helping to discern nonparallel transport effects. In three separate series of experiments (gut perfusion, intravenous injection, and organ culture), lysine had produced nonparallel transport of the two enzymes that were measured. These effects could not have been distinguished statistically merely by comparing average period-by-period values for secretion, or even by comparing average ratios, because of the variability in secretion from animal to animal and tissue to tissue.

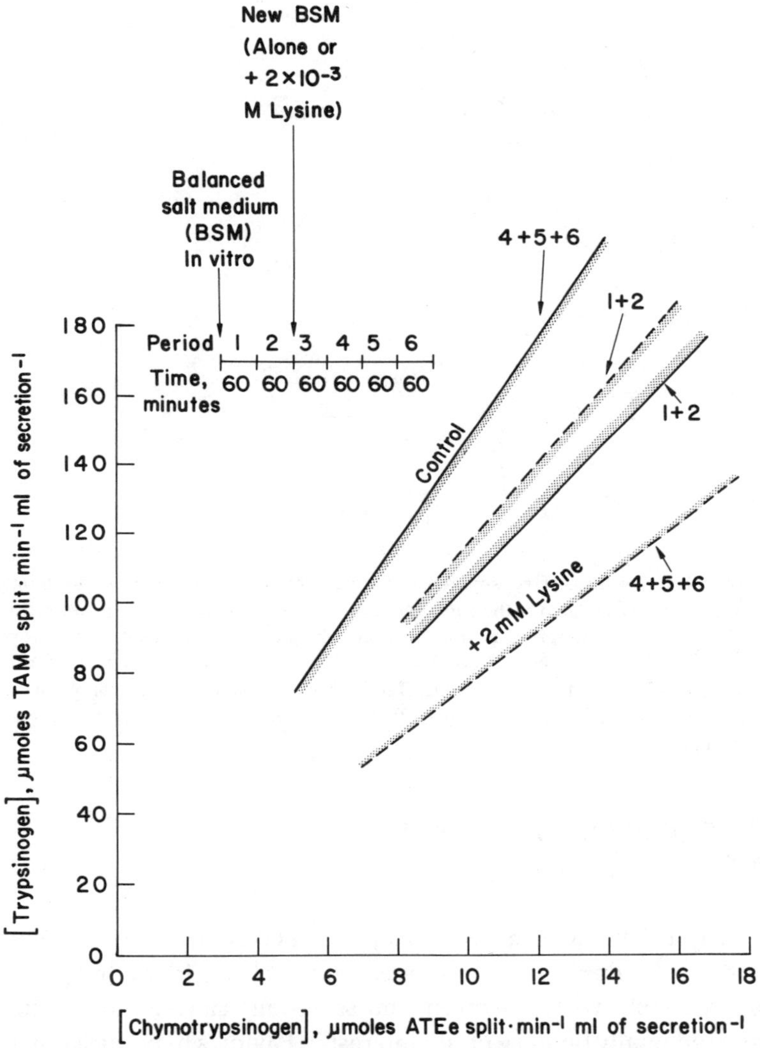

Figure 11 Effect of lysine on relative concentrations of trypsinogen and chymotrypsinogen in secretion collected from the duct of the rabbit pancreas in organ culture. Design is shown at top of figure. Solid lines represent data from experiments in which bathing solution (BSM) was changed at 2 hr, but no amino acid was added (Control). Dashed lines are from lysine experiments in which 2×10^{-3} M lysine was added when bath was changed at 2 hr. Stippled areas represent either + or − SE of estimate around slope (only + or − is shown). Solid lines, periods 1 and 2, $n = 10, r = 0.92$; periods 4, 5, and 6, $n = 18, r = 0.91$. Dashed lines, periods 1 and 2, $n = 10, r = 0.75$; periods 4, 5, 6, $n = 18, r = 0.92$. Periods 4, 5, and 6, BSM alone versus BSM plus 2×10^{-3} *M* lysine, $p < 0.001$. (Reprinted, with permission, from Rothman, 1974.)

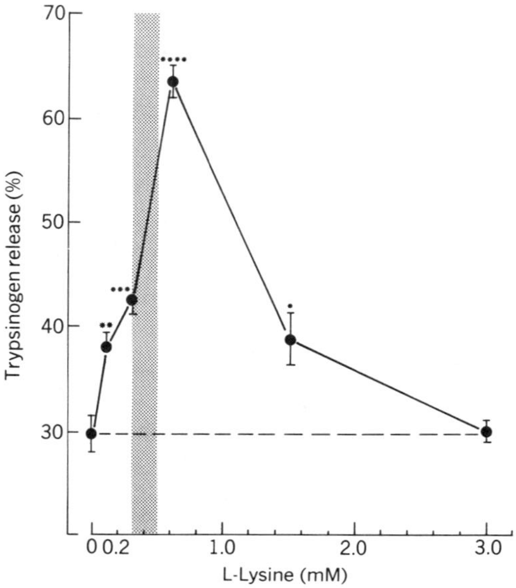

Figure 12 Percent release of trypsinogen into a postmicrosomal supernatant of homogenized rat pancreatic tissue as related to the concentration of L-lysine in the homogenization medium. Shaded bar represents range of normal plasma values of L-lysine in rat. Points represent means, and error bars represent SE. For 0 and 0.6 m*M*, $n = 10$. For all other concentrations, $n = 7$. (*) $p < 0.05$; (**) $p < 0.01$; (***) $p < 0.005$; (‡‡) $p < 0.001$. Broken line indicates mean value for trypsinogen release in absence of lysine. (Reprinted, with permission, from Grendell and Rothman, 1981.)

13.9 CHYMODENIN AND NONPARALLEL TRANSPORT

In the early 1970s Adelson purified a peptide taken from a side fraction of the CCK purification from hog duodenum (Adelson, 1975). In light of the studies on the effect of lysine on enzyme secretion that I have just discussed, and more generally because of our interest in Pavlov's hypothesis concerning the regulation of digestion, Adelson and I set out to determine whether or not the duodenal peptide that he had purified had a selective effect on the secretion of one or another digestive enzyme. Could it be a duodenal hormone that produces the selective secretion of one or more enzymes? He had already shown the peptide to be active in terms of pancreatic function, producing a modest overall enhancement of protein secretion in situ and the release of amylase from isolated zymogen granules (Adelson, 1971).

After a preliminary assessment, we settled on two enzymes for further analysis, seemingly the least and most affected by administration of the peptide in situ, lipase and chymotrypsinogen respectively.

The results of the first series of experiments (Adelson and Rothman, 1974; Adelson and Rothman, 1975), carried out in rabbits in situ, were disappointing at first glance. The peptide did not appear to have an effect on the relative rates of secretion of the two enzymes, and the peptide-treated data points fell within the range of control values when individual measurements on the pair of enzymes were plotted one against the other (Figure 13). There was considerable variance in the proportions of the two enzymes from sample to sample in the control state, even though their relative rates of secretion were clearly and significantly correlated, and values for the peptide-treated samples were not distinguishable from control values comparing average outputs parametrically. However, a nonparametric rank sum test showed that peptide-treated points had a higher chymotrypsinogen output than control points. A closer look at the data revealed that the peptide-treated points were not randomly distributed among the broad range of control values but tended to form a border of the distribution along the chymotrypsinogen axis (triangles in Figure 13). We found that if we only considered samples of secretion for which lipase output was relatively low (below 6 units in Figure 13), that is, samples taken when there was not substantial endogenous stimulation of protein secretion, then chymotrypsinogen secretion in the presence of the peptide was almost three times that in the control group when values were averaged and analyzed parametrically (Figure 14). This difference is apparent when one looks at the distribution in Figure 13 with this in mind, and it can also be realized that this large effect would be lost if one merely averaged all of the data points and compared the two groups or if differences were sought in the ratio of the two enzymes (because of the large variance in the control state).

We repeated the study in vitro using the organ preparation of rabbit pancreas and observed essentially the identical effect. Once again the peptide-treated points formed a border along the chymotrypsinogen axis of a loosely correlated control distribution (Figure 15). However, in this case, it was possible to demonstrate changes in the average rate of chymotrypsinogen secretion in the absence of changes in lipase secretion, using a period by period analysis of output values. Relative to parallel controls, chymotrypsinogen secretion had been increased by some fourfold, whereas lipase

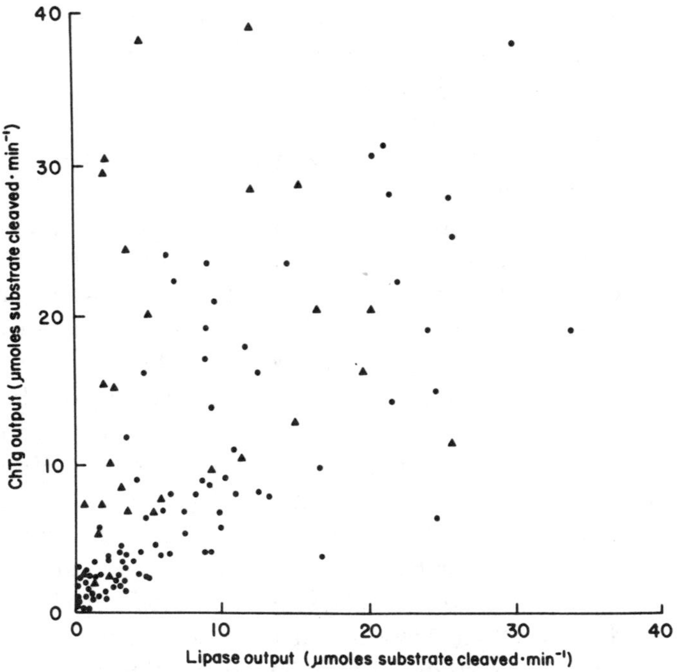

Figure 13 Total distribution of data points for individual collection periods displaying concomitant ChTg (chymotrypsinogen) and lipase outputs into the duct from rabbit pancreas in situ. Axes give micromoles substrate cleaved per minute per entire sample collected in 20 min. (▲) Chymodenin treated; (●) saline-treated controls: nine animals in each group. (Reprinted, with permission, from Adelson and Rothman, 1975.)

secretion was unaffected by the addition of the peptide (Figure 16). Thus, the peptide favored the secretion of chymotrypsinogen relative to lipase and for this reason was named chymodenin; chymo- for chymotrypsinogen, and -denin for its tissue of origin, the duodenum.

There was another aspect of the effect that was quite interesting, as well as unexpected. As I have pointed out, the control data displayed considerable variability in terms of the relative rates of secretion of the two enzymes, and the amount of one enzyme secreted was not a good indicator of the rate of secretion of the other. However, after the addition of the peptide, both in situ and in vitro, a much tighter correlation was found between the relative rates of secretion of lipase and chymotrypsinogen. In the in vitro study, for example, the control regression line had a correlation

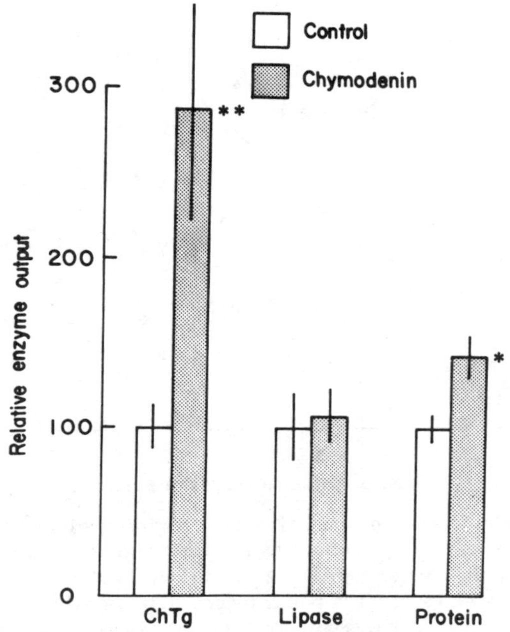

Figure 14 Effect of chymodenin on output of chymotrypsinogen (ChTg), lipase, and total protein into the duct from rabbit pancreas in situ. Mean output ± SE of mean are given. Data are normalized so that 100 equals mean output during three 20-min collection periods immediately prior to each injection of saline or chymodenin. Data are from nine control and nine chymodenin-injected animals in "basal" state. Dose of chymodenin was 10 μg/injection. Number of independent injections for saline control and chymodenin-treated were 9 and 16 for ChTg, 11 and 16 for lipase, and 10 and 15 for protein. (**) $p < 0.01$; (*) $p < 0.025$; p for ChTg was calculated using Cochran's modification of the Behrens-Fischer t test for populations of unequal variance. (Reprinted, with permission, from Adelson and Rothman, 1975.)

coefficient of 0.53; a significant but not a very tight correlation. However, the correlation coefficient increased dramatically to 0.96 merely as a result of adding the peptide to the medium (Figure 15). Thus, the variability that had been seen in the control situation had almost vanished as a result of adding the peptide. This meant two things: (1) that the variability seen in the control situation was of biological origin, and not the result of measurement or other experimental error; and (2) that the observed variability was itself variable. This was a new kind of nonparallel transport. The variability between the secretion of two enzymes was not fixed as the parallel transport hypothesis

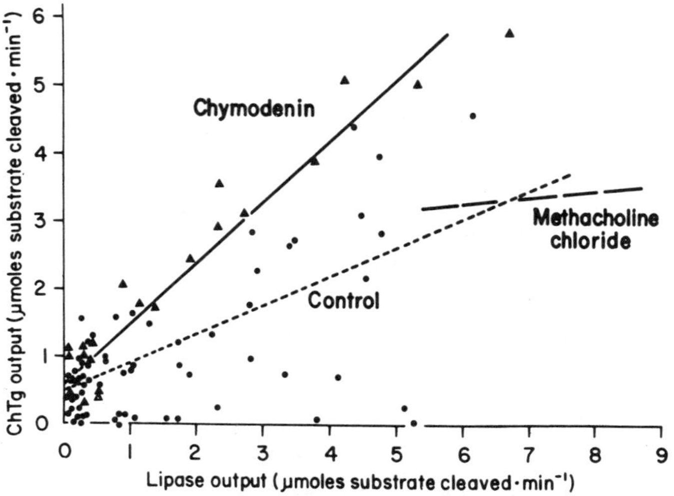

Figure 15 Data points for concomitant ChTg (chymotrypsinogen) and lipase outputs into the duct from rabbit pancreas in organ culture. (▲) Chymodenin treated; (●) control. Individual data points for methacholine treatment (1 mg/125 ml bath fluid) were beyond range of figure and are omitted as were two control points (which were used, however, in calculation of regression line). Goodness of fit was significant at $p < 0.05$ level for each regression line, and lines were significantly different from each other at $p < 0.001$ level. For chymodenin-treated pancreases, $\bar{b} = 0.898$ and $r = 0.96$; for control data, $\bar{b} = 0.395$, $r = 0.53$; for methacholine chloride, $\bar{b} = 0.111$ and $r = 0.53$. (Reprinted, with permission, from Adelson and Rothman, 1975.)

required, but rather the extent to which the amount of one enzyme secreted could predict the amount of another being secreted at the same time was itself variable.

Thus, we had found two additional cases of nonparallel transport in the effect of chymodenin on the relative rates of chymotrypsinogen and lipase secretion in vitro and in situ.* In addition, we found that the degree to which the secretion of different enzymes covary is also variable. There was an additional nonparallel transport effect observed en passant, if you will. When a cholinergic agonist was added to the medium bathing the pancreas in organ culture, there was a nonparallel response in which lipase secretion was favored relative to chymotrypsinogen (Figure 17), unlike the effect of chymodenin.

*Adelson and Miller recently reproduced these observations and reported additional examples of chymodenin-induced non-parallel transport (Adelson and Miller, Science, 1985).

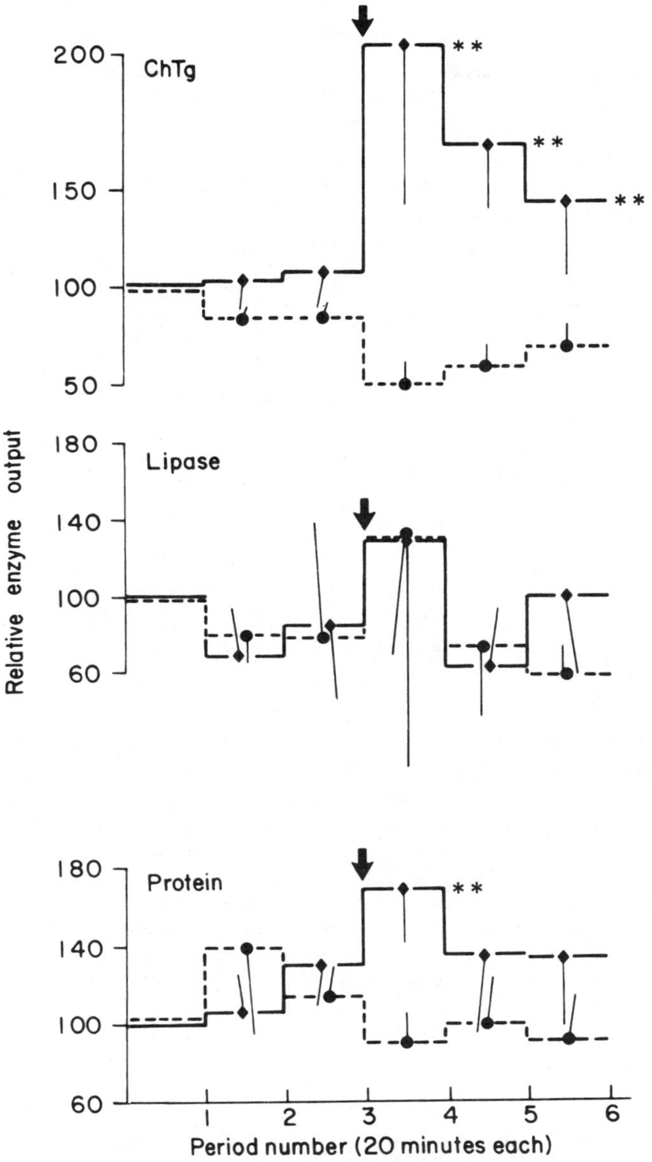

Figure 16 Effect of chymodenin on ChTg (chymotrypsinogen), lipase, and protein output into the duct from rabbit pancreas in organ culture. Data were normalized so that 100 equals output of each substance in first of three 20-min pretreatment periods. Error bars are canted from vertical for clarity. Time of treatment is indicated by arrow. (———) Chymodenin treatment; (----) untreated control; (**) p less than 0.01. Eight organs were treated with chymodenin, 11 served as controls.

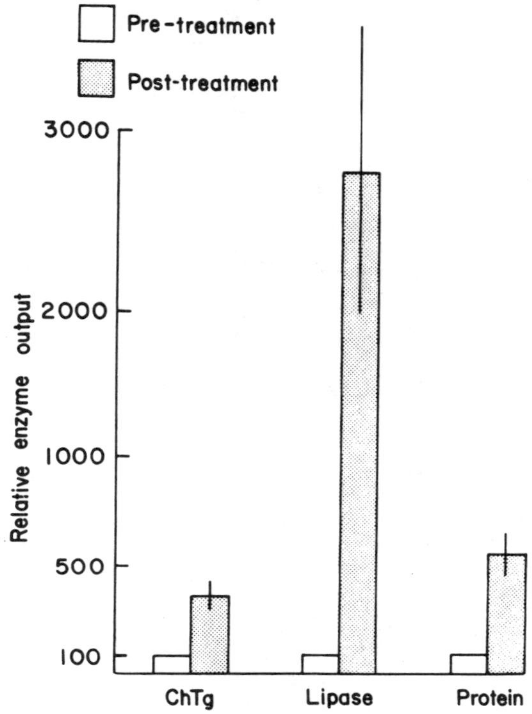

Figure 17 Effect of methacholine chloride on pancreatic secretion of ChTg (chymotrypsinogen), lipase, and protein into the cannulated duct of the gland in culture. Pancreases had been pretreated in previous hour with chymodenin as described in Figure 16. Data were paired and normalized as follows: for each pancreas, output of each substance in 20-min collection period immediately prior to treatment was determined and normalized to a value of 100. Output of same substance was then followed for two collection periods (40 min) after methacholine chloride treatment; mean of these two outputs was compared to pretreatment period for same gland in a paired *t* test. Nine pancreases were used. All differences were significant at $p < 0.01$ level comparing controls and posttreatment outputs. Increase (nonparallel) in lipase secretion was significantly greater than that in protein at $p < 0.05$, and in ChTg at $p < 0.01$. (Reprinted, with permission, from Adelson and Rothman, 1975.)

13.10 THE INDEPENDENT SECRETION OF DIGESTIVE ENZYMES

As a result of these observations, I decided to reexamine an older series of experiments in which glucose had been perfused through the duodenum of anesthetized rabbits and the secretion of trypsinogen, chymotrypsinogen, and amylase followed (Rothman, 1976). Glucose perfusion had had no effect on the overall or average rate of secretion of any of the enzymes (Figure 18), an apparently negative study. However, it was possible that the *relative* rates of secretion of the different proteins might have been altered. When a regression analysis was performed such relative changes could be demonstrated for each of the three enzyme pairs (Figures 19 and 20; Table 3). In each case the calculated slope had been altered by about two to three times as a result of glucose perfusion. Therefore, we had three more cases of nonparallel transport, this time involving three enzymes in one situation, the secretion of each varying relative to the others.

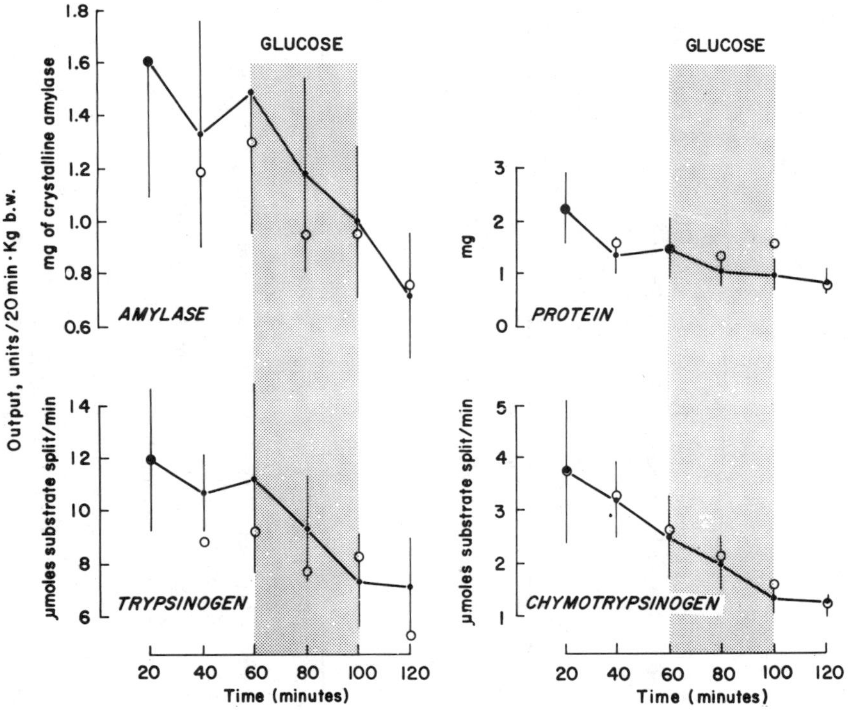

Figure 18 Secretion of amylase, trypsinogen, chymotrypsinogen, and total protein into the cannulated duct of the rabbit pancreas in situ as a function of time. Duodenum was perfused (as in Figure 8) for all six 20-min periods shown, both in the absence of glucose (•) ($n = 9$, values are means ± SE) and in its presence (○) ($n = 6$). When 5 m*M* glucose was added to perfusion medium, it was added at 60 min (period 3) and removed after 5th collection period (100 min). For the + glucose series (○) only mean values are presented. None of the + glucose points (○) for different times or measures was significantly different from time-paired controls (•). (Reprinted, with permission, from Rothman, 1976.)

TABLE 3. Effect of Duodenal Perfusion of 5 m*M* Glucose on Proportions of Three Enzymes in Secretion as Determined by Linear Regression of Enzyme Pairs[a]

	Slope, *b*					Intercept, *a*	
	−Glucose	*r*	+Glucose	*r*	Δ%*b*	−Glucose	+Glucose
Trypsinogen/ chymotrypsinogen	1.8 ±0.3[b]	0.89	5.6 ±1.2[c]	0.83	208%	5.1 ±1.3[d]	0.3 ±1.7[b]
Chymotrypsinogen/ amylase	2.0 ±0.4	0.86	0.9 ±0.1	0.96	117%	0.6 ±0.7	0.6 ±0.1[d]
Trypsinogen/ amylase	3.4 ±1.0	0.73	6.5 ±1.0	0.91	91%	6.3 ±1.8[d]	6.5 ±1.0[d]

[a]Number is 12 for each function. Goodness of fit, $p < 0.01$ for all functions; *r*, correlation coefficient; −glucose is data from periods 1 (20 min) and 2 (40 min) in Figure 18; +glucose is data from periods 5 and 6 in Figure 18; no changes in *b* or *a* were seen in time-paired controls. Taken from Rothman (1976).

[b]Standard error of the estimate.

[c]Difference (−vs. +glucose) in all slopes significant at $p < 0.001$.

[d]Different from zero intercept, significant at $p < 0.005$.

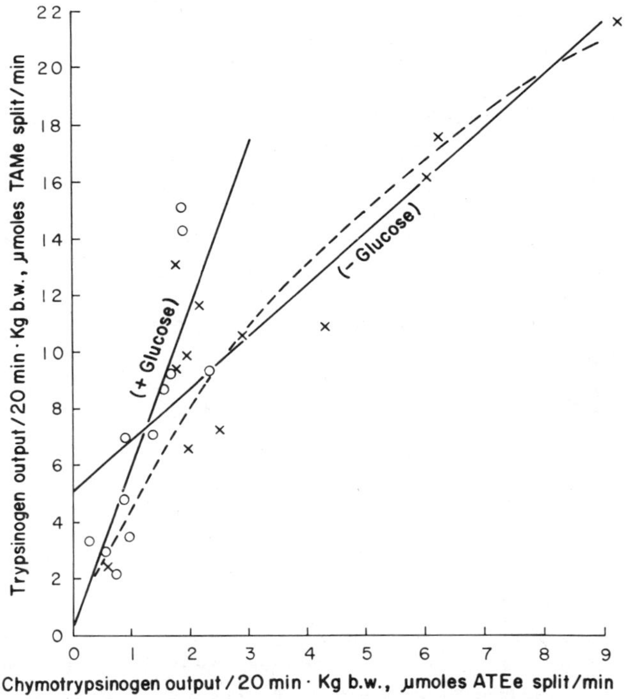

Figure 19 Plot of secretion of enzyme pair chymotrypsinogen-trypsinogen into the cannulated duct of the rabbit pancreas in situ before glucose was added to perfusion medium (X) (periods 1 and 2 in Figure 18) and after (O) (periods 5 and 6). Solid lines are calculated by linear regression analysis for both groups. Curved (----) line is drawn to show that the − glucose data fit a nonlinear function about as well as they fit a linear function (see text for discussion). (Reprinted, with permission, from Rothman, 1976.)

Yet another type of nonparallel transport became evident from these experiments. In several cases the calculated regression lines did not go through the origin but intercepted the axes at points significantly different from 0,0. This could mean one of two things. The secretion of one enzyme could occur in the complete absence of the secretion of the other. This would, of course, be evidence that the secretion of different enzymes occurred independently and was not linked obligatorily in some fashion, as in the exocytosis of granules of mixed enzyme content. Alternatively, the function describing their relative rates of secretion might be nonlinear and the fact that the line did not regress through the origin might therefore simply reflect the presence of a curvilinear function. In this case, the relative rates of secretion would vary continuously (be continuously nonparallel) as

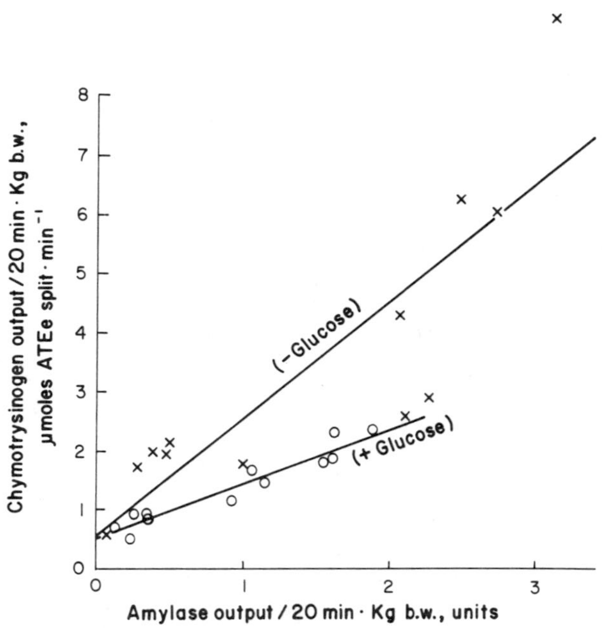

Figure 20 Plot of secretion of the enzyme pair amylase-chymotrypsinogen into the cannulated duct of the rabbit pancreas in situ before glucose was added to the perfusion medium (×) (periods 1 and 2 in Figure 18) and after (○) (periods 5 and 6). Lines are calculated linear regressions. The + glucose intercept is significantly different from origin (see Table 3). (Reprinted, with permission, from Rothman, 1976.)

a function of the overall rate of secretion. Such behavior had been documented in other cases, one of which I discussed above, and others of which I will consider below. In this case (that is, for the control trypsinogen-chymotrypsinogen function in Figure 19), it was not possible to distinguish between the two possibilities. Of course, both effects might occur, nonlinear functions and positive intercepts, either separately or in the same function (see Figure 6). Although it could not be determined whether the control function in Figure 19 was either a straight line or a curve, for the "+ glucose" line in Figure 20, in which case the data points approach the axis closely, a true positive y-axis intercept seems clear, indicating that the secretion of chymotrypsinogen can occur independently of amylase.

In addition, it was found that glucose perfusion could alter the apparent intercept as well as the variance around it. Glucose perfusion eliminated the calculated y-axis intercept for the trypsinogen-chymotrypsinogen pair, and the data now regressed through the origin (Figure 19). For amylase-chymotrypsinogen, although glucose perfusion did not alter the numerical value of the y-axis intercept, it reduced the variance around the observed intercept so that it became statistically distinguishable from the origin, unlike the control state (Figure 20, Table 3). This reflects the fact that glucose perfusion had increased the covariance of the secretion of the two enzymes, another example of a treatment that further entrains the secretion of two enzymes.

But perhaps the most interesting effect seen in this study was related to changes in the degree of covariance in the secretion of *three enzymes.* If the extent of covariance in the secretion of different enzymes was due to variability in the secretion of enzyme from granule pools of different mixed enzyme content, then changes in the variability of selection among these different granules, either increases or decreases, would occur in parallel for all enzymes. Only if the proteins were secreted independently of each other would it be possible, for example, for there to be a change in the covariance of chymotrypsinogen secretion relative to trypsinogen in one direction (more or less variable), and a change for chymotrypsinogen secretion relative to amylase in the opposite direction. This is exactly what was observed. For the first two periods after the initiation of glucose perfusion, there was a large decrease in the entrainment of the secretion of trypsinogen and chymotrypsinogen, that is, they were secreted more independently of each other (Figure 21). At the same time entrainment in the secretion of chymotrypsinogen and amylase became much greater (Figure 21). In the

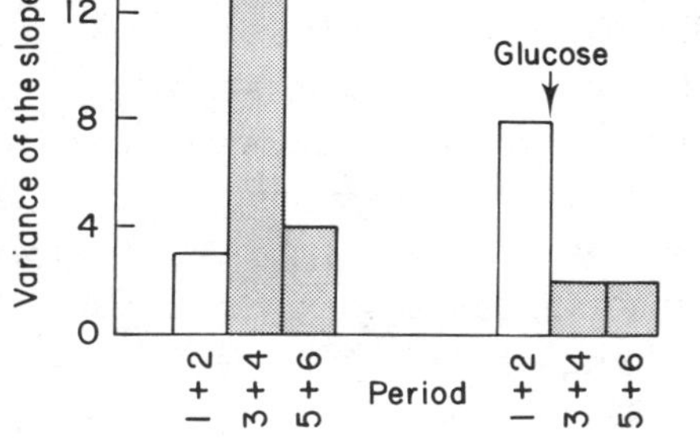

Figure 21 Variance around slope of two enzyme-pair plots, chymotrypsinogen-trypsinogen (Tg/ChTg) and amylase-chymotrypsinogen (ChTg/amylase), before (clear bars) and after (shaded bars) addition of glucose to perfusion medium. Same experimental design as Figures 18, 19, and 20. Treatment altered variance around calculated slope in both cases, increasing it transiently for Tg/ChTg (F test, $p < 0.05$) and decreasing it for ChTg/amylase ($p < 0.05$). (Reprinted, with permission, from Rothman, 1976.)

former case, the variance increased some fivefold, whereas in the latter it was decreased by about 75%. These observations presented compelling evidence for the hypothesis that the proteins were secreted independently of each other, whatever the mechanism of secretion might be.

13.11 INCREASE, DECREASE, AND NO CHANGE

Several years ago Tseng, Grendell, and I performed a series of experiments (Tseng et al., 1982) in which we attempted to repeat earlier work by a Swiss group (Dick and Felber, 1975; Felber et al., 1974) who had demonstrated that extracts of rat duodenum, taken from animals previously fed either a glucose meal or a protein hydrolysate "lactalbumin", injected directly into the intrapancreatic circulation via the celiac artery led to the enzyme-specific secretion of amylase or trypsinogen, respectively. They proposed the existence of enzyme-specific hormones in the duodenal mucosa, a class of "chymodenins," if you will.

We were unsuccessful in reproducing exactly what they observed, but were able to find a similar, although less dramatic, phenomenon that was consistent with their observation and whose characteristics were enlightening. Unlike the Swiss group's report, in our hands the injection of extracts from glucose-fed rats did not increase the secretion of amylase when period-by-period averages were compared. And although protein hydrolysate-treated extracts did increase trypsinogen secretion by about twofold, it produced a roughly parallel increase in the average rate of amylase secretion as well. We pursued the matter further by regression analysis. When the secretion of amylase was plotted against that of trypsinogen for individual samples, we found that both treatments had altered the regression lines. The function for extracts taken from fasted rats (controls) had a substantial positive y-axis intercept (trypsinogen secretion in the absence of amylase) (Figure 22). The regression line calculated for data from rats injected with the glucose-exposed extract was altered in two ways. The slope of the function was about doubled, favoring trypsinogen secretion, and the intercept became indistinguishable from one passing through the origin. The effect of the extract from protein hydrolysate-fed rats was similar in that it also both lowered the y-axis intercept toward the origin and increased the slope of the function, although it did both to a lesser degree.

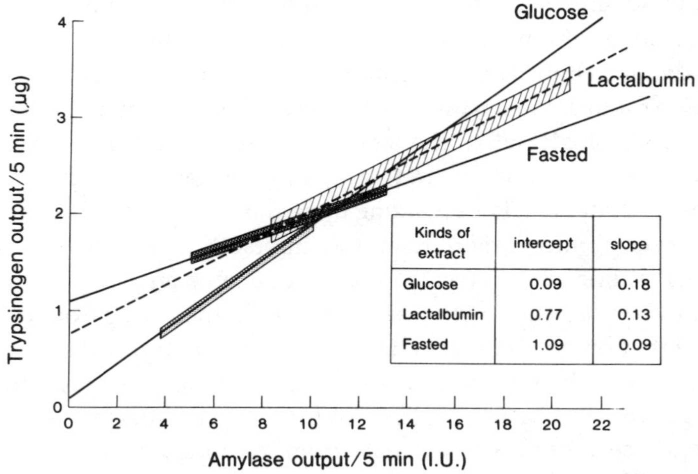

Figure 22 Effect of intraceliac injection of duodenal extracts from fasted, lactalbumin-fed, and glucose-fed rats on relative proportions of amylase and trypsinogen found in pancreatic secretion collected from the cannulated duct of anesthetized rats. Regression lines were calculated from scatter of individual pairs of enzyme measurements for all data points from time of injection (20 min) until end of the study (70 min) (see Figure 23). Glucose: $n = 60$, $r = 0.54$, $p < 0.01$; lactalbumin: $n = 60$, $r = 0.42$, $p < 0.01$; fasted: $n = 49$, $r = 0.42$, $p < 0.01$. Calculated slopes and intercepts are given in inset. For each line, shaded box represents range of amylase output for condition and SE of estimate of trypsinogen output for same group. Effect of extracts from glucose-fed rats was to produce a more amylase-dominant fluid as compared with fasted control, whereas extracts from lactalbumin-fed rats increased trypsinogen content relative to amylase as compared with fasted group (see text for further discussion). Slopes and intercepts of each function were significantly different from each other at $p < 0.001$. (Reprinted, with permission, from Tseng et al., 1982.)

Thus, although a change was observed, these results once again did not appear to agree with those of the Swiss group since both treatments seemed to have produced the same effect. However, on further inspection the functions were different in important ways. The shaded areas in Figure 22 show the range of amylase output or secretion values along one axis and the standard error of the estimate of trypsinogen output along the other. Looking at these polygons one can see that the amylase values for the two treated groups showed very little overlap. The protein hydrolysate-treated group lay outside or beyond the point of intersection between this function and the control (fasted) function, whereas the opposite was the case for the glucose-treated group, in which case almost all of the points lay closer to the origin than the point of intersection. This was very interesting. It meant that the two treatments had produced very different effects even though they had both altered the parameters of the function in the same direction.

As I said, the control and experimental functions intersected each other. The reason for this is that treatment had altered both slope and intercept. In such a case, a single set of changes in the parameters of a function can produce different effects depending upon the location of the data points along the function; that is, depending upon the overall rate of secretion. For points closer to the origin than the intersection of the control and experimental lines, the intercept effect dominates and, in our case, that favors amylase secretion; whereas for points further away from the origin than the intersection, the slope effect dominates and trypsinogen secretion would be favored. Interestingly, the data from the protein hydrolysate-treated group lay primarily outside of the point of intersection, favoring trypsinogen secretion, whereas data from the glucose-treated group lay closer to the origin, favoring amylase secretion. Thus, in the end the data did support the observations of the Swiss group.

But beyond this they also demonstrated a curious aspect of the phenomenon of nonparallel transport that is important to appreciate. If a treatment changes a function so that the new function intersects the old, this single change in a function's parameters can lead to an increase in the proportion of one enzyme in secretion, relative to another, for values outside the point of intersection, produce no change at all if the data points fall in the area of the intersection between the two lines, or produce a decrease for values closer to the origin than the intersection. Therefore, a single treatment effect can alter the relative proportions of the two enzymes in secretion in one direction, in the opposite direction, or not at all, depending upon how active secretion is overall. (N.B.: A somewhat similar effect to that seen in Figure 6 due to the nature of a curve.)

13.12 THE POWER FUNCTION

In the same study (Tseng et al., 1982), CCK was injected into anesthetized rats and the secretory response for amylase and trypsinogen measured. A dramatic nonparallel effect was produced in which the ratio of amylase to trypsinogen increased by about three times (Figure 23). (N.B.: The non-parallel effect was established within 5 min.) A similar nonparallel effect had been reported earlier by others for a different enzyme pair in response to

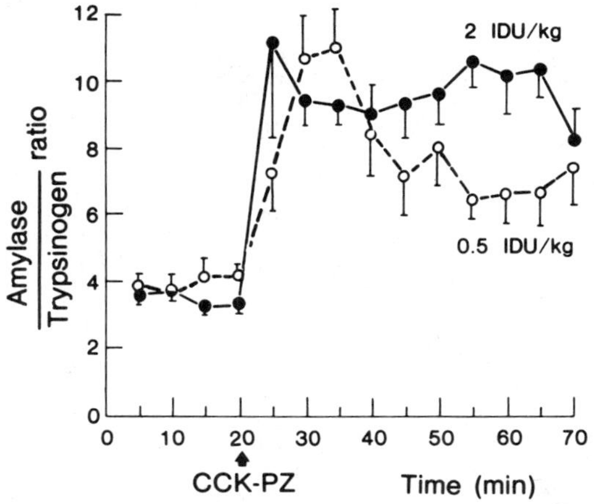

Figure 23 Ratio of amylase to trypsinogen in pancreatic secretion collected from the cannulated duct of anesthetized rats before and after intraceliac injection of cholecystokinin-pancreozymin [CCK–PZ, 2.0 Ivy dog units (IDU)/kg body wt (—●—) or 0.5 IDU/kg body wt (---○---)] at time indicated by arrow. Points, means; error bars, SE; $n = 6$ for all points. After injection of CCK–PZ, all ratios were significantly different from those seen during last preinjection period (20 min) with $p < 0.05 - 0.001$. (Reprinted, with permission, from Tseng et al., 1982.)

CCK in the rat (Dagorn, 1978). When the individual data points were plotted, amylase versus trypsinogen, it became clear that the function relating the secretion of the two enzymes was not linear, but became more and more amylase dominant as overall enzyme secretion became greater (Figure 24). That is, the proportions of the two enzymes varied continuously as a function of the overall output of enzyme in what was presumably the same state (after the injection of CCK).

This was not the first time that we had seen such a relationship. We had reported the existence of a nonlinear relationship between the secretion of two enzymes in 1974, for the effect of cholinergic stimulation on the secretion of trypsinogen and chymotrypsinogen by the rabbit pancreas in situ, as well as for the original nonparallel response to CCK, which also displayed the same type of nonlinear relationship (see Figure 6). The former (cholinergic) function became more chymotrypsinogen dominant as overall output became greater (the ratio of trypsinogen to chymotrypsinogen being 2.5 ± 0.3 SE for the highest one-third of observed values, and 5.7 ± 0.7 for the lowest third), whereas the latter (CCK) function became more trypsinogen dominant as overall output became greater (Figure 6).

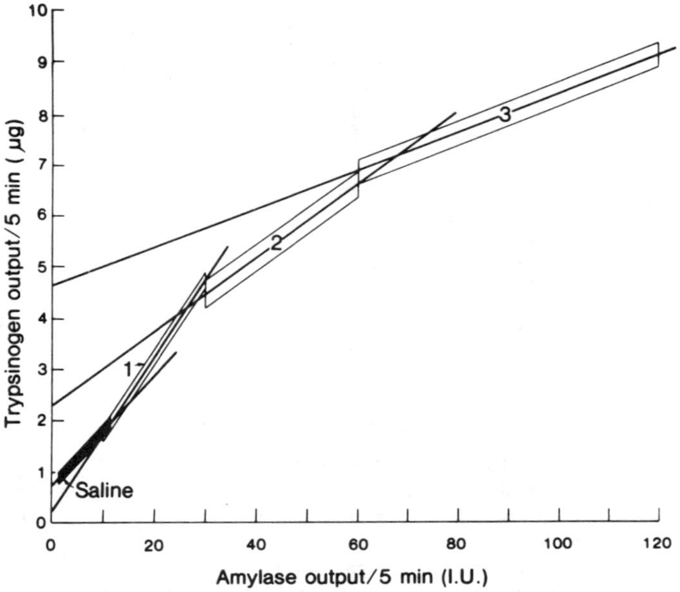

Figure 24 Effect of intraceliac injection of cholecystokinin-pancreozymin [CCK–PZ, 0.5 or 2.0 Ivy dog units (IDU)/kg body wt] on relative proportions of amylase and trypsinogen in pancreatic secretion collected from the cannulated duct of anesthetized rats as compared with values for saline-injected controls. Regression lines were calculated from scatter of individual pairs of enzyme measurements for all data points from time of injection (20 min) until end of study (70 min) (see Figure 23). Data were analyzed over three ranges of amylase output for CCK–PZ-treated group (segment 1: 0–30 IU/5 min; segment 2: 31–60 IU/5 min; and segment 3: 61–120 IU/5 min). For each linear segment, box represents range of amylase output and SE of estimate for trypsinogen output over that range (segment 1: $n = 35$, $r = 0.68$, $p < 0.01$; segment 2: $n = 44$, $r = 0.32$, $p < 0.05$; segment 3: $n = 37$, $r = 0.51$, $p < 0.01$; and saline: $n = 84$, $r = 0.59$, $p < 0.01$). Regressions for segments 1, 2, and 3 are significantly different from each other in terms of both intercept and slope at $p < 0.01$. Segment 1 and saline also are significantly different (slope at $p < 0.02$ and intercept at $p < 0.001$). (Reprinted, with permission, from Tseng et al., 1982.)

Before we move on, one additional point about the effect of CCK on amylase and trypsinogen secretion by the rat pancreas is worth noting. As I have pointed out, if control and experimental lines intersect, then a single treatment may favor the secretion of one enzyme closer to the origin than the point of intersection of the two lines, may favor neither in the region of the intersection, and may favor the other enzyme beyond it. Thus, in this case there are three different potential effects of one treatment, depending upon the overall level of secretory activity. The situation becomes even more complex if the function is a curve as in Figure 6 or 24. In this case, as the data points are farther and farther removed from the origin beyond the point of intersection of the two functions, the ratio of enzymes again begins to approximate more and more closely the control function as the curve falls off. Indeed, the slope may become less than that of the control function. This effect is produced by CCK in the rat (Figure 24). Compare the line segment in Figure 24 labeled "saline" (control) to the various parts of the CCK curve. At the intersection, CCK produces no change at all. Beyond the intersection of control and CCK functions trypsinogen secretion is favored (increased slope) and then as the function falls off at higher rates, amylase secretion becomes increasingly favored so that control ratios may again be observed or even reversed toward a more amylase-dominant slope. Thus, moving out from the origin, one treatment, CCK, produces five distinct, continuously varying effects on the ratio of the two enzymes relative to the control, depending upon the overall rate of secretion.

Although a good fit to a linear function is frequently obtained for plots of enzyme pairs close to the origin, it has become increasingly clear that a single general nonlinear function may describe the relationship between the rates of secretion of any enzyme pair for all rates of secretion. The best fit has thus far invariably been to a simple power function of the form, $y = ax^b$. This may be significant because this simple relationship connecting the rates of secretion of any two enzymes is consistent with a model for secretion in which the relative rates of secretion of the individual enzymes are independently determined as a function of concentration gradient and permeability; the parameters of the function merely reflect the different kinetic parameters at work for the different enzymes.

In two recent studies, Tseng, Grendell, and I (Tseng et al., 1984; Grendell et al., 1984) have found numerous examples of nonparallel transport in which the parameters of such power functions are altered by various physiological stimuli. For example, in the presence of CCK the injection of an effective dose of either glucose or lysine into the celiac artery of the rat leads to the selective secretion of trypsinogen or amylase, respectively (Figure 25). When these data are plotted and the rates of secretion of the two enzymes compared, the CCK curve is shifted toward the amylase axis as a result of glucose injection and toward the trypsinogen axis as a result of lysine injection (Figure 26).

In the other series of experiments, secretion of the same two enzymes was followed in the presence of either insulin or glucagon, along with CCK.

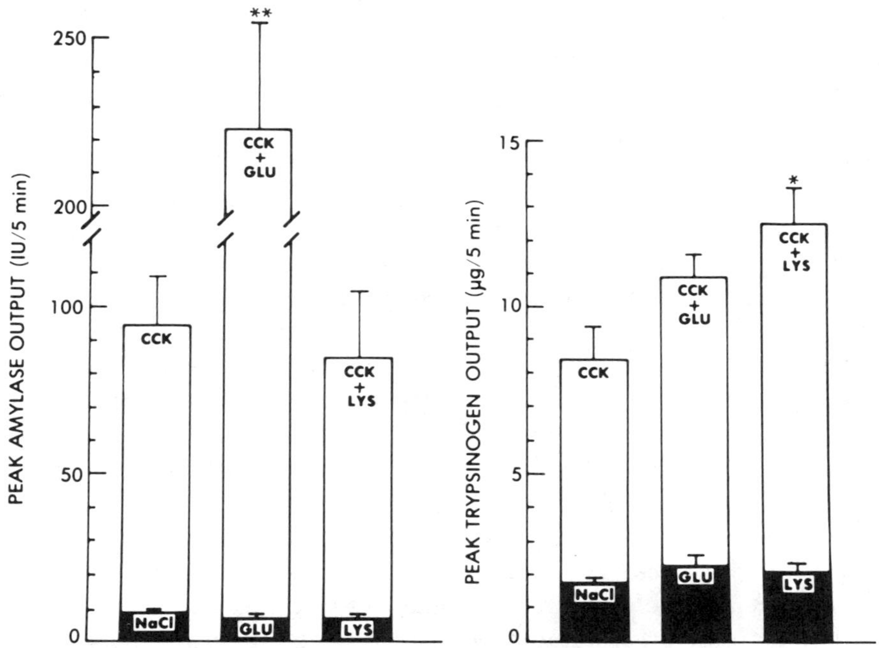

Figure 25 Peak amylase and trypsinogen output from pancreaticobiliary duct of anesthetized rats following injection into the celiac artery of 250 μl of 0.9% NaCl (NaCl), NaCl containing 15 m*M* D-glucose (Glu), NaCl containing 0.5 m*M* L-lysine (Lys), or NaCl containing 0.5 IDU/kg body wt cholecystokinin (CCK). CCK + Glu or CCK + Lys indicate combinations in NaCl at doses specified. Bars, means; error bars, SE. For NaCl, Glu, and Lys, $n = 6$; for CCK, $n = 9$; for CCK + Glu and CCK + Lys, $n = 10$. Values for enzyme output following CCK + Glu or CCK + Lys are significantly different from those observed after CCK alone as indicated by asterisks: (*) $p < 0.01$; (**) $p < 0.005$. (Reprinted, with permission, from Grendell et al., 1984.)

Both hormones increased the rate of trypsinogen secretion relative to that of amylase (Figures 27 and 28). However, they did so in opposite ways. Insulin increased trypsinogen output, whereas glucagon decreased amylase secretion (Figure 29). When the data points were plotted, the curves were again shifted; this time both treatments produced more trypsinogen dominant functions in a dose-dependent manner (Figures 30 and 31). However, points along the insulin curve for the most effective dose tended to lie further from the origin than data points from the glucagon experiments, which tended to lie closer to the origin (Figure 32). This produced the two different treatment responses, insulin increasing the amount of trypsinogen

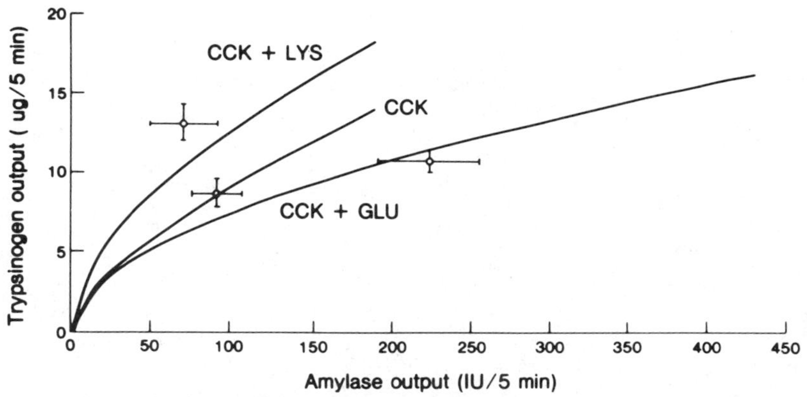

Figure 26 Regression lines calculated from scatter of individual pairs of enzyme measurements for all data points over a 50-min period following injection of CCK, CCK + Glu, or CCK + Lys as described in Figure 25. Each regression line is extended to upper limit of range of values observed for amylase output. In each case, best fit was using an equation of form $y = ax^b$: for CCK, $y = 0.44x^{0.66}$, $n = 87$, $r = 0.87$, $p < 0.01$; for CCK + Glu, $y = 0.66x^{0.53}$, $n = 97$, $r = 0.91$, $p < 0.01$; and for CCK + Lys, $y = 0.97x^{0.53}$, $n = 97$, $r = 0.70$, $p < 0.01$. Differences among regressions were assessed by log-log transformation to approximate linear functions. Each regression differed from others at $p < 0.001$. Open circles and error bars represent average peak output (mean ±SE) for condition indicated by adjacent regression line (see Figure 25). (Reprinted, with permission, from Grendell et al., 1984.)

secreted and glucagon decreasing amylase secretion. In addition, as the dose of either insulin or glucagon was increased beyond that most effective, the effect was diminished and the values and functions returned to or toward the control (Figures 30 and 31). In these experiments each and every separate curve represents two types of nonparallel transport: the first, in its own right merely by dint of being a curve, and the second, relative to the parameters of the control (CCK alone) function.

I would like to end this part of my discussion of nonparallel transport with a final example taken from recent work by Niederau, Grendell, and myself (Niederau et al., 1985). In this work, nonparallel transport was demonstrated pursuant to the inhibition of secretion, in particular, inhibition of CCK-stimulated secretion by the rabbit pancreas in organ culture by the glutaramic acid analog, proglumide. These experiments demonstrate two interesting effects. In the first place, inhibition of secretion, as well as stimulation, can occur in a nonparallel fashion. In this particular case, chymotrypsinogen secretion was preferentially reduced relative to amylase in a dose-dependent manner (Figure 33). In addition, the parameters of the curves relating the two variables were not only altered by treatment, as with

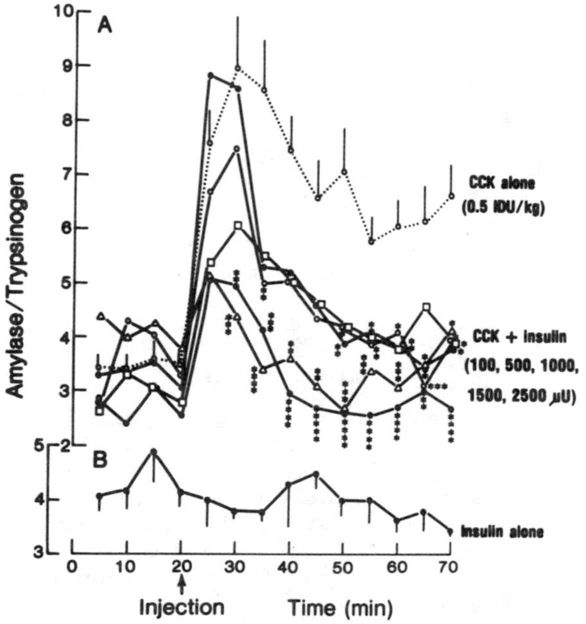

Figure 27 (A) Ratio of amylase to trypsinogen in pancreatic secretion collected from the cannulated duct of anesthetized rats prior to and after intraceliac injection (arrow) of CCK (0.5 IDU/kg body wt) with insulin (100 μU, ○; 500 μU, ◖; 1000 μU, △; 1500 μU, □; or 2500 μU, ●) or of CCK (0.5 IDU/kg body wt) alone (--- ○ ---). (B) Effect of injection of 500 μU of insulin alone. Points, means; error bars, SE. Because of overlap, error bars are omitted from all points for CCK + insulin groups before and after injection. For CCK alone, $n = 15$; $n = 6$ for all insulin doses except $n = 5$ for 1500 μU; $n = 6$ for insulin alone. Asterisks indicate significant differences between CCK alone and CCK + insulin groups for same time periods: (*) $p < 0.05$; (**) $p < 0.025$; (***) $p < 0.01$; (****) $p < 0.001$. (Reprinted, with permission, from Tseng et al., 1984.)

insulin or glucagon, but altered so that as the inhibition became more profound, the shape of the curve changed from one that increasingly favored chymotrypsinogen secretion at higher overall output, to one that favored amylase secretion (Figure 33). Thus, this nonparallel transport not only involved a change in the slope and pitch of a particular curve but a change in its direction as well. In mathematical terms this is simple enough; the exponent is merely changed from one above to one below unity. To explain such a change in terms of a vesicle mechanism requires complex models. On the other hand, in an equilibrium-based transport process, this

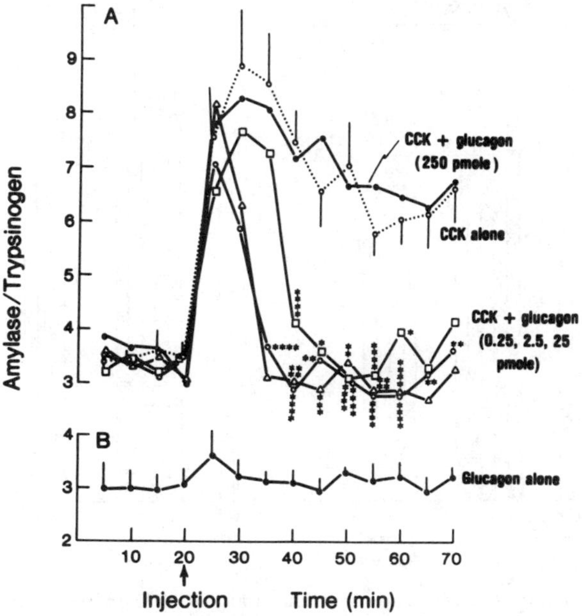

Figure 28 (A) Ratio of amylase to trypsinogen in pancreatic secretion collected from the cannulated duct of anesthetized rats prior to and after the intraceliac injection (arrow) of CCK (0.5 IDU/kg body wt) with glucagon (0.25 pmol, —□—; 2.5 pmol, —○—; 25 pmol, —△— or 250 pmol, —●—) and CCK (0.5 IDU/kg body wt) alone (---○---). (B) Effect of injection of 2.5 pmole of glucagon alone. Points, means, error bars, SE. Because of overlap, error bars are omitted from glucagon with CCK groups. Asterisks indicate significant differences between CCK alone and CCK with glucagon for same time period: (*) $p < 0.05$; (**) $p < 0.025$; (***) $p < 0.01$; (**/**) $p < 0.001$. (Reprinted, with permission, from Tseng et al., 1984.)

effect would simply reflect a change in the kinetic parameters for the independent transport of the two proteins.

13.13 EXPLANATIONS FOR NONPARALLEL TRANSPORT WITHIN THE BORDERS OF THE PARADIGM

Two approaches have been taken to deal with observations of nonparallel transport from a paradigmatic perspective. The first is that they are merely experimental artifacts, and that accurate measurements of secretion would

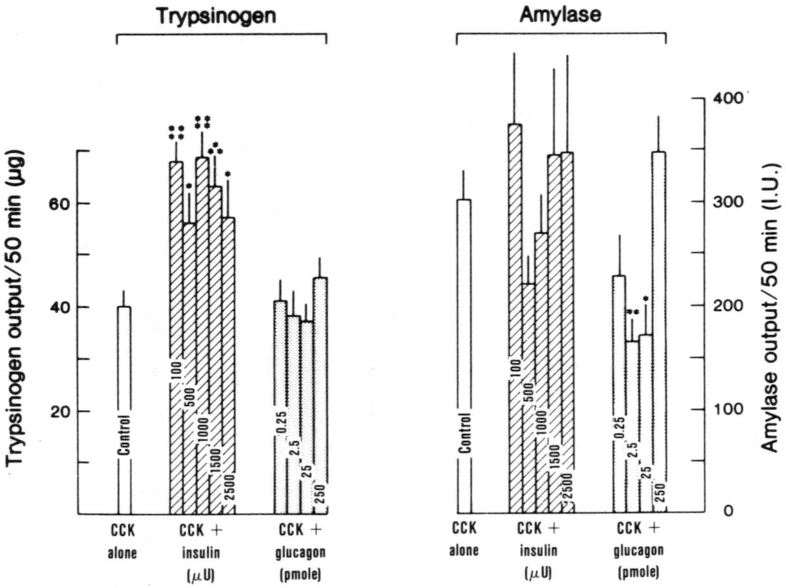

Figure 29 Effect of intraceliac injection of insulin (100, 500, 1000, 1500, or 2500 μU) with CCK (0.5 IDU/kg body wt), of glucagon (0.25, 2.5, 25, or 250 pmol) with CCK (0.5 IDU/kg body wt), or of CCK (0.5 IDU/kg body wt) alone (control) on total output of amylase and trypsinogen in pancreatic secretion collected from the cannulated duct of anesthetized rats from time of injection (20 min) until end of the experiment (70 min) (see Figures 27 and 28). Columns, means; error bars, SE. CCK alone, $n = 15$; $n = 6$ for each insulin dose except 1500 μU where $n = 5$; $n = 6$ for all doses of glucagon. Asterisks indicate a significant difference between CCK alone and CCK with either insulin or glucagon: (*) $p < 0.05$; (**) $p < 0.025$; (***) $p < 0.01$; (****) $p < 0.001$. (Reprinted, with permission, from Tseng et al., 1984.)

show it always to be parallel. Such claims of artifact are common in science and particularly so when an established idea is being questioned. However, beyond the claim, more than 15 years after our first report of nonparallel transport, during which time there have been many and varied reports of this phenomenon, there has not been a single example of a nonparallel transport observation that has been documented as being due to a particular artifactual cause. At this point it is hard to imagine that the range of observations that have been made, such as the series that I have just discussed, could possibly all merely reflect assay artifacts of one type or another. I have discussed some of these issues elsewhere in the past and refer you to this article (Rothman, 1977).

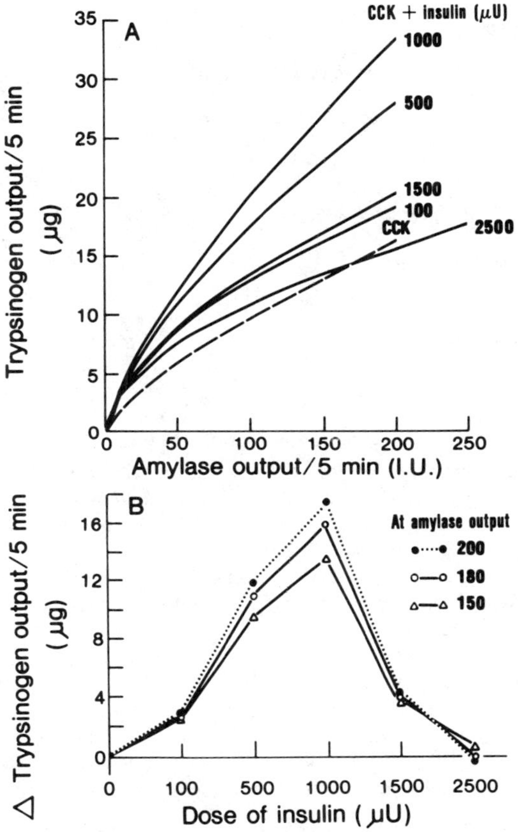

Figure 30 (A) Effect of intraceliac injection of insulin (100, 500, 1000, 1500, or 2500 μU) with CCK (0.5 IDU/kg body wt) (——) or CCK (0.5 IDU/kg body wt) alone (----) on relative proportions of amylase and trypsinogen found in pancreatic secretion collected from the cannulated duct of anesthetized rats. Regression lines were calculated from scatter of individual pairs of enzyme measurements for all data points from time of injection (20 min) until end of study (70 min) (see Figure 27). CCK plus insulin (100 μU): $n = 57, r = 0.88, p < 0.01$; CCK + insulin (500 μU): $n = 56, r = 0.89, p < 0.01$; CCK + insulin (1000 μU): $n = 60, r = 0.90, p < 0.01$; CCK + insulin (1500 μU): $n = 46, r = 0.89, p < 0.01$; CCK + insulin (2500 μU): $n = 56, r = 0.77, p < 0.01$; CCK alone: $n = 138, r = 0.88, p < 0.01$. All functions are of form that best fit data, $y = ax^b$. Each insulin function was significantly different from that for CCK alone at $p < 0.001$ as estimated by analysis of log-log transformations. (B) Mean trypsinogen output following intraceliac injection of CCK (0.5 IDU/kg body wt) plus insulin versus the dose of insulin. Values are derived from curves in Figure 30A at specified amylase output levels. Three functions are shown (at 200 IU amylase/5 min, --- ●---; at 180 IU/5 min, — ○ —; at 150 IU/5 min, — △—). △ Trypsinogen output, trypsinogen output produced by CCK with insulin minus trypsinogen output produced by CCK alone. (Reprinted, with permission, from Tseng et al., 1984.)

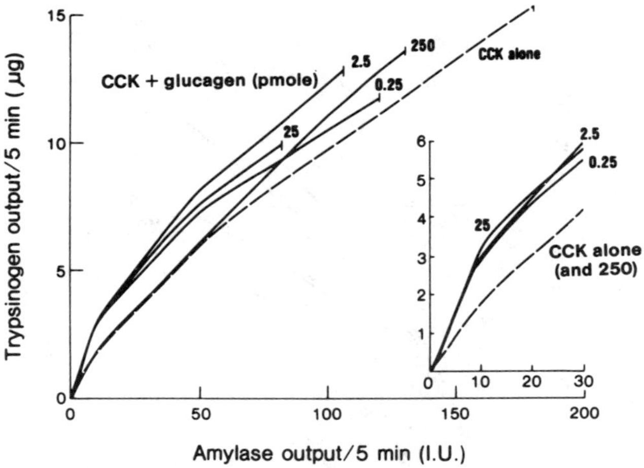

Figure 31 Effect of intraceliac injection of glucagon (0.25, 2.5, 25, or 250 pmol) with CCK (0.5 IDU/kg body wt) (——) or of CCK (0.5 IDU/kg body wt) alone (----) on relative outputs of amylase and trypsinogen found in pancreatic secretion collected from the cannulated duct of the anesthetized rat. Regression lines were calculated from scatter of individual pairs of enzyme measurements for all data points from time to injection (20 min) until end of study (70 min) (see Figure 28). Each regression line is extended to upper limit of range of values observed for amylase output. CCK + glucagon (0.25 pmol): $n = 58, r = 0.86, p < 0.01$; CCK + glucagon (2.5 pmol): $n = 56, r = 0.84, p < 0.01$; CCK + glucagon (25 pmol): $n = 56, r = 0.84, p < 0.01$; CCK + glucagon (250 pmol): $n = 60, r = 0.92, p < 0.01$. All functions are of form that best fit data, $y = a x^b$. Each glucagon function was significantly different from that for CCK alone at $p < 0.001$ as estimated by analysis of log-log transformations. Inset shows functions close to origin. Both CCK and CCK with 250 pmol of glucagon are essentially identical close to origin and are shown by a dashed line in inset. (Reprinted, with permission, from Tseng et al., 1984.)

As it has become clear that observations of nonparallel transport cannot simply be discounted, a second line of thought has emerged. This view can be stated as follows: "Although nonparallel transport occurs, it can be explained by the exocytosis model. The vesicle system that is responsible for secretion is simply somewhat more complex than originally thought, but nevertheless the model still applies in a general sense." Thus, exocytosis as a random, mass transport process by which means all of the various enzymes are released en masse without mind to the nature of particular secreted molecules, in the manner that I discussed above, must be put aside and a more complex vesicle system devised, but a vesicle system of the same general sort nevertheless.

But how could this be accomplished? Two things are required and there is at present no substantive evidence for either. The first is that different granules would have to contain different enzymes either singly or in

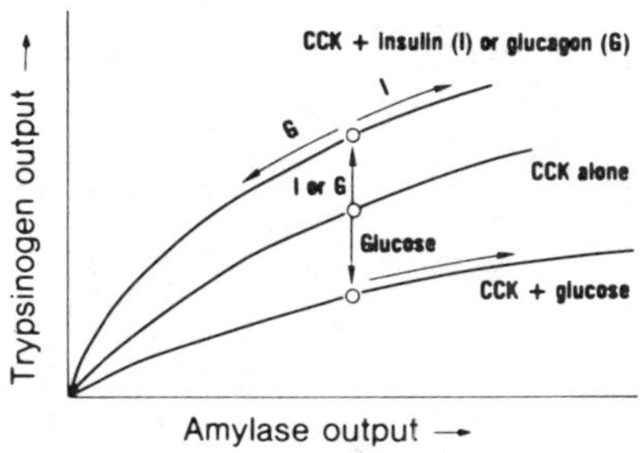

Figure 32 Idealized schema showing effect of insulin (I), glucagon (G), and glucose on relative rates of amylase and trypsinogen secretion by rat pancreas in presence of CCK. CCK augments output of both enzymes, increasingly favoring amylase secretion as overall enzyme output is increased. Both glucagon and insulin shift CCK curve (CCK alone) toward a more trypsinogen-dominant function; however, glucagon points (G) tend to fall closer to origin, whereas insulin (I) points tend to be further away. Glucose shifts CCK curve toward a more amylase-dominant function, and points are found further from the origin. See text for further discussion. (Reprinted, with permission, from Tseng et al., 1984.)

combination. The second is that the secretory mechanism would have to be able to distinguish or select between these different granules either within single cells or between cells. There is no evidence, other than the circular "evidence" of nonparallel transport itself, to suggest that individual granules contain single species of enzyme or that there are ordered differences between granules of different mixed enzyme composition. Indeed, from the very first immunocytochemical antibody studies to the most recent, all of the evidence suggests that each granule and each cell contains all of the secreted enzymes, and ordered differences in enzyme content between granules or cells have never been reported. Regional or geographic organization has not been demonstrated, although small differences have been reported between the enzyme content of acinar cells surrounding the islets of Langerhans, about 1% of the gland mass, and the remaining 99% (Malaisse-Lagae et al., 1975). Such differences probably cannot account for a measurable nonparallel effect, no less for the variety of such phenomena that have been reported, or for the varied characteristics of nonparallel transport that I have discussed.

In one reported case of nonparallel transport (Dagorn, 1978), the author proposed that the stimulant that had produced the effect elicited secretion of enzyme from a group of cells that contained one mixture of enzyme, whereas basal or unstimulated secretion was derived from another group of cells with a different mix of enzymes whose secretion was not augmented by the stimulant. Of course, it is important to remember that the existence of

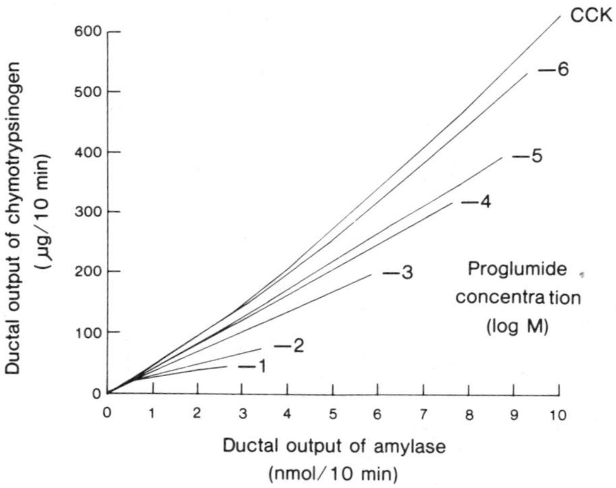

Figure 33 Regression lines calculated from the scatter of all individual pairs of enzyme outputs into the cannulated duct of the rabbit pancreas in organ culture for each different dose of proglumide (in the presence of CCK). Each regression line is extended to the upper limit of the range of values observed for amylase output. In each case the best fit resulted using an equation of the form $y = a\,x^b$ (power model):

Proglumide (log M)	Function (x = amylase; y = chymotrypsinogen)	Correlation Coefficient	Data Points (n)
0 (CCK alone)	$y = 40.4x^{1.19}$	0.94	81
−6	$y = 42.2x^{1.14}$	0.90	36
−5	$y = 39.5x^{1.06}$	0.94	45
−4	$y = 39.2x^{1.03}$	0.94	54
−3	$y = 34.0x^{1.01}$	0.97	63
−2	$y = 29.6x^{0.75}$	0.75	63
−1	$y = 26.8x^{0.54}$	0.71	27

Reprinted, with permission, from Neiderau et al., 1985.

the nonparallel effect itself did not provide evidence for this hypothesis. This explanation was an hypothesis proposed in an attempt to account for the observation within the borders of the exocytosis model for secretion.*

*Similarly, in Adelson and Miller's recent work with chymodenin (Adelson and Miller, Science, 1985), they argue that the high correlation between lipase and chymotrypsinogen secretion seen in the presence of chymodenin (see Figures 13 and 15) is proof of their "linked" secretion by exocytosis. I do not understand how a correlation between two variables, however tight, proves linkage. This is simply a hypothesis that is consistent with this isolated observation and offers an explanation for it in paradigmatic terms.

How easily can we explain such observations and those that I have discussed above in terms of a model of secretion based on ordered intercellular or intergranular differences in enzyme content? We can begin by considering the proposal in terms of the power function described above. For example, such a function could be due to the variable admixture of material from two groups of granules or cells with different enzyme content. In such a situation, the steepest portion of the function, that closest to the origin, would most closely approximate the enzyme content of one pool and the portion of the curve farthest away from the origin would most closely reflect the contents of the other pool. From some of the figures presented in this chapter we can see that such source pools would have to differ from severalfold to as great as an order of magnitude in their content of different enzymes. Such differences would probably be obvious with immunocytochemistry, but have yet to be observed, or at least reported. But there are additional problems. As I noted above, it is possible to change the direction and parameters otherwise of such curves, and this requires further regulatory modulation as well as additional enzyme pools for each pair of enzymes. The central complexity of such a model is in the number of pools required to account for all of the potential possibilities. In order to account for the coordinated variable secretion of all of the enzymes in a continuous fashion in this manner, the number of different pools containing enzyme in different ordered proportions would be of the order of 19!—the number of secreted proteins minus one factorial. This is more than the number of cells or granules in the pancreas of a rodent. But even if we consider that the number of permutations is limited to only *pairs* of enzymes, the number of different pools would still have to be in the hundreds.

We must also consider this explanation in light of other phenomena such as positive intercepts which demonstrate that one enzyme can be secreted in the absence of another, and the observation that the extent of the covariance of the secretion of two particular enzymes may increase, while at the same time the covariance between one of these enzymes and a third species may decrease. That is, these changes do not suggest mixing from fixed pools of mixed enzyme content. One could go on, but of course it is always possible to imagine some manner of means, however complex, to explain such results in vesicular terms. On the other hand, what evidence there is on hand does not support the existence of either intergranular or intercellular organization, and the sole reason for thinking that such organization exists is the underlying belief that protein secretion must occur by exocytosis. If this theoretical "crutch" is removed, then certainly the burden of proof must be on the proposal for such a complex system.

If in all of the studies that I have discussed in this chapter, only parallel transport had been seen, namely all lines regressed through the origin, no changes in intercepts or slopes of functions were observed, no changes in variance occurred, no curves were formed, and so forth, then the conclusion that parallel transport applied in general and that there was no reason to question the mass transport exocytosis model would have reasonably been drawn. The rejection of each prediction of the parallel transport model requires that the exocytosis model be substantially modified, namely made more complex, and that the negative evidence regarding the existence of ordered pools be put aside to await positive evidence, in order to save the vesicle model.

On the other hand, the many complexities of mechanism visited upon the vesicular model by nonparallel transport are simply, even trivially, explained by a model in which different enzymes move independently of each other as a result of their own kinetic energy. All of the observations can be accounted for as merely being due to changes in the transport parameters of concentration gradient and permeability. Beyond this we must remember that the mechanism underlying nonparallel transport must be considered in light of all the other evidence that I have discussed in this book.

13.14 SUMMARY

It has been many years since our first report of nonparallel transport and almost 100 years since Pavlov's original proposal. I have discussed some of the reasons why I believe that it has been hard for some investigators to find what increasingly appears to be a common phenomenon. And certainly if one wishes to find nonparallel effects, then the traditional practice of averaging values may well hide more than it reveals. But there is probably another reason for the past obscurity of this phenomenon, a behavioral one if you will. An observation that is not consistent with expectation is often lost or missed even by the most honest of observers. In a study on the inadmissibility of anomaly, Jerome Bruner, a Professor of Psychology at Harvard University, found that when he showed subjects incongruous playing cards, for example a red spade, that at first the anomaly was often not noticed, that is the spade was seen as black whatever its true color. With more trials, however, it was seen, but often in an ambiguous and uncertain fashion. With time and experience, most observers came to appreciate its existence. However, for some, it became a matter of great importance to deny its reality.

SELECTED REFERENCES

1. Adelson, J. W. Un nouveau polypeptide, extrait du duodenum porcin, possedant une activite biologique chez la rat. *Biol. Gastroenterol.* **4**:355, 1971.
2. Adelson, J. W. Chymodenin, an overview, in, *Gastrointestinal Hormones*, J.C. Thompson, Ed. Univ. of Texas Press, Austin, 1975, pp. 563–574.
3. Adelson, J. W. and S. S. Rothman. Selective pancreatic enzyme secretion due to a new peptide called chymodenin. *Science* **183**:1087–1089, 1974.
4. Adelson, J. W. and S. S. Rothman. Chymodenin, a duodenal peptide: specific stimulation of chymotrypsinogen secretion. *Am. J. Physiol.* **229**:1680–1686, 1975.
5. Babkin, B. P. *Secretory Mechanisms of the Digestive Glands.* Hoeber, New York, 1950.
6. Dagorn, J. C. Nonparallel enzyme secretion from rat pancreas: In vivo studies. *J. Physiol. (London)* **280**:435–448, 1978.
7. Dick, J. and J. Felber. Specific hormonal regulation, by food, of the pancreas enzymatic (amylase and trypsin) secretion. *Horm. Metab. Res.* **7**:161–166, 1975.
8. Felber, J. P. A. Zermatten, and J. Dick. Modulation, by food, of hormonal system regulating rat pancreatic secretion. *Lancet* **2**:185–188, 1974.
9. Glazer, G. and M. L. Steer. Requirements for activation of trypsinogen and chymotrypsinogen in rabbit pancreatic juice. *Anal. Biochem.* **77**:130–140, 1977.
10. Grendell, J. H. and S. S. Rothman. Digestive end-products mobilize secretory proteins from subcellular stores in the pancreas. *Am. J. Physiol.* **241**:G67–G73, 1981.
11. Grendell, J. H., H. C. Tseng, and S. S. Rothman. Regulation of digestion: I. Effects of glucose and lysine on pancreatic secretion. *Am. J. Physiol.* **246**:G445–G450, 1984.
12. Grossman, M. I., H. Greengard, and A. C. Ivy. The effect of dietary composition on pancreatic enzymes. *Am. J. Physiol.* **138**:676–682, 1943.
13. Iacino, D., G. A. Schele, and C. Liebow. Secretory response of the rabbit pancreas to cholecystokinin stimulation. *Am. J. Physiol.* **239**:G247–G254, 1980.
14. Malaisse-Lagae, F., M. Ravazzola, P. Robberecht, A. Vandermeers, W. J. Malaisse, and L. Orci. Exocrine pancreas: evidence for topographic partition of secretory function. *Science* **190**:795–797, 1975.
15. Niederau C., J. H. Grendell, and S. S. Rothman. The effects of proglumide on ductal and basolateral secretion of pancreatic digestive enzymes. *Am. J. Physiol.* **249**:G100–G107, 1985.
16. Pavlov, I. P. *The Work of the Digestive Glands.* Griffin, London, 1910.
17. Rothman, S. S. "Nonparallel Transport" of enzyme protein by the pancreas. *Nature* **213**:460–462, 1967.
18. Rothman, S. S. The molecular regulation of digestion: Bond-specific and short-term. *Am. J. Physiol.* **226**:77–83, 1974.
19. Rothman, S. S. The independent secretion of different digestive enzymes by the pancreas. *Am. J. Physiol.* **231**:1847–1851, 1976.
20. Rothman, S. S. The digestive enzymes of the pancreas, a mixture of inconstant proportions. *Ann. Rev. Physiol.* **39**:373–389, 1977.

21. Rothman, S. S. To transport parallel or not to transport parallel? *Am. J. Physiol.* **240**:G253, 1981.

22. Rothman, S. S. and H. Wells. Selective effects of dietary egg white trypsin-inhibitor on pancreatic enzyme secretion. *Am. J. Physiol.* **216**:504–507, 1969.

23. Rothman, S. S. and H. Wilking. Differential rates of digestive enzyme transport in the presence of cholecystokinin-pancreozymin. *J. Biol. Chem.* **253**:3543–3549, 1978.

24. Scheele, G. A. and G. E. Palade. Studies on the guinea pig pancreas. Parallel discharge of exocrine enzyme activities. *J. Biol. Chem.* **250**:2660–2670, 1975.

25. Steer, M. L. and G. Glazer. Parallel secretion of digestive enzymes by the in vitro rabbit pancreas. *Am. J. Physiol.* **231**:1860–1865, 1976.

26. Steer, M. L. and T. Manabe. Cholecytokinin-pancreozymin induces the parallel discharge of enzymes from the in vitro rabbit pancreas. *J. Biol. Chem.* **254**:7228–7229, 1979.

27. Tseng, H. C., J. H. Grendell, and S. S. Rothman. Food, duodenal extracts, and enzyme secretion by the pancreas. *Am. J. Physiol.* **243**:G304–G312, 1982.

28. Tseng, H. C., J. H. Grendell, and S. S. Rothman. Regulation of digestion. II. Effects of insulin and glucagon on pancreatic secretion. *Am. J. Physiol.* **246**:G451–G456, 1984.

CHAPTER 14

A Matter of Concentration

The exocrine pancreas has two major secretion products: the digestive enzymes, and water with its associated electrolytes. The discovery of the first hormone, secretin, was based on the observation that the intravenous injection of extracts of duodenum increased the secretion of the water and electrolyte component by the pancreas (Bayliss and Starling, 1902a, b). Subsequently, duodenal extracts were found to contain another polypeptide, now known as cholecystokinin (CCK), that preferentially stimulated the secretion of the digestive enzymes (Harper and Raper, 1942). The capability of dissociating changes in the rate of fluid and electrolyte secretion, a secretin or "hydrelatic" effect, from changes in protein secretion, a CCK or "ecbolic" effect (Thomas, 1950), indicated that different mechanisms were responsible for the two secretions.

The ability to dissociate the two processes led to the realization that the bulk of the material secreted by the pancreas was due to the secretion of water. In some species and circumstances it was possible to increase protein secretion to maximal rates with minimal alterations in the bulk or volume of the secreted material, and conversely, the volume of secretion could be varied from virtually zero to maximal values by altering electrolyte secretion without greatly altering protein secretion (Harper, 1967). More recently, with the availability of pure and synthetic CCK and secretin, this seemingly clear distinction has broken down somewhat. In some species and circumstances, CCK increases fluid and electrolyte secretion and secretin enhances protein secretion at least to some extent, whereas in others, such as the rabbit, essentially complete dissociation can be found.

Nevertheless, in this context, if protein secretion occurred by exocytosis, then the expulsion of the highly concentrated contents of secretion granules into a large volume of fluid would result in the extensive dilution of the secreted proteins. This dilution effect would presumably apply during highly augmented protein secretion, as well as during unstimulated or so-called "basal" secretion, because the concentration of protein in the duct lumen can be greatly increased without significantly altering the bulk of secreted fluid.

This view of the secretory process conflicts with another, in which the concentration of protein in the duct lumen of pancreatic tissue, previously stimulated to secrete protein at a highly augmented rate with secretagogues, would to be approximately the same as that in the zymogen granule. In this case, ductal contents would in great part be comprised of the contents of the granules and the dilution of secreted proteins by fluid secretion would be minimal at least at the acinar level. (Dilution presumably would still occur as the material travels through the duct system.) Indeed, if the concentration were the same in the duct lumen as in the zymogen granule, then all of the contents of the duct lumen would be derived from granules if secretion occurred by exocytosis. If the concentration were half that in the granule, then half of the volume would be derived from this source and so forth. Electron microscopic pictures showing electron-opaque duct lumina may be viewed as evidence for this hypothesis (Case, 1978) (Figure 1).*

Thus, there are two seemingly contradictory lines of thought that we can consider: One that proteins are greatly diluted upon their secretion even at high rates, and another that little dilution occurs during augmented secretion and that protein secretion accounts for the bulk of material secreted into the duct system under these circumstances at the acinar level. In a series of experiments Ho and I compared the protein concentration in fluid collected from the major pancreatic duct of the rabbit during augmented protein secretion to the concentration of these proteins in zymogen granules extracted from pancreatic tissue, in order to provide a maximal estimate of the extent of dilution that actually takes place as well as to correlate changes in protein concentration to changes in the bulk of secreted material.

*Also see Blomfield and Settree, *Exp. and Mol. Path.*, 1983.

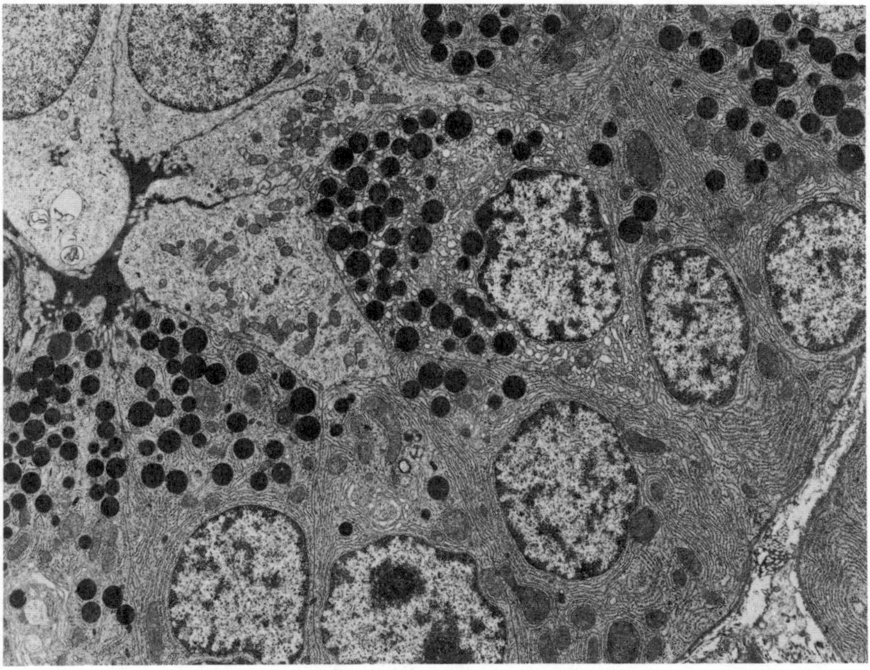

Figure 1 Electron micrograph of a human pancreatic acinus showing a number of acinar and centro-acinar cells around an acinar lumen. Note that the material in the lumen is only slightly less electron opaque than that in the zymogen granules. (This micrograph is from Dr. Horst Kern, and is reproduced with permission from Reference 3.)

Our best estimate of granule protein concentration ranged from approximately 135–270 mg of protein per milliliter of granule volume, or about 4–9 m*M* for an estimated composite protein molarity (Ho and Rothman, 1983). Direct measurements on populations of zymogen granules isolated from rabbit pancreas after an overnight fast gave an average value of approximately 269 mg of protein/ml of granule volume and an observed range of about 140–450 mg/ml (Figure 2).

How does this value compare with the concentration of protein in secretory fluid in the unstimulated state (after an overnight fast), as well as after stimulation of protein secretion with a cholinergic agonist? Basal or unstimulated secretion contained protein at an average concentration of 18 mg/ml of secretory fluid (Figure 2) or about 7% of the protein concentration in the granule. Thus, in terms of a mass or bulk transport model, the dilution effect on granule contents was more than tenfold in the unstimulated state.

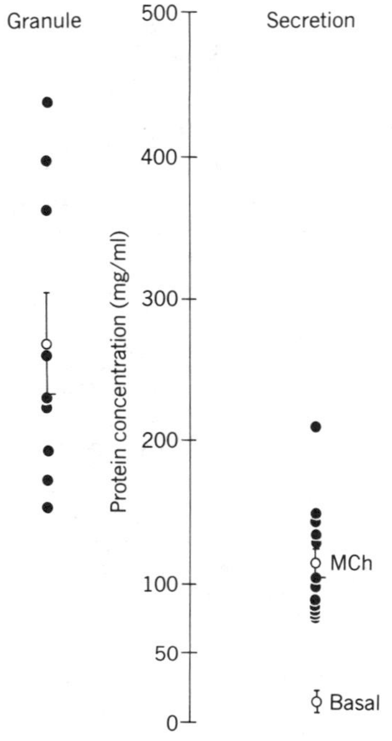

Figure 2 Comparison of the protein concentration of rabbit pancreatic zymogen granules, of secretion produced by the rabbit pancreas in situ in response to an injection of the secretagogue acetyl-β-methylcholine chloride (MCh), and of secretion produced by the rabbit pancreas in the absence of exogenous secretagogues (Basal). Secretion was collected from the cannulated duct of the rabbit pancreas in situ. Values for granule preparations isolated from nine separate groups of fasted rabbits are shown (●, Granule). The maximal protein concentration achieved in response to the administration of MCh (0.5 mg/kg body weight, subcutaneous) for each experiment is also shown (●, Secretion). Basal is the average protein concentration of secretion during the period immediately preceding the administration of the secretagogue. The mean protein concentrations ± SEM (ᴏ̣) were as follows: granules, 269 ± 35 mg/ml ($n = 9$); MCh-stimulated secretion, 117 ± 11 mg/ml ($n = 12$); and basal secretion, 18.3 ± 2.3 mg/ml ($n = 12$).

A cholinergic agonist increased the protein concentration of secretion to an average of 117 mg/ml at the peak or to 43% (117/269) of the estimated protein concentration in the zymogen granule (Figure 2). That is, approximately 43% of the bulk of secretion on the average would be of granule origin if an exocytosis-type system were operating. The highest value observed was 209 mg of protein/ml of secretion or 80% of the average granule concentration.

The maximal protein concentration in secretion under these experimental conditions can be estimated from the y-axis intercept of a plot of the reciprocal of secretory protein concentration versus the reciprocal of protein output (Figure 3). This value was 214 mg/ml, or again about 80% of the estimated protein concentration in the zymogen granule.

Therefore, cholinergic stimulation increased the concentration of protein in secreted fluid from about 7% of that found in the zymogen granule to an average of 43%, and as high as 80%, of that value. Thus, it seems that secreted material is not greatly diluted when protein secretion is most active, even after its passage down the duct system. This means that under these conditions, protein secretion should account for a large percentage of

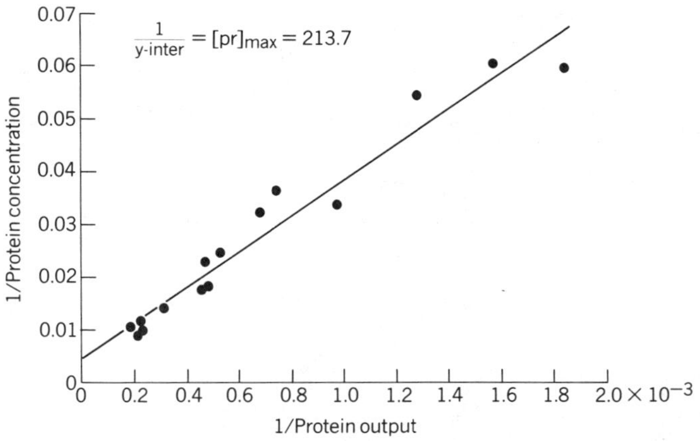

Figure 3 Reciprocal of protein concentration in secretion fluid versus the reciprocal of protein output. Acetyl-β-methylcholine chloride (0.5 mg/kg body weight) was injected subcutaneously into rabbits in situ and secretion collected at 5-min intervals from the cannulated duct of the pancreas. The mean protein concentration (mg/ml) and output (mg/5 min) were determined for each 5-min time period for 75 min (see Figure 4), and the reciprocal of each value was plotted against the other measure for the same period (●). The line was constructed by linear regression. The reciprocal of the y intercept is 213.7 mg/ml.

the bulk of secreted fluid, averaging about 43%, and as high as 80% of the total bulk of secreted material. Because this would reflect an increase from only 7% of the volume during unstimulated secretion, there should be large increases in the bulk of material secreted by the gland as a result of cholinergic stimulation, approximately a doubling on the average (to account for the increase from 7 to 43%) and as great as a fourfold increase (to account for the 80% estimate). Not only should there be a large increase in secreted bulk, but that increase should be correlated to the increase in the protein concentration in secreted fluid, that is, the greater the increase in concentration, the greater the increase in the volume of secreted material. Moreover, as the concentration of protein falls over time, as augmented secretion abates, so should the volume increment that protein secretion brings to the bulk. Hence, the volume of secretion should fall substantially as well.

Were these events seen? No significant change in average flow (or the volume of secreted material) was observed upon administration of the cholinergic agonist to rabbits in situ coinciding with the changes in protein concentration. Nor was there a reduction in volume when augmented protein secretion decreased (Figure 4). For the period of peak protein

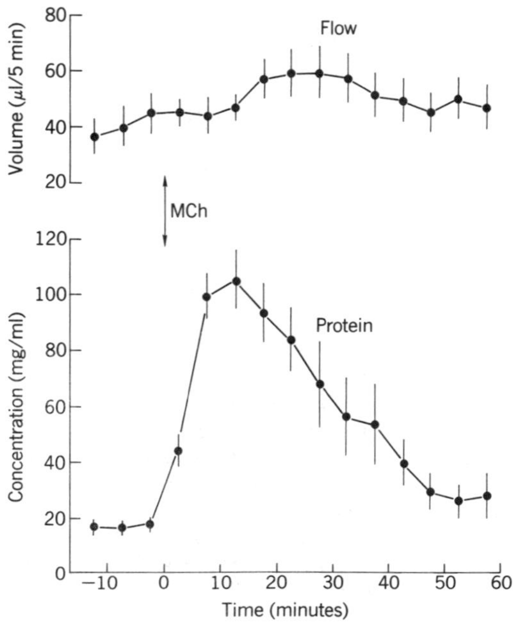

Figure 4 Flow rate and protein concentration as a function of time. Acetyl-β-methylcholine chloride (MCh, 0.5 mg/kg body weight) was administered subcutaneously (arrow). The points and bars are the means ± SEM of 12 experiments. No significant increases in flow occurred coincident with increased protein secretion, nor were decreases in flow seen coincident with the subsequent decline in protein secretion.

concentration, that is, when the increase in volume should have been greatest, volume flow increased in about half of the studies and decreased in the other half (Figure 5). Thus, although protein concentration in ductal fluid during highly augmented rates of protein secretion approached that in the zymogen granule and accordingly would have been expected to produce a substantial increase in the volume of secreted material, decreases in volume were as commonly observed as increases, and there was no correlation between protein concentration and fluid flow in kind, in degree, or in time course (Figures 4 and 5).

If we view secretion in terms of the mass transport or exocytosis model, this presents us with a paradox. The secretion of protein appears to account for a substantial portion of the bulk of secreted material, and yet when we significantly increase its secretion, there is no resultant increase in secretory bulk. As I have discussed, the alternative equilibrium model views protein secretion as a diffusion-like process in which individual protein molecules

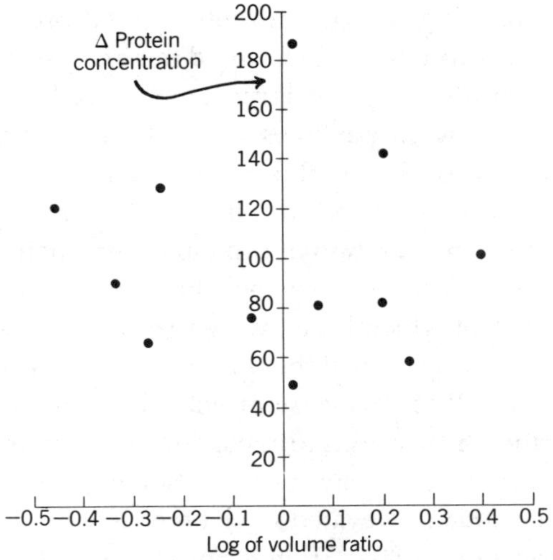

Figure 5 Relationship between secretagogue-induced changes in protein concentration of secretion and flow rate. Acetyl-β-methylcholine chloride (MCh, 0.5 mg/kg body weight) was administered subcutaneously to the rabbit in situ. The 5-min period in which the secretory response (mg/ml) was greatest was determined for each experiment as was the flow rate for that period. Each was compared to the protein concentration and flow rate for the 5-min period immediately prior to the administration of MCh in the same animal. Protein concentration is shown as the difference between the peak concentration (after stimulation) and the concentration immediately preceding injection of the secretagogue. Flow rate is expressed as the log of the ratio of flow rate during the period of maximal protein concentration to flow for the period immediately prior to the administration of the stimulant. Each point (•) shows the relationship for a different animal. There is no correlation between the change in protein concentration and the change in flow rate ($n = 12$).

leave the cell as a result of their independent movement across granule and then plasma membrane. In a membrane transport process, solute can move from one compartment to another without the volumes of either compartment necessarily being changed. If we consider these observations in terms of this hypothesis, we no longer are confronted by a paradox. It is possible for ductal fluid to have a high concentration of protein, even equal to that in the zymogen granule, without there being any significant increase in the bulk of secreted material.

This analysis is based on the view that the degree of dilution of protein from zymogen granule to distal ductal orifice reflects the extent of admixture, or net influx, of fluid subsequent to protein secretion by the acinar cell, and applies as long as the net result of protein transit through the duct system is its dilution (or if its concentration is unaltered, i.e., it is not diluted), whatever fluctuations might occur en route. Only if protein were concentrated by its passage through the duct would this analysis not be applicable. It is widely held that the duct system is responsible for the major secretion of water and electrolytes by the pancreas, and indeed it is thought that it may be responsible wholly for the net flux of water across this epithelium, and therefore dilution is to be expected. Nevertheless, although there is no indication that a net concentration of protein occurs, neither is it clear that protein is in fact diluted during its passage through the duct system. That is, there is no evidence of which I am aware that demonstrates that protein concentration is actually greater at the proximal end of the duct system, and indeed there is some that is consistent with there being little change in protein concentration from acinus to distal ductal orifice. In the rabbit when the pancreatic duct is blocked, the concentration of protein within the duct system is not substantially elevated over time, as would be expected if a concentration gradient existed from acinus to distal orifice (or if the converse situation applied and fluid were extracted in substantial quantities from the duct lumen concentrating protein therein) (Rothman, 1980). Indeed, in the rabbit no significant flow-dependent alteration in protein concentration, for over an order of magnitude range in flow rate, is seen using a variety of modalities to alter flow. Nor is there any indication of a correlation between spontaneous flow rate and protein concentration.

One could argue that because it is not clear exactly how the concentration of protein found in secretion collected from the pancreatic duct is related to that secreted into the acinar lumen, that we cannot exclude the possibility that the concentration at that site is indeed much lower than in secreted fluid. Nonetheless, regardless of how we view the likelihood of such an event, the development of a high concentration of protein in secreted fluid, similar to that in the zymogen granule, in the absence of parallel changes in the bulk of secreted material can never be viewed as evidence in support of the exocytosis hypothesis.

SELECTED REFERENCES

1. Bayliss, W. M. and E. H. Starling. On the causation of the so-called 'peripheral reflex secretion' of the pancreas. *Proc. Roy. Soc.* **69:**352–353, 1902a.
2. Bayliss, W. M. and E. H. Starling. The mechanism of pancreatic secretion. *J. Physiol. (Lond.)* **28:**325–353, 1902b.
3. Case, R. M. Synthesis, intracellular transport, and discharge of exportable proteins in the pancreatic acinar cell and other cells. *Biol. Rev.* **83:**211–354, 1978.

4. Harper, A. A. and H. S. Raper. Pancreozymin, a stimulant of the secretion of pancreatic enzymes in extracts of the small intestine. *J. Physiol. (Lond.)* **102:**115–125, 1942.
5. Thomas, J. E. *The External Secretion of the Pancreas.* C.C. Thomas, Springfield, Illinois, p. 62, 1950.
6. Harper, A. A. In *Handbook of Physiology, Section 6: Alimentary Canal, Vol. III, Secretion,* C.F. Code, Ed. American Physiological Society, Washington, D.C., pp. 969–995, 1967.
7. Ho, J. J. L. and S. S. Rothman. Protein concentration in the pancreatic zymogen granule. *Biochim. Biophys. Acta* **755:**457–466, 1983.
8. Rothman, S. S. Passage of proteins through membranes—old assumptions and new perspectives. *Am. J. Physiol.* **238:**G391–G402, 1980.

CHAPTER 15

A Consideration of the General Concept of Protein Permeability

There are two central reasons why scientists invent hypotheses and theories: as tools to design experiments to permit orderly observation, and to attempt to describe and explain how things occur and are organized in our perceived world with as much accuracy, precision, and completeness as possible. The first type of hypothesis, if proven correct, is not usually claimed by its inventor or used by him or her to provide a detailed explanation of how a process is organized or occurs, even though its affirmation may well help define the system, ruling in or out particular explanations for its characteristics. On the other hand, when the second type of hypothesis is supported by sufficient experimental or theoretical evidence, its inventors make the claim that their theory provides a reasonably accurate accounting of the actual mechanism at work. Often, and perhaps invariably, a hypothesis of the first kind, if successful, leads to one of the second kind. One can dispute, as a philosophical or practical matter, whether the second type of hypothesis can ever really be proven true in a rigorous fashion and provide an accurate and detailed description of the physical world. Nevertheless, there are many mechanistic hypotheses that are successful in that they are good predictors of behavior, even if they cannot be rigorously proven.

The cisternal packaging–exocytosis or vesicle theory of secretion is a theory of the second or mechanistic type. Even though the theory has not been fully formulated in all of its details or has in some aspects even been coherently developed, nor, as we have discussed in this book, has it been a particularly good predictor of behavior, it does attempt to provide a detailed model for how molecules are moved within and out of the cell and outlines for us the machinery that might carry out such transport. The equilibrium or diffusion-based hypothesis, on the other hand, has primarily been a hypothesis of the first type. It is not mechanistic, at least not in the same sense. It was originally conceived of as a tool, an alternative to the vesicle model, to help test claims that had been made for it. Might another model, indeed an alternative that is in many ways antithetical to the vesicle model, fit observation as well? Such a comparison not only tests the predictability of an hypothesis more rigorously than inductive inference merely applied to it in its own terms, but it also tells us how satisfactory our experiments are in terms of their ability to distinguish between contrasting ideas. More broadly, the equilibrium hypothesis, as all such diffusion-based hypotheses, provides a general and deep conceptual framework within which to learn about and describe the phenomenological qualities of transport. As such, it makes no claim about the mechanistic details of the transport process, although it may enlighten us as to what is possible and what is not possible. That is, it does not deal,in any specific or explicit manner, with, or otherwise consider exactly what chemical or physical events between which particular molecules account for the passage of proteins through membranes. Unfortunately, this is occasionally misunderstood and it is thought incorrectly that because we talk of diffusion, that the hypothesis proposes that the specific physical mechanism that accounts for transport is simple diffusion through pores in biological membranes.

What we have learned about the movement of proteins, or at least those whose movement we have studied into and out of the acinar cell of the pancreas, is that it appears to have the same general characteristics as the movement of virtually all types of molecules whose transport across biological membranes have been studied heretofore. Protein movement, just as that of sodium, glucose, glycine, and so forth, appears to be responsive to chemical gradients, is reversible, and the molecules appear able to move individually. Thus, their transport has many of the characteristics of a simple diffusive equilibrium, whatever particular mechanism may in fact account for specific transport events.

This is not the same as saying that simple diffusion through pores accounts for transport. Transport may involve chemical reactions between the transported molecule and other molecules, proteins and lipids, embedded in the membrane, or it indeed may be purely physical. And of course, transport in one direction may be favored, it may be active or passive, and chemical reactions or bonding, if they occur, may or may not require the expenditure of metabolic energy. These possibilities are not distinguished by the hypothesis. Testing it merely enlightens us as to what behavior a successful mechanistic model must account for. It does not offer such a model for our consideration, even though it may in the end lead us to one, as the first type of hypothesis evolves into the second.

If there had been no historical or other reason for viewing protein transport as sui generis (from transport processes as generally conceived), then the commonality in the behavior that we have observed for protein movement and that seen for other molecules that have been studied would be reassuring. It would tell us that the movement of atoms and molecules of whatever kind across biological membranes adheres, at least in a general sense, to the same rules although they are not of course accounted for by a single mechanism or even a single type of mechanism. Moreover, such movement is of the same general nature as that seen in purely physical systems over short distances, namely, movement is produced by the kinetic motion of individual atoms and molecules.

However, even though it increasingly seems that the movement of individual protein molecules through biological membranes is a ubiquitous event, found throughout the phyla, even if it is not the only means by which these molecules cross membranes, such a broadly conceived view of protein transport has generally been avoided, even eschewed. Rather, evidence has been sought for *the* particular special (usually chemical) mechanism that accounts for transport, vesicle models of secretion aside. This seems to be because protein transport is viewed as an extraordinary, not an ordinary, transport event; an exception to a general rule of impermeability, rather than an example of permeability. For this reason it has been common to seek, and seek at the outset, that unique or particular molecular or chemical event that makes it possible for a particular protein to cross a particular membrane without that membrane's fabric being torn.

The usual pattern of going from a general characterization of transport in terms of a diffusion-based model, to a more detailed analysis of movement, to derive inferences about the kinetic, thermodynamic, and compartmental nature of transport and then on this basis to consider models for mechanism, has not been generally followed. This view, in which an essentially impermeable membrane is seen as having to be broached in some special manner for transport to occur, has not only led to the development of numerous specific, even detailed, models for mechanism (a hypothesis of the second kind) in the absence of the usual foundation of information about the kinetic, thermodynamic, and compartmental characteristics of the transport process, but it has also led to the proposal of very detailed *general* models for protein transport, most notably the signal hypothesis, that attempt to explain all protein movement across or into membranes within a single particular specific mechanistic framework. The signal hypothesis is rather remarkable in that it proposes at the outset such a manifold array of specific events to carry out transport. I believe that the construction of such detailed, highly specific models to account for all protein transport across all membranes is likely to fail if a successful general characterization of protein transport is the goal.

And certainly, the signal hypothesis cannot be considered to be a successful general model at this point. There are far too many exceptions to its postulates; exceptions that for the most part have no clear raison d'être. In addition, as I have discussed in Chapter 7, too much additional complexity has had to be added to the original model, in the form of ad hoc and post hoc hypotheses, to explain observations that deviate from prediction. There seems to be no reason to suppose that only elongated chains can cross membranes, or only nascent or forming chains, or that covalent modification is a necessary general requirement, or that at least some formed proteins do not spontaneously enter and cross membranes as a function of special anatomical and chemical features of the protein, the membrane, and the aqueous environment, and so forth.

I see no valid reason not to use the most general and deepest hypothetical framework available, that is, a diffusion-based hypothesis. If a single unique (complex, chemical, and so forth) mechanism indeed accounts for all protein transport, then the diffusion-based hypothesis will lead us to that conclusion. If protein transport is diverse, then it will clearly establish that fact. As I have said, the reluctance of many investigators to approach the subject in this most general fashion seems to stem in great part from the a priori assumption that membranes are fundamentally impermeable to proteins and that our task in all cells that we study is to elucidate the special mitigating process. This no longer seems a feasible position.

In order to retain the notion that a single type of process, such as the model proposed in the signal hypothesis, accounts for all protein transport and is the particular means used by cells to overcome the essential impermeability of membranes to this group of molecules, we must discard in one fashion or another all observations that do not fit this model either by considering them exceptions to the general rule or artifacts of one sort or another. However, as the transport of more and more proteins is studied across a greater and greater variety of membranes, the more difficult such attempts appear to have become and the more likely that the existence of diverse modus operandi for protein transport will be established as the natural fact. In any event, the accumulation of exceptions and modifications to current mechanistic models for protein transport suggests that model building thus far may be occurring at the wrong level of abstraction.

The existence of a great many types or varieties of protein transport should not surprise us, even if our task would be made easier if there were only one special mechanism to account for everything. We know that there are numerous modes of transport even for the, relatively speaking, simple sodium ion. For sodium, differences in transport modality exist from cell type to cell type and from membrane type to membrane type, and it is even common for more than one means of transmembrane movement to exist for a particular membrane, from simple diffusion to active carrier-mediated transport. Why should proteins, the most complex and structurally diverse of all molecules, cross membranes in one and only one way? Indeed, might we not well expect the diversity of transport mechanism to reflect the diversity of protein structure and function? Thus, there may be many different types of protein transport, and indeed the movement of every individual protein across each specified membrane may reflect the formal individuality required by a diffusive paradigm. One cannot even presume that the same molecule crosses two different membranes in the same manner, or a given membrane in just one way. There is sufficient reason to reject such notions for simple molecules, no less for proteins.

If "diversity" is the keynote, then how are we to gain a general understanding of such processes, because after all that is our ultimate goal? It is the existence of such diversity that must lead us back to the security of a diffusive paradigm. In my view only in this way can we, in the final analysis, be sure to develop a rational and scientifically rigorous notion of protein transport from which we can generalize broadly.

15.1 SIMPLE DIFFUSION

One reason that protein transport has been considered a subject apart from the study of transport processes more generally has been the belief that these large molecules, unlike smaller molecules, cannot pass out of or into cells by simple diffusion through aqueous channels of one sort or another, even to a modest extent. That is, it has been thought that such channels pose an absolute barrier to macromolecules. This view has its experimental basis in the measurement of average pore radius, as I discussed in Chapter 4, as well as in assumptions about cells that are thought to exclude this possibility a priori.

It has been demonstrated that the average pore radius for many, although not necessarily a majority of cells, is too small to accommodate most proteins. However, as I have pointed out, such measurements of pore radius are necessarily average values and do not preclude the presence, even in cells with rather small average pore radii, of some pores with a sufficiently large radius through which proteins might pass. Whether or not this is the case, what the upper limit or range of upper limits are for the radius of an individual channel in a given cell or, how such values might vary from cell type to cell type is not known, but the possibility that large pores exist cannot be excluded on the basis of current experimental knowledge.

As for the a priori exception to the notion of large pores, one can argue that they cannot exist because, if they did, the integrity of the cell would be compromised; that is, large pores are not compatible with a viable cell. They would allow cellular proteins not properly destined for export to leak out, and thus over time the cell would lose its protein contents and become nonfunctional. But of course, all that the cell really requires is that loss, if it occurs, be such that new synthesis can balance it and that a steady concentration of the protein can be maintained within the cell.

It is important to measure and directly characterize the release of various intracellular proteins that occurs normally from cells and not to assume that all such release is merely from broken, sloughed, or otherwise comprised cells. Are intracellular proteins lost normally? At what rate? Are some lost more readily than others? If so, how do we account for this? And so forth. It has recently been shown in the pancreas that many lysosomal enzymes appear in secretion (Rinderknecht et al., 1979). Not only that, but when digestive enzyme secretion is augmented by a stimulant, so is the secretion of these "nonsecreted" proteins. The secretion of some is increased to the same extent as the digestive enzymes themselves, whereas others are increased to a lesser degree. In any event, since such things can be measured with relative ease, we should not make the assumption that the cell membrane presents an absolute barrier to the release of any substance, even proteins.

It also may be argued that large pores in membranes would act as low-resistance electrical shunts and prevent the normal development of "excitable" transmembrane potential phenomena, or would otherwise make the maintenance of ion and electrical gradients difficult, if not impossible. Of course, in this light one might expect that neurons and muscle cells might be particularly devoid of large pores as compared with, say, secretory cells; but even so, left to their own devices all gradients would wind down in short order even if only small ion and water-accepting channels existed, without the presence of biological pumps or other gradient generating processes. Moreover, it is these small pores that represent the major conductance pathway. Even if large pores exist in the cell membrane, they would only have to account for a very small percentage of the total pore area, and carry a small percentage of the total current across the membrane, to account for substantial protein transport. The extent to which the apical membrane of the acinar cell would have to be punctuated with large pores in order for passage through them to account for the release of digestive enzyme, even in the most highly stimulated case, turns out to only require a minute fraction of the cell membrane, as well as a small percentage of the total number of pores and the total pore area (see Chapter 4) (Rothman, 1980).

Of course, the fact that the existence of such pores has not been ruled out does not mean that we can assume that they exist, no less that they account for protein transport across membranes in the pancreas or elsewhere. The point is merely that the existence of large pores would allow for the simplest type of transport mechanism: simple physical diffusion. As such it should be considered; indeed, it should be the favored hypothesis until and unless it is explicitly excluded by the kinetic and thermodynamic evidence, regardless of our preconceived notions about such a possibility.

15.2 Q_{10}

One way of excluding simple diffusion as the sole mechanism of transport is to measure the temperature dependence of the transport process. Simple diffusion is a rather temperature-insensitive event compared with reaction-dependent processes such as enzyme-mediated transport. Its rate changes only slightly in the range of the body temperature of homeotherms to ambient temperature. For temperature changes of 10°C, free diffusion is altered by about 1.0 to 1.3-fold. This is called the Q_{10} of the process. For reaction-mediated transport in general, Q_{10} values tend to run between 2–3.

We have made measurements of this sort for protein secretion by the pancreas and for the concentration-dependent release of digestive enzyme from zymogen granules. In the latter case, Liebow found that release was quite temperature sensitive at the initial rate, although the steady or equilibrium state was insensitive to changes in temperature (Liebow, 1985). For chymotrypsinogen, at the initial rate of release for the most temperature sensitive range, values for Q_{10} from 10–20 were seen (Figure 1). A substantial temperature dependence indeed! Thus, it seems that the release of digestive enzyme from zymogen granules, or at least release of the enzymes that we measured, cannot be explained by simple diffusion.

This information is more difficult to obtain for transport across the cell membrane. In this case, thus far we are only able to measure the temperature dependence of transport in the intact cell and that poses significant problems. Isenman measured the temperature dependence of amylase secretion by the whole rabbit pancreas in organ culture and found Q_{10} values of a little over 2.0 (Rothman, 1980). At first glance, this suggests, that once again we are not dealing with simple diffusion. However, protein secretion in this situation is dependent upon the presence of fluid and

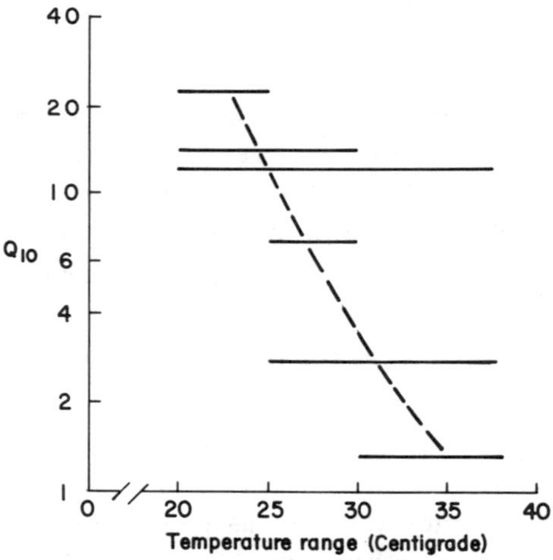

Figure 1 Effect of temperature on the rate of enzyme release from zymogen granules. Shown on the ordinate are the values for Q_{10} for chymotrypsinogen release found using 30-min release times. Shown on the abcissa is the temperature range in degrees centigrade. Ordinate values are determined by measuring the incremental release over a 30-min period (release after 30 min minus initial release) at two different temperatures and dividing the ratio of these two release rates by the difference in the two temperatures divided by 10 [that is, $(R_2/R_1)/((T_2 - T_1)/10)$]. This value is shown by the horizontal bar beginning at the lower temperature (T_1) and extending to the higher temperature (T_2) at the appropriate ordinate height. Zero time relase and release at the stipulated temperature were computed from four samples each. The dashed line is an approximation showing the decrease in temperature sensitivity of release going from 20 to 37C. (Reproduced, with permission, from Liebow, 1985.)

electrolyte transport, a temperature-sensitive process itself, and therefore the measured temperature dependence could reflect that of ion, and not protein, transport. When she measured the temperature sensitivity of electrolyte secretion, she found that it had almost the identical Q_{10} value over the temperature range from 20 to 37°C (see Figure 9 in Chapter 10). This either meant that both processes had approximately the same temperature dependence or that the observed value reflected the temperature sensitivity of ion transport and that protein movement, in its own right, had a Q_{10} value of the order of 1.0, as in a diffusive process. The latter view was supported by the observation that protein secretion in a given state was essentially wholly dependent upon fluid flow for its expression; that is, it varied in direct proportion to variations in the volume of secretion (Rothman, 1980) (see Chapter 10).

However, the evidence also indicated that the transport process might be close to equilibrium across the membrane (concentration in ductal fluid did not increase with decreased flow; see Chapter 10), and thus the temperature sensitivity of the equilibrium state and not the process of release itself, best estimated at the initial rate of release, was probably being measured. One cannot avoid this problem by measuring protein release from isolated cells, because in this case release occurs into the medium across both basolateral and apical cell surfaces, and only a composite temperature dependence is measured. In this regard, it is possible to measure in isolation the temperature dependence of release across the basolateral membrane of the cell for the gland in organ culture, also an equilibrating process. Its Q_{10} at the initial rate was found to be about 2.3 (Isenman and Rothman, 1977; see Chapter 10).

Thus, it appears that we can exclude simple diffusion as being able to account for protein movement across granule membrane; however, it has not yet been excluded as being responsible for transport across the apical cell membrane itself. I will consider this further below.

15.3 THE LIPID BILAYER

In addition to the belief that proteins cannot leave cells through pores, it has also been thought that they cannot cross the membrane via its lipid layer. The reason for this belief is the view that secretory proteins are spheres whose surfaces are randomly studded with polar amino acids most of whose charge would have to be neutralized or otherwise masked in order for the molecule to enter the lipid phase. Such charge neutralization would be energetically costly (although the high Q_{10} values for the release of protein from granules should not be missed), and as a result the protein would be expected to partition predominantly in the favorable aqueous environment. However, this qualitative view of the problem is too narrowly conceived. For example, it does not consider the potential energy of the concentration gradient across the membrane, nor the meaning of the partitioning of a small percentage of the protein in the membrane phase in regard to its transport. Moreover, it envisions resolutely polar bodies entering a resolutely nonpolar bulk lipid phase, and it is not clear that this is a satisfactory description of the situation. For example, is the entrance of a protein into a lipid bilayer identical to its solvation in a bulk lipid solution? Of course, it is really more accurately akin to the protein penetrating the surface of a lipid solution a meager distance of two lipid molecules, or about 70Å. This view of protein permeation also assumes that the lipid is solely organized as a homogenous bilayer at all points on the membrane. However, nonbilayer configurations may exist that provide a different, perhaps more favorable, arrangement for the passage of proteins, such as the hexagonal array found by Cullis and DeKruijff during the passage of cytochrome c across artificial lipid vesicle membrane (Cullis and DeKruijff, 1979; DeKruijff and Cullis, 1980), either induced by membrane proteins or the transported protein itself.

It is also interesting to consider the possibility that the surface of a protein is not merely a random or disordered charged surface, but that surface amino acid residues can be ordered in ways that make transport more likely. For example, rather sharp changes in the relative hydrophilicity and hydrophobicity of the protein may be possible. In this regard, it is important that membrane lipids include substantial amounts of amphipathic phospholipids whose polar ends face the surrounding aqueous medium. Interactions between these polar groups and charged moieties on the surface of proteins, forming polar pairs, may act to make the protein more hydrophobic, as well as alter the stability and form of the lipid layer. That is, it is possible that lipoprotein complexes could be formed between proteins and membrane lipids that might account for transport (Figure 2A).

We have found that some digestive enzymes, most notably chymotrypsinogen and trypsinogen, avidly form complexes with certain amphipathic phospholipids in vitro, and indeed chymotrypsinogen can be quantitatively extracted from solution in the presence of small amounts of the phospholipid phosphatidyl inositol (PI) in stoichiometrically constant complexes in which a mere 3 moles of phospholipid is able to extract a mole

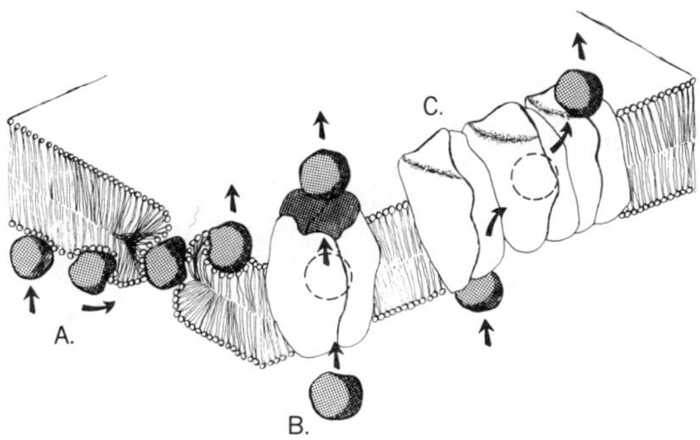

Figure 2 Artist's conception of a protein passing through a membrane in three ways: (A) secretory protein–membrane lipid complex; (B) passage of protein through a pore (pore could either be a constant feature of membrane or one produced only as the result of permeable protein interacting with a membrane component—a fugacious pore); and (C) secretory protein-membrane protein complex. (Reproduced, with permission, from Rothman, 1980.)

of protein from solution (Rothman, 1980). The reaction is calcium sensitive and has other characteristics reminiscent of the secretion process (Rothman, 1978) (Figure 3).

None of this is to say that this is the mechanism, or even a mechanism, by which protein transport occurs in the pancreas, or anywhere for that matter, although there is some evidence in intact cells that is consistent with the idea. We simply ought not exclude the possibility that proteins, even water-soluble proteins, can cross the lipid portion of the membrane. This is not only possible, but at least certain proteins, often ones that are known to cross membranes otherwise, such as cytochrome *c*, can penetrate and cross even simple artificial lipid bilayers (Quinn and Dawson, 1969; Steinemann and Lauger, 1971). Notions of lipid–protein interactions, beyond solution in bulk phases, should be considered and evaluated in situ.

15.4 PROTEIN–PROTEIN INTERACTIONS

If aqueous channels and passage across the lipid layer are excluded from consideration, then we are left to explain protein transport in terms of specific interactions between the transported protein and a protein or proteins embedded in the membrane. I say "left" because in a diffusive paradigm, this is ostensibly the most complex possibility of those that we have considered. However, complex or not, protein transport by means of such "carrier" complexes is known to occur, including curiously enough the

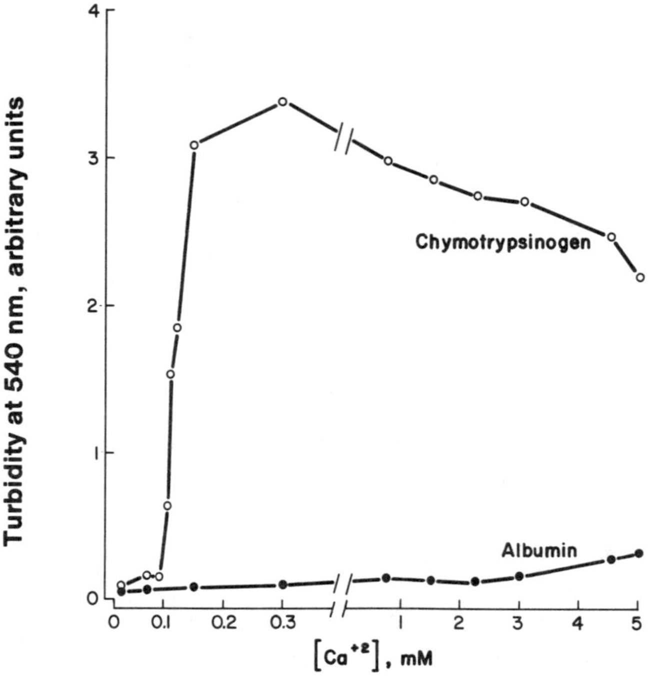

Figure 3 The effect of calcium concentration on the formation of inositol phosphatide–chymotrypsinogen and inositol phosphatide–albumin complexes as estimated from the turbidity of suspensions. Chymotrypsinogen (○) concentration was 41 μM and albumin (●) concentration was 15 μM. Inositol phosphatide concentration was 1.0 μM for both albumin and chymotrypsinogen cases. Values are the mean of a minimum of three observations at each point for both chymotrypsinogen and albumin. (Reproduced, with permission, from Rothman, 1978.)

uptake of bacterial toxins by host cells that are destroyed as a result. Moreover, this type of mechanism for protein transport is consistent with our knowledge of the transport of small organic molecules, such as sugars and amino acids, by carrier-mediated processes. As I have already noted, the diversity and complexity of proteins as a class of molecules recommends the existence of such transport processes. Nevertheless, we should not assume that such special means account for protein transport in any given case until and unless we are able to exclude, by knowledge of the behavior of the transport process, alternative simpler means of transport (as just discussed) that may conflict with our natural biases, but that nevertheless can and should be excluded by appropriate experimental test before being discarded.

15.5 A CLUE

We have one important clue regarding the nature of protein transport across the apical membrane of the acinar cell, and it, interestingly enough, reinforces the view that this process may be the result of the passage of protein molecules through aqueous, or at least polar, channels, which is of course the simplest possibility. It was reported several years ago that the pancreatic epithelium has a substantial transepithelial permeability to substances such as mannitol, inulin, and sucrose, and as such appears to be classified as a "leaky epithelium," perhaps the leakiest that has been described thus far (Jansen et al., 1979). In accordance with the view that these molecules do not enter cells readily, if at all, and that the leakiness of epithelia in general is the result of the paracellular passage of fluid and ions, the investigators who made this discovery assumed that they were dealing with a paracellular phenomenon. However, stimulants of protein secretion also produced large increases in the permeability of the tissue to these substances. To explain this observation in terms of paracellular shunts, one had to propose additionally that stimulants of protein secretion also produce an increase in the permeability of paracellular shunts. Although this certainly was possible, there was no convincing reason why this should be expected. Such an effect would not seem to have any positive purpose in promoting secretion and might even cause the loss of substantial amounts of secreted fluid into the interstitium. Based on questions of this kind, Mélèse and I attempted to distinguish between cellular and paracellular passage in a series of experiments (Mélèse and Rothman, 1983a, b, c) in which we tested a variety of different water-soluble molecules using the system in which the original observations had been made, the rabbit pancreas in short-term organ culture. All of our results indicated the same thing: Cellular passage was involved and the increased permeability seen as a result of stimulation was due to changes in the permeability of the apical membrane of the cell.

It seemed unlikely that the transcellular movement of these molecules occurred by means of a vesicular process (endocytosis, directed transcellular movement, and exocytosis) because (Mélèse and Rothman, 1983a, b, c): (1) the concentrations of the different substances found in ductal fluid, relative to their concentration in the medium bathing the gland, differed widely (they were not the same; as one would expect for "neutral" nonelectrolytes in a vesicular process); (2) the concentration of some of the substances in secretion were as great as 50–70% of the medium concentration, and therefore a vesicle shuttle would have to account for a like percentage of the total fluid secreted by the pancreas; and (3) the stimulant produced the same increase in the secretion of a substance even when it was absent from the bathing medium if it was present in the cell.

The change in permeability with stimulation was two to four times for molecules ranging from phosphate ions, to hexitols, to inulin. This means that there is a substantial increase in the permeability of the apical membrane of the acinar cell to water-soluble (that is, polar) compounds that occurs coincident with the increase in protein secretion. Moreover, this includes molecules as large as inulin (and perhaps accounts for the effect of stimulants noted above on the secretion of lysosomal enzymes). Thus, a major reorganization of the membrane appears to take place when protein secretion is stimulated by hormones.

This increase in permeability is, however, not tied to protein secretion, even though its appearance is correlated to its onset, or more specifically, its rate of onset (that is, the greater the rate of protein secretion initially, the greater the rate at which permeability increases) (Figures 4, 5, and 6) (Mélèse and Rothman, 1983c). The independence of the two processes is demonstrated by the fact that the concentration of nonelectrolytes in secreted fluid remains elevated over time in the continued presence of the stimulant (and the nonelectrolyte), whereas protein secretion itself falls off and returns to or toward the unstimulated rate in a relatively short period of

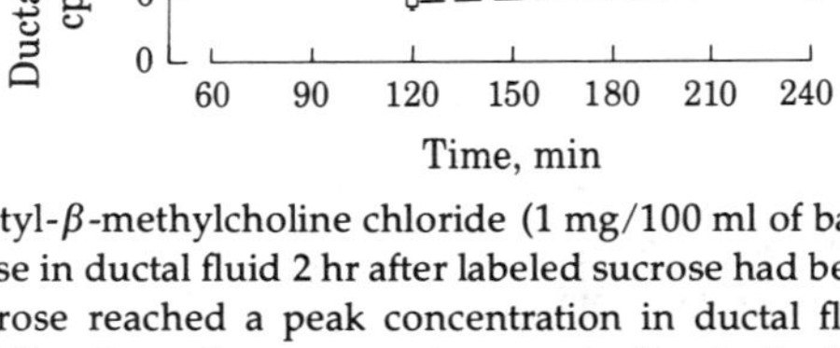

Figure 4 Effect of acetyl-β-methylcholine chloride (1 mg/100 ml of bath fluid) (bar) on the concentration of sucrose in ductal fluid 2 hr after labeled sucrose had been removed from the bathing medium. Sucrose reached a peak concentration in ductal fluid at five times the prestimulus value and five times its concurrent concentration in the bathing medium. The concentration of [^{14}C] sucrose in the bath immediately before the addition of the stimulant and at the end of the experiment is shown (□). Data are expressed as mean ± SEM; n = six preparations. $p < 0.01$ without acetyl-β-methylcholine chloride (at 120 min) versus with acetyl-β-methylcholine chloride (at 140 min). (Reproduced, with permission, from Mélèse and Rothman, 1983c.)

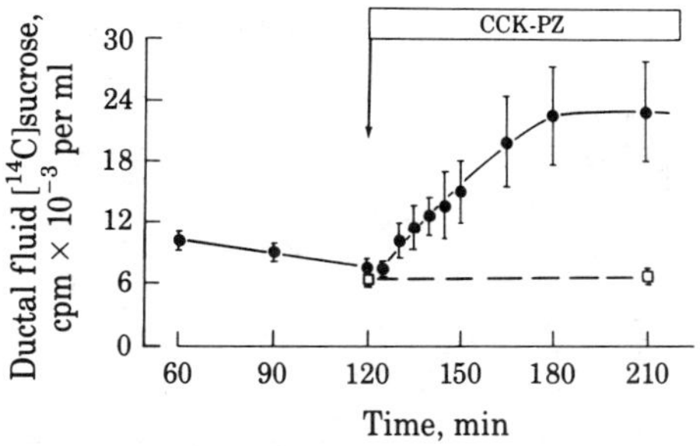

Figure 5 Effect of CCK–PZ (6.0 Ivy dog units/100 ml of bath fluid) (bar) on the concentration of sucrose in ductal fluid 2 hr after labeled sucrose was removed from the medium. The circumstances are the same as those in Figure 4. In the presence of the stimulant, the concentration of sucrose in ductal fluid rose, reaching an apparent steady-state value at approximately three times the prestimulus level and at approximately three times that present in the bathing medium. The concentration of $[^{14}C]$sucrose in the bath immediately before the addition of the stimulant and at the end of the experiment is shown (□). Data are expressed as mean ± SEM; n = six preparations. $p < 0.01$, without CCK–PZ (at 120 min) versus with CCK–PZ (at 210 min). (Reproduced, with permission, from Mélèse and Rothman, 1983c.)

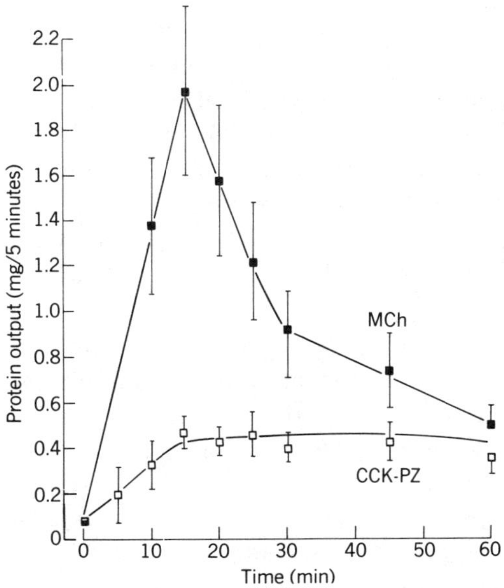

Figure 6 Effect of acetyl-β-methylcholine chloride (1.0 mg/100 ml of bath fluid) (■) and CCK–PZ (6.0 Ivy dog units/100 ml of bath fluid) (□) on protein output from rabbit pancreas in short-term organ culture. Data are expressed as mean ± SEM. Acetyl-β-methylcholine chloride, n = six preparations; CCK–PZ, n = six preparations. (Reproduced, with permission, from Mélèse and Rothman, 1983c.)

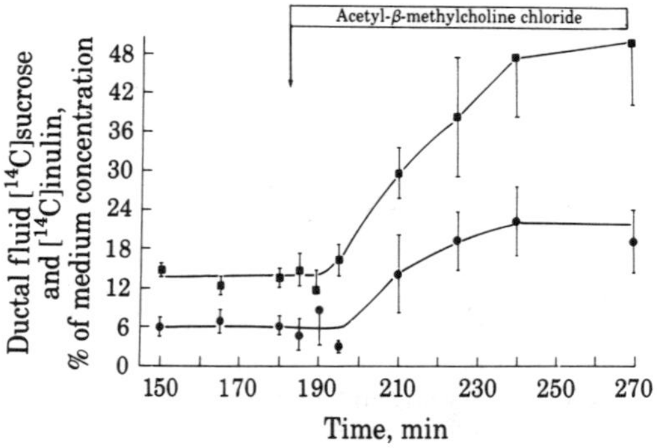

Figure 7 Concentration of sucrose (■) and inulin (●) in ductal fluid in unstimulated and stimulated states. Values are expressed as percent of bath concentration. The biological preparation was the same as that described in Figures 4, 5, and 6. The steady-state concentration of sucrose and inulin in ductal fluid rose in the presence of the stimulant acetyl-β-methylcholine chloride (1 mg/100 ml of bath fluid) (bar) to approximately three to five times that seen in the unstimulated state. Data are expressed as mean ± SEM. Sucrose, $n = 12$ preparations; inulin, $n = 12$. $p < 0.01$ for sucrose and $p < 0.01$ for inulin, without acetyl-β-methylcholine chloride (at 180 min) versus with acetyl-β-methylcholine chloride (at 270 min). (Reproduced, with permission, from Mélèse and Rothman, 1983c.)

time (Figures 6 and 7). This suggests that although membrane reorganization is related to protein secretion, it is not a direct result of the transport of the digestive enzymes. That is, it is an independent, related phenomenon. This eliminates, among other possibilities, the notion that the effect could be due to the insertion of new (leakier) membrane into the apical surface of the cell (as in exocytosis), since in this case the two processes would be linked, and a fall-off in secretion would lead to a fall-off in the permeability of the membrane and to the reestablishment of the initial state.

In any event, relatively large water-filled or other polar channels are apparently formed in the membrane during protein secretion. This supports the idea that proteins exit through such channels, although it is not excluded that they are merely formed coincident to a reorganization of the membrane that accounts for protein transport in other ways. Thus, despite preconceptions one might harbor that such things do not occur, the evidence lends support to the hypothesis that large channels or polar lacunae exist in the apical membrane of the acinar cell that account for the release of protein from these cells. This process, then, may be different from the highly temperature-sensitive release of protein across zymogen granule membrane. Moreover, these observations reinforce, and are consistent with, those that I discussed earlier on the flow dependence of secretion, as well as other evidence for the equilibrium nature of the secretion process across the apical plasma membrane.

15.6 CELL THEORY AND THE PROBLEM OF PROTEIN TRANSPORT

When the vitalist-reductionist debate ended with vitalism the clear loser, at least in the original sense of the term, the idea became dominant that cells and organisms do not have unique properties that distinguish them from the inanimate world. It seemed that biological systems were merely remarkably complex versions of what one sees outside of living things. Even if our metaphysics leaves us with many worrisome questions about this notion, as a practical scientific matter our knowledge of biological systems leaves little other choice at this time but to accept it.

We have been reduced to defining the unique qualities of life in terms of the specific nature of the proteins and nucleic acids within cells and the reactions that they carry out. The cell differs from its environment and from the inanimate world in general because it contains these special compounds that engage in special reactions that carry out the unique functions of cells, such as motility, irritability, reproduction, and so forth. It is these characteristics that give the cell its place apart from the inanimate world. In a sense, this definition of living affords us with a meager remnant, "a remainder biscuit," of vitalism, namely, the special nature of the chemical contents and reactions of cells.

It is in this context that one must envision the hypothesis that cell membranes are permeable to protein molecules in a general sense and not merely in particular instances. That is, in this context such a permeability can be viewed as being akin to the loss of the stuff of life. Of course, as I have discussed, the issue is really whether or not the cell can maintain itself and the internal concentration of a substance whatever loss may occur. In most ways that we have considered cells in modern times, they appear to be systems capable of maintaining disequilibrium steady or stationary states, not absolute differences between themselves and their environments. Certainly transport processes in general reflect this fact. But the modern notion of vitalism excludes proteins and other macromolecules from this rubric. Intracellular proteins are thought to be housed absolutely within the confines of the cell. Only if the cell is damaged or a special process exists for its exit can it leave. Introducing the notion that cell membranes are permeable to proteins in a more general sense crosses this last vitalistic barrier. This is not to say that such a permeability in fact exists, or that if it does that it would be substantial, or the same for different molecules, or specific or nonspecific, but merely that the existence of a permeability of cells to proteins in general is a hypothesis that cannot be excluded a priori on the basis of current knowledge, and that it is an hypothesis that should be tested. If it turns out, as I think it will, that absolute impermeability, even for proteins, cannot be assumed, then perhaps it will be appropriate to state that the separation between cell and environment is relative in all of its aspects and absolute in none.

SELECTED REFERENCES

1. Cullis, P. R. and B. DeKruijff. Lipid polymorphism and the functional roles of lipids in biological membranes. *Biochim. Biophys. Acta* **559**:399–420, 1979.
2. DeKruijff, B. and P. R. Cullis. Cytochrome *c* specifically induces non-bilayer structures in cardiolipin-containing model membranes. *Biochim. Biophys. Acta* **602**:477–490, 1980.
3. Isenman, L. D. and S. S. Rothman. Transport of amylase across the basolateral membrane of the pancreatic acinar cell. *Proc. Natl. Acad. Sci. USA* **74**:4068–4072, 1977.
4. Jansen, J. W. C. M., J. J. H. M. DePont, and S. L. Bonting. Transepithelial permeability in the rabbit pancreas. *Biochim. Biophys. Acta* **551**:95–108, 1979.
5. Liebow, C. Protein transport through zymogen granule membrane: reconsideration after a decade, in, *Nonvesicular Transport,* S. S. Rothman and J. J. L. Ho, Eds. John Wiley & Sons, New York, 1985.
6. Mélèse, T. and S. S. Rothman. Increased phosphate efflux from acinar cell during protein secretion. *Am. J. Physiol.* **245**:C121–C124, 1983a.
7. Mélèse, T. and S. S. Rothman. Distribution of three hexose derivatives across the pancreatic epithelium: Paracellular shunts or cellular passage? *Biochim. Biophys. Acta* **763**:212–219, 1983b.
8. Mélèse, T. and S. S. Rothman. Pancreatic epithelium is permeable to sucrose and inulin across secretory cells. *Proc. Natl. Acad. Sci. USA* **80**:4870–4874, 1983c.
9. Quinn, P. J. and R. M. C. Dawson. Interactions of cytochrome c with monolayers of phosphatidyl choline, phosphatidic acid and cardiolipin. *Biochem. J.* **115**:65–75, 1969.
10. Rinderknecht, H., I. G. Renner, and H. H. Koyama. Lysosomal enzymes in pure pancreatic juice from normal healthy volunteers and chronic alcoholics. *Dig. Dis. Sci.* **24**:180–186, 1979.
11. Rothman, S. S. Chymotrypsinogen-inositol phosphatide complexes and the transport of digestive enzyme across membranes. *Biochim. Biophys. Acta* **509**:374–383, 1978.
12. Rothman, S. S. Passage of protein thorugh membranes—old assumptions and new perspectives. *Am. J. Physiol.* **238**:G391–G402, 1980.
13. Steinemann, A. and P. Lauger. Interaction of cytochrome c with phospholipid monolayers and bilayer membranes. *J. Membr. Biol.* **4**:74–86, 1971.

Epilogue

One should not construct a model too near to reality, lest he mistake the model for the thing itself.

Attributed to A.V. Hill

As an experienced investigator I have seen not a few instances during my career that have revealed the possibilities of mistakes even after the most careful precautions have been taken. All venturesome scientists are aware of the many chances of going astray as they enter a new field. Michael Faraday, a prince of experimenters, testified "that I may be largely wrong I am free to admit—who can be right altogether in physical science which is essentially progressive and corrective?" If Faraday could feel thus toward experimentation in the realm of physics, how much readier to acknowledge the possibility of error should be an investigator who labors in the more complex and difficult realm of biology.

Walter Cannon

Anomalous observations, that is, observations that do not appear to be consistent with current, widely held scientific theory, are frequently reacted to initially in one of two ways. Either the observation is thought to be incorrect, an artifact of one sort or another, or alternatively, it is seen as being consistent with the paradigmatic view of the process, incorporated into that view by modifying the general theory to accommodate the discordant observation, and alternative explanations are viewed either as unnecessary or as reflecting antiparadigm bias on the part of those who propose them. These methods of dealing with the anomalous often occur in sequence. At first, an anomalous observation may be thought to be the result of experimental artifact, an imprecise or improperly controlled observation. When it becomes clear that the observation itself is correct as the result of a clearer understanding of its nature, other evidence or replication by others, then it is fit into the broader paradigm.

Many of the observations that my colleagues and I have made, including some of those that I have discussed in this book that bear on the validity of the cisternal packaging–exocytosis paradigm as an adequate comprehensive explanation for the process of secretion, have been reacted to in this way. This is not to say that this or that observation was actually shown to be artifactual or indeed explicable within paradigmatic borders. Rather, it was, and in some cases still is, argued that the observations might conceivably be artifactual even though one may not know exactly in what fashion or that this is in fact the case. Or, it may be argued that one can imagine ways in which the observation might fit into the paradigm even though experimental evidence that this is indeed so is lacking. I have discussed many of these alternative explanations in this book to give the reader a sense of what kinds of alternatives to the equilibrium model have been suggested to account for anomalous observations that have been made, as well as to account for the theoretical problems inherent to the vesicle paradigm.

Supporters of the paradigm may feel that for particular anomalous observations, such criticism, even if unsupported by evidence in its own right, is justified by other positive evidence in support of their view. However justified or unjustified that may be, we cannot merely consider each anomalous observation in isolation. It is in the aggregate that the problems that they raise for the vesicle model become truly imposing, one observation reinforcing another. One of my major purposes in writing this book was to bring together in one place and one contextual framework a variety of observations, many from different theoretical and experimental perspectives, that raise questions about the vesicle paradigm's adequacy as a satisfactory model to account fully for protein secretion and transport.

To retain the vesicle transport model as the sole potential explanation for the process of secretion in the pancreas in face of these discordant observations does not make sense to me. One must account not merely for one or another anomalous observation within paradigmatic borders but for them all; not only those that I have selected for discussion here either, but for the whole body of such anomalous observations that are to be found in the literature. It is my belief that to do so is a substantial task. Although it is conceivable that all of the anomalous observations may in the end be explained within paradigmatic borders, it seems to me that the questions that these observations raise are sufficient to warrant that alternative hypotheses, particularly those that are simpler, more general, and deeper, such as the equilibrium theory, be explored with energy. Whatever the future holds, for the moment the paradigm cannot explain the anomalous because evidence that it does is not available, and therefore it cannot be viewed as a successful theory at present.

In this light, it is important to appreciate that none of the observations that I have discussed "refute" or prove the exocytosis or vesicle model false in any general sense. That is, they do not prove that such processes do not exist in general or even in the pancreas in particular. They merely indicate that vesicle models, as they have been envisioned in the past, do not provide a sufficient basis to account for the process of secretion and that a general alternative exists that should be explored. Of course, it should be remembered that a vesicle mechanism accounting wholly for the natural process of secretion of a product and its simple existence in a particular cell are different matters that require different experimental and theoretical support. Conversely, the existence of vesicle mechanisms does not prove that nonvesicular transport processes for proteins or other cell products do not occur. Of course, in the end both types of process, vesicular and nonvesicular, may be shown to occur and function either in concert or in different situations, and for this reason experimental support should be sought for both.

More important than which model is closer to the truth and I believe the central point to be gleaned from this book, is that the first priority in the proper application of the scientific method is that theories for processes be developed from the observation of actual events in real space-time; their kinetic, thermodynamic, compartmental, and other characteristics, and not merely be constructed as cartoons from static evidence, weak a priori assumptions about what is or is not possible, and the logically inappropriate transposition of existential evidence from system to system. This observational data base is essential for a successful theory of whatever kind.

If the paradigm proves to be incorrect or at least incomplete, this should in no way be viewed as diminishing the historical and creative significance of the contributions of those whose work gave it life. Without their model and the evidence they brought forth in its support, there would be no alternative, and progress would not have been made. A scientific model should not be judged solely, or even primarily, in terms of whether or not it turns out to be correct, but rather in terms of its usefulness in helping us learn about natural processes.

As I said at the outset, criticism, even the outright rejection of older ideas and notions, is the foundation upon which scientific progress is built. The central contributions of our most esteemed forebearers are almost invariably based on their questioning the common views of their time, as the names Galileo, Lavoisier, Pasteur, Semmelweis, Darwin, and Einstein attest. Yet these innovators, as well as innumerable others whose discoveries were not as significant, faced great difficulty in convincing their peers not merely of the correctness of their views, but even of their right to question the traditional perception. Thus although criticism lies at the heart of scientific progress, it is rarely applauded in practice. As a result, new theories and the evidence that supports them are often clouded by controversy when they are first put forth. In the present case, the danger is not that experimental efforts to find support for the vesicle paradigm will prematurely flounder due to arguments such as those that I have put forth in this book. If experience is any indication, it is more likely that exploration of the alternative will be the victim of controversy.

For myself, I think that good science, like good political democracy, must allow for, even welcome, differences of opinion as well as the confusion and uncertainty that such differences produce. The alternative is scientific autocracy and intellectual stagnation: rigid, not rigorous, science.

Index